MAKE YOUR OWN COMPILER THIS WAY

Targeting for PIC16 and STM8 MCUs

Mengjin Su

diycompiler@gmail.com

GitHub: /mengjin-su/diy-c-compiler

Table of Contents

Preface

Compilers are also a type of software that plays an important role in computer systems and software development. Daily programming of software engineers/technicians in the IT industry often involves using a compiler to translate a high-level language (C, C++, C#, …) into runnable code that runs directly on the target machine.

Most people might not be aware of how a compiler is designed or how it works. Indeed, the colleges will offer a class on the compiler system. But it usually stays at the theoretical level, avoiding the details of actual design practice, or it offers only incomplete examples. It leaves students feeling lost as a result.

The process of designing a compiler is similar to that of programming. It will use the compiler on the users' computers. The situation is somewhat like that of a blacksmith, who forges a small hammer using a big hammer from his toolbox.

It is for sure that compiler design is not easy. It requires designing and manipulating complex data structures and often involves deep recursive function calls. But it is worth the effort to see the results generated by your own compiler running on your target machine. Also, understanding compiler principles will be beneficial for practical programming, resulting in more efficient and reliable code.

Targeting two well-known 8-bit MCUs (PIC16F and STM8) as the running platforms, this book presents the complete procedure for designing a C compiler. The development procedure is accomplished using the GNU GCC toolchain (C and C++), running in terminal/console mode, on both Windows and Linux. All source code (parser, assembler, linker) and test examples are available on GitHub.

Moreover, the development will use the tools or commands flex and bison, which are included in the GNU GCC package dedicated to compiler/parser design. It is not an exaggeration to say that these tools are crucial for designing and building the parser. Actually, users can design or create a new language with these tools.

PART-I:
Introduction and Warm-up

This chapter of the book outlines the details of designing a C compiler for PIC16 8-bit microprocessors from Microchip. Microchip introduced the first generation of 8-bit PIC (34 instructions) microprocessors in the early 90s of last century. After the year 2000, the new generation of PIC16 microprocessors (enhanced PIC16) came to the market, which kept the same RISC structure as the older PICs, with an enhanced instruction set (56 instructions).

The enhanced PIC processor family runs faster and delivers better performance. Moreover, it gives better support for high-level languages (C language). This book focuses on the design of the C compiler for the new generation PIC16 microprocessors.

The C compiler includes the following executable files(commands):

- `cpp1.exe` – C language preprocessor – expands and inserts all included files.
- `cc16e.exe` – C language parser – generate PIC16 assembly output
- `as16e.exe` – PIC16 assembler – generate intermedia object code (float code)
- `lk16e.exe` – PIC16 linker – generate the HEX file (binary code) for the target MCUs.

Chapter-1
Select Designing Tool and Preparation

Simply speaking, a compiler is a set of programs or executable files that are used to compile for target machines or MCUs. Like other software, compilers are designed or developed using a design tool. In this book, all the designs are designed/developed using the GNU GCC Tool, which was developed under the UNIX/Linux operating system and later ported to the Windows environment.

1.1 GNU GCC Tool Select

There are several GNU GCC toolchain have been ported to Windows. All of them are free to users.

- `MinGW (Minimalist GNU for Windows)`
  This appears to be the most widely used C/C++ compiler tool running in the Windows environment today. It runs under Windows as the native commands. It contains all the functions and features to cover all the needs in this book. It can run on both 32-bit and 64-bit versions of Windows. Additionally, new updates continue to be released and are available to users.

- `DJGPP (DJ's GNU Programming Platform)`
  This C/C++ tool package appears to be very easy to use. But it can only run under 32-bit environment. That's why it's out of phase today.

- `Cygwin (Cygnus and Windows)`
  This C/C++ tool package essentially creates a Linux/Unix-like environment on Windows. That means its behavior and results differ from those of Windows.

This book illustrates the whole process of compiler design using the MinGW toolchain under the Windows environment. MinGW is available for free download at Source Forge:

https://sourceforge.net/projects/mingw

1.2 Flex and bison – Parser Design Tools

Flex and **bison** are the specific tools/commands, coming with GNU toolchain, for parser design. Flex is a variant of **lex** that originated in UNIX, and bison is derived from **yacc**.

- **flex** – a tool for designing a lexer that converts the input character stream of source files or C files to the syntax tokens;
- **bison** – a tool for designing a parser at the language syntax level. It takes the tokens from the lexer as the native inputs. Users use so-called parsing rules (or production rules) to identify or match the grammar of the target language and generate a parsing tree as output.

Flex and Bison come with the MinGW toolchain; no extra installation is needed under Windows. However, under Linux, they are not installed by default, unlike g++. The user needs to install them manually as follows, in administrator mode:

```
$sudo apt-get install flex

$sudo apt-get install bison

$sudo apt-get install g++
```

1.3   Set Up Target Compiler Project Folder

As with other software development, it is appropriate to set up a folder for it. As an example, in this book, the project folder is located on the hard disk 'F:', named as "**p16ecc**", as shown in Figure-1.

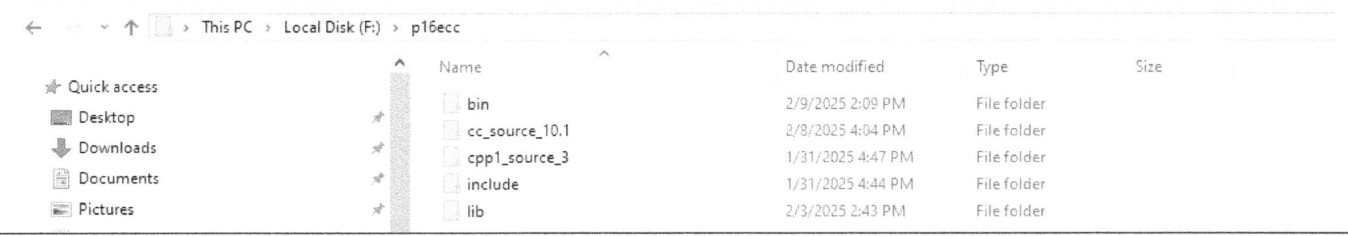

Figure-1

In the project folder **/p16ecc** above, there are several sub-folders or directories that are mandatory:
- `/bin`   contains all the executable files or commands of the target compiler;
- `/include`   contains header files for PIC16 microprocessors;
- `/lib`   contains compiler system library source files.

To make the target compiler working as other terminal commands, the user needs to add the project folder to the system PATH, or simply type the command as follows from the terminal:

```
PATH=F:\p16ecc\bin;%PATH%
```

In a Linux environment, things are a bit more complicated. To make the target compiler work like other terminal-level commands, the user needs to add the configuration to the system. Assume the project folder is located at /home; the user needs to modify the system configuration file '.bashrc', which is usually found under /home. Open the file with the editor vi command, add the new configuration at the end of the file:

```
export PATH=$PATH:$HOME/p16ecc/bin
```

Chapter-2

C Preprocessor Design

When lunching the compiler, the first thing it does is to run the C preprocessor(**cpp1**). The preprocessor is a text processor that processes the statements that are leading with letter '#' in the source file (`#ifdef`, `#ifndef`, `#endif`, `#define`, `#include`, etc.). This chapter presents a simple preprocessor that only processes the `#include` statements in the input source file. The C preprocessor is invoked by the **cc16e** compiler. It can also run alone.

There are two types of `include` statements for addressing the files to be included from different sources:

- `#include` *<filename>*   involves the file specified by *filename* from **/include** directory (system file).

- `#include` *"filename"*   involves the file specified by *filename* from user defined directory or The project directory (user file).

The preprocessor's operation is simply reading the file contents to replace the statements, then generating a temporary file after expansion. Obviously, the output file could be much larger than the original.

The operation of the preprocessor seems simple and straightforward. But there are a few considerations that should be taken care:

1. Nested or chained inclusions. This happens like *file1* → *file2* → *file3* …

2. Cross/recursive inclusion between files (this will cause endless operation).

3. The output of the preprocessor, **cpp1**, may contain a number of source files. There should be a way to identify which source file each line belongs to.

2.1   Preprocessor cpp1 in C/C++

```
Project directory: /p16ecc/cpp1_source_1
Source files: main.cpp, cpp1.cpp, cpp1.h, makefile
Output: cpp1.exe
```

2.1.1   Start of the cpp1 project

As a project, it is natural to make project script, make file, for it:

```
CC = g++
RM = rm
CP = cp

EXE = cpp1.exe
```

```makefile
OBJ = main.o cpp1.o

OPTIONS= -c -Wall -Os

$(EXE): $(OBJ) makefile
        $(CC) -static $(OBJ) -o $(EXE)
        $(CP) $(EXE) ../bin

%.o: %.cpp makefile
        $(CC) $(OPTIONS) $<

clean:
        $(RM) *.o
        $(RM) $(EXE)
```

2.1.2 Design of cpp1

cpp1 can be launched by typing a command from the terminal (or inside of **cc16e**) in the form:

    cpp1 *file*

main.cpp is the start of **cpp1**, shown as code-2.1:

```cpp
1   #include <stdio.h>
2   #include <string.h>
3   #include <string>
4   #include "cpp1.h"
5
6   #define CC_FOLDER_NAME      "p16ecc"
7   #define CC_FOLDER_LEN       strlen(CC_FOLDER_NAME)
8   #define IS_SLASH(c)         ((c) == '/' || (inWindows && (c) == '\\'))
9   #define IS_SEPERATOR(c)     ((c) == '=' || (inWindows && (c) == ';') || \
10                              (!inWindows && (c) == ':'))
11
12  int main(int argc, char *argv[], char *env[])
13  {
14      bool inWindows = false;
15      std::string incPath;
16
17      if ( argc < 2 )
18      {
19          printf("missing input file!\n");
20          return -1;
21      }
22
23      for (; env && *env; env++)   // search for "PATH=" setting
24      {
25          if ( memcmp(*env, "OS=Windows", 10) == 0 )
26              inWindows = true;
27          else if ( strncasecmp(*env, "path=", 5) == 0 )
28          {
29              char *p = strstr(*env, CC_FOLDER_NAME);
30              int len = CC_FOLDER_LEN;
31
32              if ( p && IS_SLASH(p[-1]) && IS_SLASH(p[len]) &&
33                        memcmp(&p[len + 1], "bin", 3) == 0            )
34              {
35                  while ( !IS_SEPERATOR(p[-1]) ) p--, len++;
36
37                  incPath.assign(p, len);
38                  incPath += "/include/";
39              }
40          }
41      }
42
43      std::string output = argv[1];
44      output += "_";
45
46      return cpp1(&incPath, argv[1], (char*)output.c_str());
47  }
```

/p16ecc/cpp1_source_1/main.cpp   Code-2.1

- #25~26: identify its running in a Windows environment.
- #27~40: form the string for the system **/include** directory.

- #43~44: generate output file name.
- #46: start to parse the input file.

The function, claimed in cpp1.h

int cpp1(std:string *inc_path, char *file_in, char *file_out, bool sys_file=false);

is to parse the input, shown in code-2.2a and code-2.2b.

```
1    #include <stdio.h>
2    #include <string.h>
3    #include <string>
4    #include "cpp1.h"
5
6    typedef struct File_t_ {
7        std::string     name;
8        FILE            *fhandle;
9        int             line_no;
10       bool            sys_file;
11       struct File_t_  *next;
12   } File_t;
13
14
15   static File_t *fileStack = NULL;
16   static FILE   *output = NULL;
17
18   static int  lexer(File_t *file, std::string *inc_path);
19   static int  readLine(FILE *fin, std::string *line_buf);
20   static char *skipSP(char *line);
```
/p16ecc/cpp1_source_1/cpp1.cpp   Code-2.2a

- #6~12: data structure of stack for tracking input files being processed.
- #14~20: local data and functions.

```
23   int cpp1(std::string *inc_path, char *file_in, char *file_out, bool sys_file)
24   {
25       if ( file_out )
26       {
27           output = fopen(file_out, "w");
28           if ( output == NULL )
29           {
30               printf("can't open output file '%s'!\n", file_out);
31               return -1;
32           }
33       }
34
35       // elliminate duplicated file scaning...
36       for (File_t *fp = fileStack; fp; fp = fp->next)
37           if ( fp->name == file_in && fp->sys_file == sys_file ) return 0;
38
39       FILE *fd = fopen(file_in, "r");
40       if ( fd == NULL )
41       {
42           printf("can't open file '%s'!\n", file_in);
43           return -1;
44       }
45
46       File_t *file      = new File_t;
47       file->fhandle     = fd;
48       file->name        = file_in;
49       file->line_no     = 0;
50       file->sys_file    = sys_file;
51       file->next        = fileStack;
52       fileStack         = file;
53
54       int rtcode = lexer(file, inc_path);
55       fclose(file->fhandle);
56       delete file;
57
58       file = fileStack = fileStack->next;
59
60       if ( fileStack == NULL )     // end of cpp1, close the output.
61           fclose(output);
62
63       return rtcode;
64   }
```
/p16ecc/cpp1_source_1/cpp1.cpp   Code-2.2b

- #25~33: open the output file (for the first time running cpp1() function).
- #36~37: avoid cross/recursive operation.

- #39~44: open the input file.
- #46~52: push the input file information into the stack (don't forget initialize the line number #49).
- #54~58: start/end of parsing.

In the function lexer(), it reads the input file line by line and either expands the #include statement by calling `cpp1()` recursively or writes the line to the output directly, as shown in code-2.3 below.

```cpp
66  static int lexer(File_t *file, std::string *inc_path)
67  {
68      for (bool line_mark = true;;)
69      {
70          std::string line_buf = "";
71          int length = readLine(file->fhandle, &line_buf);
72          if ( length <= 0 ) { fclose(file->fhandle); return 0; }
73
74          file->line_no++;
75
76          char *p = skipSP((char*)line_buf.c_str()), *p1;
77          if ( memcmp(p, "#include", 8) == 0 )
78          {
79              int rtcode;
80              std::string file_path = inc_path->c_str();
81
82              p = skipSP(p + 8);
83              switch ( *p++ )
84              {
85                  case '"':    // user source file
86                      if ( (p1 = strchr(p, '"')) != NULL )
87                      {
88                          *p1 = '\0';
89                          rtcode = cpp1(inc_path, p, NULL);
90                          if ( rtcode < 0 ) return rtcode;
91                          break;
92                      }
93                      return -1;
94                  case '<':    // system source file
95                      if ( (p1 = strchr(p, '>')) != NULL )
96                      {
97                          *p1 = '\0';
98                          file_path += p;
99                          p = (char*)file_path.c_str();
100                         rtcode = cpp1(inc_path, p, NULL, true);
101                         if ( rtcode < 0 ) return rtcode;
102                         break;
103                     }
104                 default:
105                     return -1;
106             }
107             line_mark = true;
108         }
109         else
110         {
111             if ( line_mark )
112             {
113                 if ( file->sys_file )
114                     fprintf(output, "#LINE %d %s\n", file->line_no, file->name.c_str());
115                 else
116                     fprintf(output, "#line %d %s\n", file->line_no, file->name.c_str());
117             }
118
119             fprintf(output, "%s", line_buf.c_str());
120             line_mark = false;
121         }
122     }
123 }
```

/p16ecc/cpp1_source_1/cpp1.cpp   Code-2.3

- #70~74: read a line from the input file and then increase line number (#74). End the execution when reaching the end of the file (#72).
- #76~108: when input line is a `#include` statement, expand the indicated file by calling **cpp1()**.
- #109~121: for other lines, append them to the output. Note, it might need to output a line marker first (#111~117).

Compile the project from the terminal:

```
F:\p16ecc\cpp1_source_1>make
g++ -c -Wall -Os main.cpp
g++ -c -Wall -Os cpp1.cpp
g++ -static main.o cpp1.o -o cpp1.exe
cp cpp1.exe ../bin
```

The following tables (Table-2.1 and Table-2.2) show the test results:

Exp-2.1: execute command `cpp1 ch2_t1.c`:

ch2_t1.c	ch2_t1.h	Output: ch2_t1.c_

`#include "ch2_t1.h"`
 	
`int func()`
`{`
`   return ONE;`
`}`
	`#define ONE 1`	`#line 1 ch2_t1.h`
		`#define ONE 1`
		`#line 2 ch2_t1.c`
		
		`int func()`
		`{`
		`   return ONE;`
		`}`

Table-2.1

Exp-2.2: execute command `cpp1 ch2_t2.c`:

ch2_t2.c	ch2_t2.h	ch2_t3.h	Output: ch2_t2.c_

`#include ch2_t2.h"`
 	
`int func()`
`{`
`   return ONE + TWO;`
`}`
	`#include "ch2_t3.h"`
	
	`#define ONE 1`
		`#include "ch2_t2.h"`
		
		`#define TWO 2`
			`#line 2 ch2_t2.h`
			
			`#define ONE 1`
			`#line 2 ch2_t3.h`
			
			`#define TWO 2`
			`#line 2 ch2_t2.c`
			
			`int func()`
			`{`
			`   return ONE + TWO;`
			`}`

Table-2.2

2.2　Preprocessor cpp1 in Flex

```
Project directory: /p16ecc/cpp1_source_2
Source files: main.cpp, cpp1.l, cpp1.h, makefile
Output: cpp1.exe
```

The design of preprocessor **cpp1** in Section 2.1 seems simple and easy to understand. But it has limited functions and features. This section, it will add more operations to meet the needs in this book:

- Remove the comments in the input (source) files. For example:

  Before:
  ```
        p = "/* this is not a comment */"; /* this is a comment */
  ```
  After:
  ```
        p = "/* this is not a comment */";
  ```

- Merge the text to a single line for #define statements that take multiple lines. For example:

  Before:
  ```
        #define ASSIGN(a,b)  a = 10; \
                             b = 20;
  ```
  After:
  ```
        #define ASSIGN(a,b)  a = 10; b = 20;
  ```

With those enhancements, the compilation process later will be easier. Obviously, it is a lot more complicated for only C/C++ to fulfil the job. Hence, the lexer tool, **flex**, is used in this section. **Flex** is a lex-like UNIX tool with enhancements, used to generate lexers. Simply put, the user needs to write a script in the form of a regular expression to describe the strings to match or search, along with the C code inserted to handle the situations when they are encountered. **Flex** converts the script into its responding C language output (lex.yy.c as the default) for compiling using **gcc**.

2.2.1　Introduction of regular expressions

Regular expression is a rule to describe the pattern of strings to be matched. It can be considered an extension of character/string expressions in C/C++. That means, it can simply be the ASCII characters or string format that appeared in C. Plus, it can use some 'special' characters (symbols), based on their appearing positions and combinations, to describe a set of strings. The following table, Table-2.1, lists the most commonly used special characters/symbols and their usage.

Special symbols	
.	any single character, except the newline character `'\n'`.
*	zero or multiple times of the preceding pattern.
[ ]	a character set, any character in the brackets.
^	start of a line.
$	end of a line.
{ }	two usages: (1) indicates how many times the preceding pattern is allowed to repeat; (2) reference to a macro definition.
\	used to escape metacharacters, same as that in C.

+	one or more times of the preceding pattern.
?	zero or one occurrence of the preceding pattern.
|	either the preceding or the followed pattern.
"... ..."	indicate a character string, as that in C.
/	Match the preceding pattern but only followed by the following pattern.
( )	group a series of regular expressions to a new regular expression.

Table-2.1

Examples of regular expression:

```
[0-9]+                        match an integer number.
-?[0-9]+                      match a positive or negative integer number.
[a-zA-Z]+                     match any length of alphabet string
[_a-zA-Z][_a-zA-Z0-9]*        match an identifier of a function or variable name.
```

Note: for more details about regular expressions, please refer to «**lex & yacc**», by *John R. Levine, Tony Mason & Doug Brown*

2.2.2   Script for flex

The script is input and edited by user using any type text editing tool, and it can be divided into four sections, separated by special marks or delimiters, shown in List-2.1:

```
%{

Section-1: normal C code head part (inclusions, macro definitions, …)

%}

Section-2: lexer macro definitions and user defined states for the lexer

%%

Section-3: lexer rules: the patterns, in the form of regular expression, plus the actions in C

%%

Section-4: all other C code (functions and data)
```

List-2.1

What flex does can be briefly described as follows:
- *Section-1* and *Section-4*, as well as the C codes in other sections, will be simply copied to the output, lex.yy.c, without any change by flex.
- *Section-2 and Section-3* are the key part that constructs a lexer after converted by flex.
- *Section-3* lists the sequence of rules. The sequential order of the rules matters. A change to the sequential order might alter the lexer behavior.
- The lexer generated by **flex** basically is a state machine. It starts by calling `yylex()`, and then it will keep looping until reaching a `return` statement or terminated by calling `yyterminate()`.

11

- There are some functions and variables generated by **flex** for manipulating the lexer from outside. Their names are leading with "`yy`" or "`YY_`". The most frequently used of them are:

`yyin`	The input file (handle) for the lexer.
`yytext`	The pointer that indicates the buffer to hold the string matched.
`yyleng`	String length indicated by `yytext`.
`yylineno`	The current line number of the current input file being read (`yyin`)
`yylex()`	Start running the lexer.
`yyterminate()`	Stop running the lexer.

2.2.3 Design CPP1 using flex

File `main.cpp` remains the same as in Section 2.1. In file cpp1.l, the function `cpp1()`, located int the section-3 of the script, is similar to that in the last section. The difference is that it will start running the lexer by calling `yylex()`, shown as Code-2.4:

```
128   int cpp1(const char *inc_path, char *file_in, char *file_out, int sys_file)
129   {
130       File_t *fp;
131
132       if ( file_out )
133       {
134           incPath = inc_path;
135
136           fout = fopen(file_out, "w");
137           if ( fout == NULL )
138           {
139               printf("can't open output file '%s'!\n", file_out);
140               return -1;
141           }
142       }
143
144       // elliminate duplicated file scaning...
145       for (fp = fileStack; fp; fp = fp->next)
146           if ( strcmp(fp->fname, file_in) == 0 && fp->sys_file == sys_file ) return 0;
147
148       FILE *fd = fopen(file_in, "r");
149       if ( fd == NULL )
150       {
151           printf("can't open file '%s'!\n", file_in);
152           return -1;
153       }
154
155       fp = malloc(sizeof(File_t));
156       fp->fname = malloc(strlen(file_in)+1);
157       strcpy(fp->fname, file_in);
158       fp->sys_file = sys_file;
159       fp->next    = fileStack;
160       fileStack = fp;
161
162       lastChar = 0;
163       lineMark = 1;
164
165       if ( file_out )       // root entry
166       {
167           yyin      = fd;
168           yylineno = 0;
169           yylex();
170
171           fclose(fout);
172           return error_count;
173       }
174
175       // save current status for yylex state machine
176       fp->yyin      = yyin;
177       fp->yylineno = yylineno;
178       fp->yy_state = YY_CURRENT_BUFFER;
179       yy_switch_to_buffer(yy_create_buffer(fd, YY_BUF_SIZE));
180
181       yyin      = fd;  // init. file scaning...
182       yylineno = 0;
183       return 0;
184   }
```

/p16ecc/cpp1_source_2/cpp1.l   Code-2.4

- #155~160: create an item for saving current input file information, pushed in stack then.
- #165~173: if it's called from `main()`, then it starts the lexer (#169).
- #176~179: otherwise, save the current status of the lexer into the stack for recursive running.

The section-1 of the script is shown in Code-2.5, which lists the C head part:

```
 1   %{
 2   #include <stdio.h>
 3   #include <string.h>
 4   #include <stdlib.h>
 5   #include <ctype.h>
 6   #include "cpp1.h"
 7
 8   typedef struct _File_t {
 9       char              *fname;
10       int               sys_file;
11       FILE              *yyin;        // for saving nested status...
12       int               yylineno;
13       YY_BUFFER_STATE yy_state;
14
15       struct _File_t   *next;
16   } File_t;
17
18   static FILE *fout = NULL;
19   static const char *incPath = NULL;
20   static File_t *fileStack = NULL;
21
22   static int  lineMark;
23   static char lastChar;
24   static int  define_state= 0;
25   static int  error_count = 0;
26
27   static void cpp0(char *in_file);
28   static void outputSrc(char *str);
29   static char *skipSP(char *s);
30   static File_t *popStack(void);
31
```

/p16ecc/cpp1_source_2/cpp1.l   Code-2.5

- #8~16: stack item data structure. It contains the input file information.
- #18~30: lexer local data and function claims.

In section-2 of the script, shown in Code-2.6:

```
32   %}
33
34   escape_code                [ntbrfv\\'"a?]
35   escape_character           (\\{escape_code})
36
37   c_char                     ([^'\\\n]|{escape_character})
38   c_char_sequence            ({c_char}+)
39   character_constant         (\L?\'{c_char_sequence}\')
40
41   s_char                     ([^"\\\n]|{escape_character})
42   s_char_sequence            ({s_char}+)
43   string_constant            \L?\"{s_char_sequence}?\"
44   lib_inc_file               <{s_char_sequence}>
45
46   SP                         [ \t\v\f\015]
47   newline                    [\n]
48
49
50   %x  COMMENT
51   %x  INCLUDE
52   %x  DEFINE
53   %x  DEFINE_COMMENT
54
```

/p16ecc/cpp1_source_2/cpp1.l   Code-2.6

- #34~47: macro definitions in regular expression.
- #50~53: user defined states for `yylex()`.

`yylex()` operations are described in section-3 of the script, shown in Code-2.7a and Code-2.7b, it shows the patterns to match and actions.

```
56
57   "//".*                          { /* C++ style comments, remove it. */ }
58   "/*"                            { BEGIN(COMMENT); }
59   <COMMENT>"*/"                    { BEGIN(INITIAL); lineMark = 1; }
60   <COMMENT>{newline}               { yylineno++; }
61   <COMMENT>.                       { /* C style comments, remove it. */ }
62
63   ^{SP}*"#include"{SP}+            { BEGIN(INCLUDE); }
64   <INCLUDE>{string_constant}.*{newline}   |
65   <INCLUDE>{lib_inc_file}.*{newline}      {
66                                     yylineno++;
67                                     BEGIN(INITIAL);
68                                     cpp0(yytext);
69                                   }
70
71   ^{SP}*"#define"{SP}+  |
72   ^{SP}*"#pragma"{SP}+             { BEGIN(DEFINE);  outputSrc(yytext); define_state = -1;}
73
74   <DEFINE>"//".*{newline}          { BEGIN(INITIAL); outputSrc("\n"); yylineno++; }
75   <DEFINE>{string_constant}         { outputSrc(yytext); }
76   <DEFINE>[ \t]+                   { if ( define_state ) outputSrc(yytext); }
77   <DEFINE>"/*"                      { BEGIN(DEFINE_COMMENT); }
78   <DEFINE>.                        { if ( define_state == 0 )
79                                      {
80                                        outputSrc(" ");
81                                        define_state = 1;
82                                      }
83                                      outputSrc(yytext);
84                                   }
85   <DEFINE>{SP}*"\\".*{newline}     { yylineno++; define_state = 0; }
86   <DEFINE>{newline}                { BEGIN(INITIAL); outputSrc(yytext); yylineno++;
87                                      if ( define_state >= 0 ) lineMark = 1;
88                                   }
89
90   <DEFINE_COMMENT>"*/"             { BEGIN(DEFINE); }
91   <DEFINE_COMMENT>.                ;
92   <DEFINE_COMMENT>{newline}        { yylineno++; }
93
94   {newline}                        { outputSrc(yytext);
95                                      yylineno++;
96                                   }
97   {string_constant}    |
98   .                                { outputSrc(yytext); }
```

/p16ecc/cpp1_source_2/cpp1.l   Code-2.7a

- #57: start of C++ style comment, ignore the following contents until the end of line (no output).
- #58: start of C style comment, enter state COMMENT (no output).
- #59: in COMMENT state, end of C style comments reached, reset state to INITIAL.
- #60: in COMMENT state, increase line number when line end character reached.
- #61: in COMMENT state, ignore all other characters.
- #63: "#include" statement starts, enter state INCLUDE.
- #64~69: in INCLUDE state, it gets the file name and starts the expanding by calling cpp0(). NOTE, [1] it will go back to INITIAL state, then; [2] yylex() is still active and running.
- #71~72: "#define" or "#pragma" statement starts, enter state DEFINE.
- #74~92: in DEFINE state, merge the contents of the statement into a single line, to output. Note, it will remove the comments if there is any.
- #94~98: in default state, INITIAL, output the contents in yytext, and increase the line number if and of line reached.

```
100    <<EOF>>                            { fclose(yyin);              // close input file
101
102                                         if ( lastChar != '\n' )
103                                           outputSrc("\n");          // terminate with CR
104
105                                         File_t *fp = popStack();
106
107                                         // restore status...
108                                         yy_delete_buffer(YY_CURRENT_BUFFER);
109                                         yy_switch_to_buffer(fp->yy_state);
110                                         yyin = fp->yyin;
111                                         yylineno = fp->yylineno;
112
113                                         free(fp);
114
115                                         if ( fileStack == NULL ) // end of running.
116                                         {
117                                           yyterminate();
118                                         }
119                                         else
120                                         {
121                                           lineMark = 1;
122                                           lastChar = 0;
123                                         }
124                                       }
125    %%
```
/p16ecc/cpp1_source_2/cpp1.l   Code-2.7b

- #100~124: when reaching end of file, pop the item to restore the former file's status (#105~111) from the stack. Ending the operation of yylex() when stack is empty (#117); or continue the former input file parsing (#121~122).

Function cpp0(), shown in Code-2.8, will get the file name from yytext, as well as the file source type, and then call cpp1() recursively.

```
187    static void cpp0(char *in_file)
188    {
189        char file_name[4096];
190        char c = (*in_file == '"')? '"' :
191                 (*in_file == '<')? '>' : 0;
192
193        switch ( c )
194        {
195            case '"':   // include a user file...
196                strcpy(file_name, skipSP(in_file+1));
197                *strchr(file_name, c) = '\0';
198                cpp1(NULL, file_name, NULL, 0);
199                break;
200
201            case '>':   // include a lib file...
202                sprintf(file_name, "%s%s", incPath, skipSP(in_file+1));
203                *strchr(file_name, c) = '\0';
204                cpp1(NULL, file_name, NULL, 1);
205                break;
206
207            default:
208                error_count++;
209        }
210    }
```
/p16ecc/cpp1_source_2/cpp1.l   Code-2.8

- #190~191: find out it's either a user file or system head file.

2.2.4   Test Example

Compile the project from terminal:

```
F:\p16ecc\cpp1_source_2>make
g++ -c -Os main.cpp
flex cpp1.l
gcc -c -Os lex.yy.c
g++ -static main.o lex.yy.o -o cpp1.exe
cp cpp1.exe ../bin
```

Run command from terminal:

cpp1 ch2_t3.c

ch2_t3.c:

```
#define ASSIGN(a,b)  a = 10; \
 b = 20;

char *p = "/* this is not a comment */"; /* this is a comment */
```

Ch2_t3.c_:

```
#line 1 ch2_t3.c
#define ASSIGN(a,b)  a = 10; b = 20;
#line 3 ch2_t3.c

char *p = "/* this is not a comment */";
#line 4 ch2_t3.c
```

Chapter-3

Basic Concepts and Practices of Compiler Design

The last chapter illustrates the use of **flex** to generate a lexer. Lexer works at the language keywords or the individual token level. To parse an input file at the syntax or grammar level, a parser is required. The parser takes output from a lexer and tries to match parsing rules based on language grammars. Obviously, parsing rules are targeting the language being processed and are context-oriented, unlike the lexer. This chapter will introduce a tool, **Bison**, for constructing a parser at the grammar level for the compiler. **Bison** is originally from **yacc** in UNIX, and its design script adapts the form of BNF, *Backus-Naur Form*, for users to describe the parsing rules.

Flex and Bison are usually used in pairs for parser design and work together. They both have their own design scripts, named cc.l and cc.y, as described in the following chapters. The compiler (**cc16e.exe**) design workflow is illustrated in Figure-3.1.

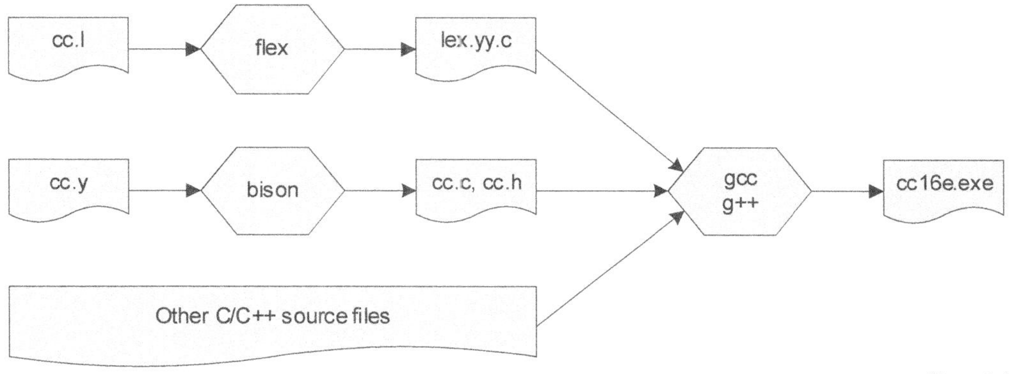

Figure-3.1

The following sections will illustrate the basic concepts or procedures of how to build a parser section by section.

3.1 Start with a Simple Parser (flex only)

```
Project directory: /p16ecc/cc_source_3.1
Source files: main.cpp, cc.l, makefile
Output: cc16e.exe
```

This section illustrates the design of a simple parser at lexer level. That means, it will only recognize the C language words (or tokens) in the input source file.

3.1.1 The start of the project – main.cpp

The **main.cpp** is similar to that in the last section, shown in Code-3.1.

```
1   #include <stdio.h>
2   #include <stdlib.h>
3   #include <string>
4   extern "C" int _main(char *filename);
5
6   int main(int argc, char *argv[])
7   {
8       for (int i = 1; i < argc; i++)
9       {
10          std::string str = "cpp1 ";
11          str += argv[i];
12
13          if ( system(str.c_str()) == 0 )
14          {
15              str = argv[i]; str += "_";
16
17              _main((char*)str.c_str());
18  //          remove(str.c_str());
19          }
20      }
21      return 0;
22  }
```

/p16ecc/cc_source_3.1/main.cpp   Code-3.1

- #4: _main() function is in C format, and located in cc.l file.
- #10~13: run the C preprocessor, **cpp1**, generating the output *file*.c_.
- #17: start parsing *file*.c_.

3.1.2 The lexer parser

The parser is generated only with **flex**, as described in **cc.l**. The first part of **cc.l** is shown in Code-3.2:

```
1   %{
2   #include <stdio.h>
3   #include <string.h>
4   #include <stdlib.h>
5   #include <ctype.h>
6
7   int  errors=0;
8   char *currentFile;
9
10  %}
11
12  letter                  [A-Za-z]
13  digit                   [0-9]
14  nonzero_digit           [1-9]
15
16  underscore              "_"
17  following_character     ({letter}|{digit}|{underscore})
18
19  identifier              ({letter}|{underscore}){following_character}*
20  dec_constant            ({digit}|({nonzero_digit}{digit}+))
21
22  SP                      [ \t\v\f\015]
23  newline                 [\n]
24
25  %%
```

/p16ecc/cc_source_3.1/cc.l   Code-3.2

- #12~23: macro definitions in the form of a regular expression
  - identifier regular expression for function and variable names (#19)
  - decimal integer regular expression (#20)

The lexer rules (match patterns and the actions) are shown in Code-3.3:

```
25  %%
26
27  ^"#line"{SP}+{dec_constant}.*{newline} {
28                              int i = 0;
29                              while ( !isdigit(yytext[i]) ) i++;
30                              yylineno = atoi(yytext+i);
31                          }
32
33  {dec_constant}          { printf("VALUE: %d\n", atoi(yytext)); }
34  "void"                  { printf("KEY_WORD: %s\n", yytext); }
35  "int"                   { printf("KEY_WORD: %s\n", yytext); }
36  {identifier}            { printf("ID: %s\n", yytext); }
37  [;{}=()]                { printf("%c\n", *yytext); }
38
39  {SP}+                   ;
40
41  {newline}               { yylineno++; }
42
43  .                       { printf("illegal character! '%c'(0x%02X) in line #%d\n",
44                                  *yytext, *yytext, yylineno);
45                          }
46
47  %%
```
/p16ecc/cc_source_3.1/cc.l   Code-3.3

- #27~31: a line mark reached (generated by **cpp1**).
- #33: an integer in decimal reached, print it out.
- #34: key word **void** reached, print it out.
- #35: key word **int** reached, print it out.
- #36: an identifier reached, print it out.
- #37: other characters that are acceptable for the parser.
- #43~45: any other characters not accepted by the parser, report error.

3.1.3 Start and run the parser

As mentioned above, _main (*file*), located in cc. , is the function to start the parser, shown in Code-3.4:

```
50  int _main(char *filename)
51  {
52      currentFile = filename;
53      yyin = fopen(filename, "r");
54
55      if ( yyin == NULL )
56      {
57          printf("can't open file - %s!", filename);
58          exit(-1);
59      }
60
61      yylineno = 1;
62      yylex();
63      return errors;
64  }
```
/p16ecc/cc_source_3.1/cc.l   Code-3.4

- #53~59: open the input file, and assign it to yyin.
- #62: start the parser.

Compile the project from the terminal:

```
F:\p16ecc\cc_source_3.1>make
g++ -c -Os main.cpp
flex cc.l
gcc -c -Os lex.yy.c
g++ -static main.o lex.yy.o -o cc16e.exe
cp cc16e.exe ../bin
```

The following, Table-3.1, is a test for the parser.

Input file: ch3_t1.c	Output file: ch3_t1.c_	Run the test from the terminal

``
void foo()
{
   int a;
   a = 100;
}
``
	``
	#line 1 ch3_t1.c
	void foo()
	{
	   int a;
	   a = 100;
	}
	``
		``
		F:\p16ecc\tests>cc16e ch3_t1.c
		KEY_WORD: void
		ID: foo
		(
		)
		{
		KEY_WORD: int
		ID: a
		;
		ID: a
		=
		VALUE: 100
		;
		}
		``

Table-3.1

3.2 A Simple C Parser Using Flex & Bison

```
Project directory: /p16ecc/cc_source_3.2
Source files: main.cpp, cc.l, cc.y, makefile
Output: cc16e.exe
```

The parser in section 3.1 can only work at the lexer level, to catch the individual words or tokens, because only **flex** is used. This section introduces an additional tool, **bison**, that works with **flex** to construct the parser at language syntax level. Similar to **flex**, **bison** will convert its script, cc.y, to a syntax parser (or parser in simple). The parser is started by calling `yyparse()` from outside, and it will start `yylex()` inherently. That means, **bison** has the dependency on **flex**, and `yyin` needs to be assigned for the input file before starting `yyparse()`.

3.2.1 Brief introduction of bison

Similar to cc.l, cc.y is also divided into four sections, as shown in List-3.1:

```
%{

Section-1: normal C code head part.

%}

Section-2: definitions: token types; names of intermediate results (or nonterminal).
```

```
%%

Section-3: parsing rules: syntax patterns plus the actions in C.

%%

Section-4: all other C code (functions and data).
```

List-3.1

- Token type – it represents the tokens from the lexer that belong to the same group. For example,

    - IDENTIFIER indicates any identifier recognized by the lexer.

    - INTEGER indicates any integer recognized by the lexer.

Note: token types are a set of integers; output them to **cc.h** as an enumeration list. And the token's actual values are usually assigned to the data union `yylval`.

- Intermediate result – it's the result generated from any parsing rule. It can be called **nonterminal** (in a parsing tree).

- Parsing rule – it can also be called **generation**. When the inputs (tokens and/or intermediate results) match the parsing pattern that complies with the syntax, it generates or converges to the result accordingly.

- Parsing rule – it's in the form of BNF notation:

```
result    :        pattern        {        action        }        ;
```

  or in the form if multiple patterns that generate the same result:

```
result    :        pattern1       {        action1       }
          |        pattern2       {        action2       }        ;
```

Among the items, the *action* part is optional. When the whole input can converge to the start of parsing (root of parsing tree) through rules, the parsing is successful, or the input complies with the syntax.

3.2.2 Script contents of cc.y

The first part of cc.y, *section-1* and *section-2*, is shown in Code-3.5:

```
 1   %{
 2   #include <stdio.h>
 3   #include <stdarg.h>
 4
 5   extern int  yyerror(char *);
 6   extern int  yylex(), yylineno;
 7   extern FILE *yyin;
 8
 9   #define YYDEBUG 1
10
11   %}
12
13   %token CHAR INT VOID CONSTANT IDENTIFIER
14
15   %start program
16
17   %%
```


- #5~9: functions and variables defined by **flex**.
- #15: names of token types.
- #17: start of parsing rule (root of parsing tree).

The rest of **cc.y** and the parsing rules are shown in Code-3.6. The list of rules that parse or match the syntax of C code, piece by piece. Note that the rules in the list cover only a subset of C syntax.

```
19   program
20        : external_definitions          { printf("parsing finished.\n"); }
21        ;
22   external_definitions
23        : function_definition
24        | external_definitions function_definition
25        ;
26   function_definition
27        : type_specifier
28          function_declarator
29          compound_statement           { printf("#%d: a function defined\n", yylineno);    }
30        ;
31   identifier
32        : IDENTIFIER
33        ;
34   primary_expr
35        : identifier
36        | CONSTANT
37        | '(' primary_expr ')'
38        ;
39   assignment_statement
40        : identifier
41          '=' primary_expr ';'         { printf("#%d: an assignment statement\n", yylineno);   }
42        ;
43   func_data_declaration
44        : type_specifier declarator_list ';'
45        ;{ printf("#%d: define variable\n", yylineno); }
46   declarator_list
47        : identifier
48        | declarator_list ',' identifier
49        ;
50   type_specifier
51        : CHAR
52        | INT
53        | VOID
54        ;
55   function_declarator
56        : identifier '(' ')'
57        ;
58   statement
59        : compound_statement
60        | assignment_statement
61        ;
62   compound_statement
63        : '{' '}'
64        | '{' statement_list '}'
65        ;
66   general_statement
67        : statement
68        | func_data_declaration
69        ;
70   statement_list
71        : general_statement
72        | statement_list general_statement
73        ;
74   %%
```

- #20~21: print the prompt when parsing is finished.
- #26~30: identify a function definition, then print the prompt.
- #39~42: identify an assign statement, then print the prompt.
- #43~45: identify a variable definition statement, then print the prompt.

3.2.3 Script contents of cc.l

cc.l is similar to that in section 3.1. At the beginning of cc.l, shown in Code-3.7, it adds a line to reference the cc.h (#4), generated by bison:

```
1    %{
2    #include <stdio.h>
3    #include <ctype.h>
4    #include "cc.h"
5    extern int yyparse();
```
/p16ecc/cc_source_3.2/cc.l

Code-3.7

As shown in Code-3.8, in cc.l, the token types (defined in cc.h) are returned instead of their values when identified from the input.

```
10   %}
11
12   letter                    [A-Za-z]
13   digit                     [0-9]
14   nonzero_digit             [1-9]
15
16   underscore                "_"
17   following_character       ({letter}|{digit}|{underscore})
18
19   identifier                ({letter}|{underscore}){following_character}*
20   dec_constant              ({digit}|({nonzero_digit}{digit}+))
21
22   SP                        [ \t\v\f\015]
23   newline                   [\n]
24
25   %%
26
27   ^"#line"{SP}+{dec_constant}.*{newline} {
28                                 int i = 0;
29                                 while ( !isdigit(yytext[i]) ) i++;
30                                 yylineno = atoi(yytext+i);
31                             }
32
33   {SP}+                       ;
34   {newline}                   { yylineno++; }
35
36   {dec_constant}              { return CONSTANT; }
37   "void"                      { return VOID; }
38   "int"                       { return INT; }
39   "char"                      { return CHAR; }
40   {identifier}                { return IDENTIFIER; }
41   [;{}=()]                    { return *yytext; }
42
43   .                           { printf("illegal character! '%c'(0x%02X) in line #%d\n",
44                                     *yytext, *yytext, yylineno);
45                             }
46
47   %%
```
/p16ecc/cc_source_3.2/cc.l

Code-3.8

- #36~40: return the token types to the syntax parse, yyparse(), when matching the inputs.
- #41: return a character to yyparse().

The function `cpp1()` in **cc.l**, shown in Code-3.9, has been modified accordingly.

```
50    int _main(char *filename)
51    {
52        currentFile = filename;
53        yyin = fopen(filename, "r");
54
55        if ( yyin == NULL )
56        {
57            printf("can't open file - %s!", filename);
58            exit(-1);
59        }
60
61        yylineno = 1;
62        if ( yyparse() != 0 )
63        {
64            printf("yyparse stopped at #line %d\n", yylineno);
65            return -1;
66        }
67        return fatalError;
68    }
```
/p16ecc/cc_source_3.2/cc.l   Code-3.9

- #53: open input file, assigned it to `yyin`.
- #62: start the parser defined in **cc.y** by calling `yyparse()`.

3.2.4 Test result

Compile the project from the terminal:

```
F:\p16ecc\cc_source_3.2>make
g++ -c -Os main.cpp
bison -d cc.y -o cc.c
flex cc.l
gcc -c -Os lex.yy.c
gcc -c -Os cc.c
g++ -static main.o lex.yy.o cc.o -o cc16e.exe
cp cc16e.exe ../bin
```

The same test file, **e1.c**, is used for the test; the result will show on the terminal as in Table-3.2:

Input file: ch3_t1.c	Run the test and terminal output

```void foo()```
 ```{```
 ``` int a;```
 ``` a = 100;```
 ```}```
	```F:\p16ecc\tests>cc16e ch3_t1.c```
	 ```#3: define variable```
	 ```#4: an assignment statement```
	 ```#5: a function defined```
	 ```parsing finished.```

Table-3.2

The parsing tree is shown below:

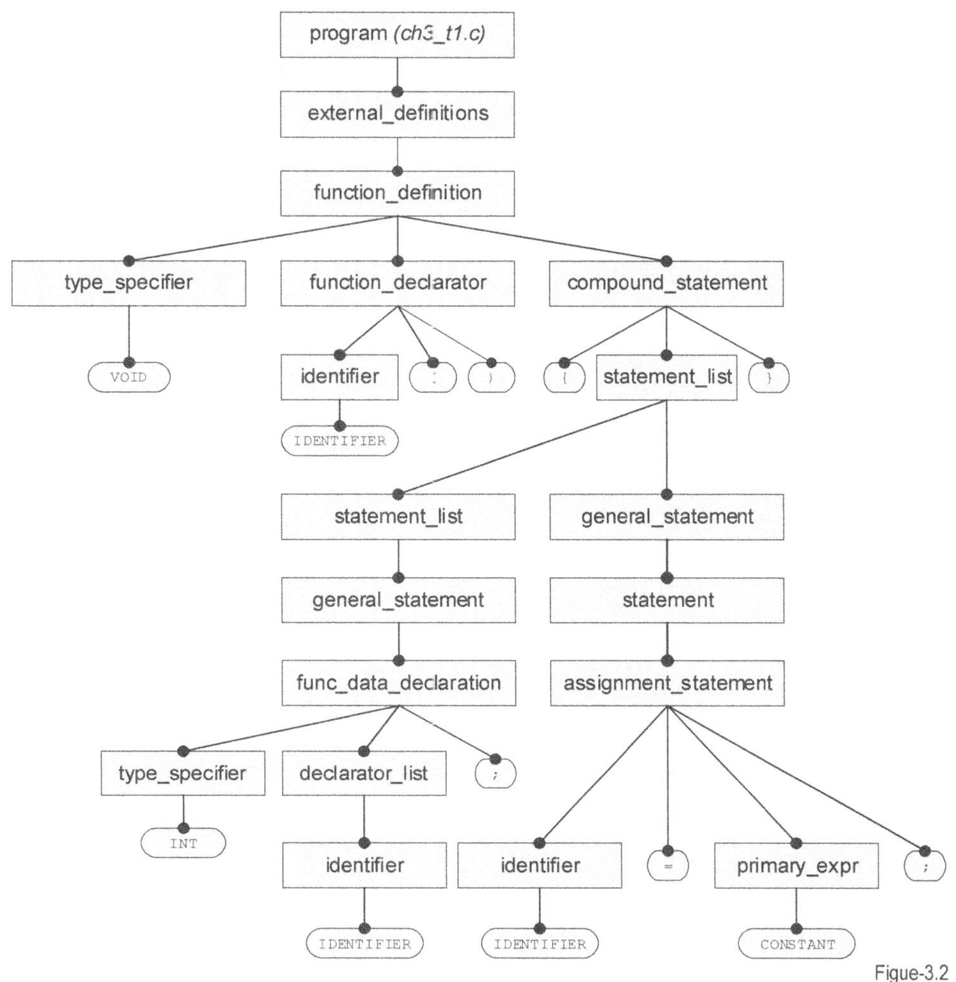

Figue-3.2

3.3 Passing Token Values

<u>Project directory</u>: /p16ecc/cc_source_3.3
<u>Source files</u>: main.cpp, cc.l, cc.y, common.h, common.c, makefile
<u>Output</u>: cc16e.exe

This section illustrates how to pass the token values and their types from the lexer to the parser.

3.3.1 Define the value carrier in cc.y

In the last section, both IDENTIFIER and CONTANT represent the type of tokens, but their real values are not present. As is known, an IDENTIFIER implies a character string, and a CONSTANT represents an integer (positive or negative). The parser needs to get the values in some way, besides their types.

As shown in Code-3.10, *Section-2* in cc.y defines a data union structure for the value carrier, yylval, that covers both types of token values. In addition, it also represents the attributes of tokens.

```
 1   %{
 2   #include <stdio.h>
 3   #include "common.h"
 4
 5   extern int  yyerror (char *);
 6   extern int  yylex (), yylineno;
 7   extern FILE *yyin;
 8
 9   #define YYDEBUG 1
10
11   %}
12
13   %union {
14       char    *s;  // string value
15       int      i;  // integer value
16   }
17
18   %token CHAR INT VOID
19   %token <i> CONSTANT
20   %token <s> IDENTIFIER
21
22   %start program
23
24   %%
```

/p16ecc/cc_source_3.3/cc.y   Code-3.10

- #13~16: define a union for `yylval`.
- #19: binding CONSTANT value attribute to the integer 'i' in `yylval`.
- #20: binding IDENTIFIER value attribute to the char pointer 's' in `yylval`.

3.3.2 Write the token value to `yylval`

```
49   {dec_constant}          { yylval.i = atoi(yytext);    return CONSTANT; }
50   "void"                  { return VOID; }
51   "int"                   { return INT; }
52   "char"                  { return CHAR; }
53   {identifier}            { yylval.s = dupStr(yytext); return IDENTIFIER; }
54   [;{}=()+-]              { return *yytext; }
```

/p16ecc/cc_source_3.3/cc.l   Code-3.11

- #49: obtain the integer value and assign it to `yylval` (the integer member), and then return the token type (CONSTANT).
- #53: obtain the identifier (make its copy) and assign it to `yylval` (the char string member), and then return the token type (IDENTIFIER). Note, the function `dupStr()` is defined in common.h/common.c.

3.3.3 Access the toekn value in `yylval`

In *Section-3* of cc.y, shown in Code-3.12, when either CONSTANT or IDENTIFIER appears, the action part can get its values from their `yylval` accordingly.

```
26    program
27        : external_definitions        { printf("parsing finished.\n"); }
28        ;
29    external_definitions
30        : function_definition
31        | external_definitions function_definition
32        ;
33    function_definition
34        : type_specifier
35          function_declarator
36          compound_statement        { printf("#%d: a function defined\n", yylineno);   }
37        ;
38    identifier
39        : IDENTIFIER              { printf("#%d: id = %s\n", yylineno, $1); }
40        ;
41    primary_expr
42        : identifier
43        | CONSTANT                { printf("#%d: value = %d\n", yylineno, $1); }
44        | '(' primary_expr ')'
45        ;
46    assignment_statement
47        : identifier
48          '=' primary_expr ';'    { printf("#%d: an assignment statement\n", yylineno);   }
49        ;
50    func_data_declaration
51        : type_specifier declarator_list ';' { printf("#%d: define variable\n", yylineno); }
52        ;
```

<table>
<tr><td>/p16ecc/cc_source_3.3/cc.y</td><td>Code-3.12</td></tr>
</table>

- #39: get the identifier value.
- #43: get the integer value.

Note, the symbol $1 (as well as $2, $3, …) is the yylval for the *n*th item in the rule; and $$ is the yylval for the result.

3.3.4 Test result

Run the project on the same testing code as before:

```
F:\p16ecc\tests>cc16e ch3_t1.c
#1: id = foo
#3: id = a
#3: define variable
#4: id = a
#4: value = 100
#4: an assignment statement
#5: a function defined
parsing finished.
```

Table-3.3

3.4 Generate Parsing Tree

Project directory: /p16ecc/cc_source_3.4
Source files: main.cpp, cc.l, cc.y, common.h, common.c, makefile
Output: cc16e.exe

The parser defined in **cc.y** has to physically build up a parsing tree (shown in Figure-3.2) when parsing is finished. The parsing tree must deliver all the information contained in the input (the source code being parsed) for further processing. Several data types need to be created and bound to the items (tokens and intermediate results) to build up the passing tree. As a result, all the intermediate results, which can be called nodes in the tree, will be bound to their own data types as well.

3.4.1 Create new data types

Several new data types are created in common.h, shown in Code-10.13a ~ Code-10.13b.

```
 8     // attribution of a node
 9     typedef struct {
10         int        type;           /* type specifier */
11     } attrib;
12
13     // node types
14     enum {
15         NODE_CON=1, NODE_STR, NODE_ID, NODE_OPR, NODE_LIST
16     };
17
18     /* constants */
19     typedef struct {
20         int          type;         /* type: NODE_CON */
21         long         value;
22     } conNode_t;
23
24     /* string */
25     typedef struct {
26         int          type;         /* type: NODE_STR */
27         char         *str;         /* string value */
28     } strNode_t;
29
30     /* identifiers */
31     typedef struct {
32         int          type;         /* type: NODE_ID */
33         attrib       *attr;
34         char         *name;
35         union node_t *parp;        /* parameter list      */
36     } idNode_t;
37
38     /* operators */
39     typedef struct {
40         int          type;         /* type: NODE_OPR */
41         attrib       *attr;
42         int          oper;         /* operator */
43         int          nops;         /* counts of operands */
44         union node_t *op[1];       /* operands (expandable) */
45     } oprNode_t;
46
47     /* list */
48     typedef struct {
49         int          type;         /* type: NODE_LIST */
50         int          nops;         /* length of list */
51         union node_t *ptr[1];      /* parameters list (expandable) */
52     } listNode_t;
```

/p16ecc/cc_source_3.4/common.h   Code-3.13a

- #9~11: object (node, constant, …) attribution.
- #14~52: five types nodes defined to apply to different objects:

  conNode_t   - for integer constants
  strNode_t   - for strings
  idNode_t   - for identifiers (function names, data variable names)
  oprNode_t   - for any kind of operations
  listNode_t   - for a sequence of nodes

```
55  typedef union node_t {
56      int          type;      /* type of node */
57      conNode_t    con;
58      idNode_t     id;
59      oprNode_t    opr;
60      listNode_t   list;
61      strNode_t    str;
62  } node;
63
64  extern node *progUnit;
65
66  void *MALLOC(int size);
67  char *dupStr(char *s);
68
69  attrib *newAttr(int atype);
70
71  // routines for 'node'
72  node *idNode  (char *s);
73  node *conNode (int v);
74  node *listNode(int length);
75  node *strNode (char *s);
76  node *oprNode (int type, int cnt, ...);
77
78  node *mergeList (node *l1, node *l2);
79  node *appendList(node *lp, node *np);
80  node *makeList (node *np);
81
82  node *cloneNode(node *np);        // clone a node
83  void delNode   (node *np);
```
/p16ecc/cc_source_3.4/common.h   Code-3.13b

- #55~62: a universal node type by binding all types of nodes.
- #64: the parsing result pointer (root node)
- #66~83: all service functions (create and delete node, …)

3.4.2 Assign data type to nodes

In Section 2 in cc.y, shown as Code 3.14, the yylval structure needs to be expanded, and all nodes (intermediate results) need to be bound to their yylval types.

```
19  %}
20
21  %union {
22      char   *s;  // string value
23      int     i;  // integer value
24      node   *n;  // a node
25      attrib *a;  // attr
26      int    *p;  // pointer
27  }
28
29  %token CHAR INT VOID FUNC_DEF FUNC_DEC FUNC_DATA
30  %token <i> CONSTANT
31  %token <s> IDENTIFIER
32
33  %start program
34
35  %type <n> program external_definitions function_definition identifier
36  %type <a> type_specifier
37  %type <n> function_declarator func_data_declaration
38  %type <n> statement general_statement compound_statement primary_expr
39  %type <n> assignment_statement statement_list declarator_list
40
41  %%
```
/p16ecc/cc_source_3.4/cc.y   Code-3.14

- #21~27: expand types in yylval.
- #35~39: bind the data types to nodes.

3.4.3 Generate parsing tree

With the expansion described above, Section 3 in cc.y now can build the parsing tree as shown in Code 3.15.

```
 43   program
 44       : external_definitions      { progUnit = $1; }
 45       ;
 46   external_definitions
 47       : function_definition       { $$ = makeList($1); }
 48       | external_definitions
 49         function_definition       { $$ = appendList($1, $2); }
 50       ;
 51   function_definition
 52       : type_specifier
 53         function_declarator
 54         compound_statement        { $$ = oprNode(FUNC_DEF, 2, $2, $3);
 55                                       $$->opr.attr = $1;
 56                                     }
 57       ;
 58   identifier
 59       : IDENTIFIER                 { $$ = idNode($1); free($1); }
 60       ;
 61   primary_expr
 62       : identifier                 { $$ = $1; }
 63       | CONSTANT                   { $$ = conNode($1); }
 64       | '(' primary_expr ')'       { $$ = $2; }
 65       ;
 66   assignment_statement
 67       : identifier
 68         '=' primary_expr ';'       { $$ = oprNode('=', 2, $1, $3);   }
 69       ;
 70   func_data_declaration
 71       : type_specifier declarator_list ';' { $$ = oprNode(FUNC_DATA, 1, $2);
 72                                       $$->opr.attr = $1;
 73                                     }
 74       ;
 75   declarator_list
 76       : identifier                 { $$ = makeList($1); }
 77       | declarator_list ',' identifier { $$ = appendList($1, $3); }
 78       ;
 79   type_specifier
 80       : CHAR                       { $$ = newAttr(CHAR); }
 81       | INT                        { $$ = newAttr(INT);  }
 82       | VOID                       { $$ = newAttr(VOID); }
 83       ;
 84   function_declarator
 85       : identifier '(' ')'         { $$ = oprNode(FUNC_DEC, 1, $1); }
 86       ;
 87   statement
 88       : compound_statement         { $$ = $1; }
 89       | assignment_statement       { $$ = $1; }
 90       ;
 91   compound_statement
 92       : '{' '}'                    { $$ = oprNode('{', 0); }
 93       | '{' statement_list '}'     { $$ = oprNode('{', 1, $2); }
 94       ;
 95   general_statement
 96       : statement                  { $$ = $1; }
 97       | func_data_declaration      { $$ = $1; }
 98       ;
 99   statement_list
100       : general_statement          { $$ = makeList($1); }
101       | statement_list general_statement { $$ = appendList($1, $2); }
102       ;
103   %%
```

`/p16ecc/cc_source_3.4/cc.y`   Code-3.15

- #44: generate the whole parsing tree. Note, `progUnit` is a global variable. Thus, the tree referenced by `progUnit` stores all the information of the input file.
- #59: string buffer needs to be released after use (to prevent memory leak).

Note, **bison** will check and confirm the data type for results (noted as '`$$`') generations.

3.5 Take a Peek of the Parsing Tree

<u>Project directory</u>: /p16ecc/cc_source_3.5
<u>Source files</u>: main.cpp, cc.l, cc.y, common.h, common.c, display.cpp, display.h, makefile
<u>Output</u>: cc16e.exe

In the last section of this chapter, it is a good time to take a look at the generated parsing, printing it to the terminal. Also, this section will add a valuable feature to the parser: inserting the input (source code to be parsed) into the result (the parsing tree).

3.5.1 Expand the data structure for nodes

The input source code to be parsed will be split during parsing and inserted into the corresponding nodes. That means the node structures need to have a string buffer for that. The following, Code-3.16, shows the changes in common.h.

```
18   typedef struct {
19       char *fileName;
20       int  lineNum;
21       char *srcCode;
22   } src_t;
23
24   /* constants */
25   typedef struct {
26       int         type;        /* type: NODE_CON */
27       src_t       *src;
28       long        value;
29   } conNode_t;
30
31   /* string */
32   typedef struct {
33       int         type;        /* type: NODE_STR */
34       src_t       *src;
35       char        *str;        /* string value */
36   } strNode_t;
37
38   /* identifiers */
39   typedef struct {
40       int         type;        /* type: NODE_ID */
41       src_t       *src;
42       attrib      *attr;
43       char        *name;
44       union node_t *parp;      /* parameter list      */
45   } idNode_t;
46
47   /* operators */
48   typedef struct {
49       int         type;        /* type: NODE_OPR */
50       src_t       *src;
51       attrib      *attr;
52       int         oper;        /* operator */
53       int         nops;        /* counts of operands */
54       union node_t *op[1];     /* operands (expandable) */
55   } oprNode_t;
56
57   /* list */
58   typedef struct {
59       int         type;        /* type: NODE_LIST */
60       src_t       *src;
61       int         nops;        /* length of list */
62       union node_t *ptr[1];    /* parameters list (expandable) */
63   } listNode_t;
```
/p16ecc/cc_source_3.5/common.h   Code-3.16

- #18~22: new data structure that holds source code piece.
- #27, #34, #41 #50 and #60: places that hold source code for nodes.

3.5.2 Collect and fetch source code

Two functions added in cc.l, shown in Code-3.17, that handle the job.

```
47    %%
48
49    ^"#line"{SP}+{dec_constant}.*{newline} {
50                              yytext[yyleng-1] = '\0';
51                              updateSourceInfo(yytext);
52                          }
53
54    {dec_constant}          { addSrcCode(); yylval.i = atoi(yytext);    return CONSTANT; }
55    "void"                  { addSrcCode(); return VOID; }
56    "int"                   { addSrcCode(); return INT; }
57    "char"                  { addSrcCode(); return CHAR; }
58    {identifier}            { addSrcCode(); yylval.s = dupStr(yytext); return IDENTIFIER; }
59    [;{}=()+-]              { addSrcCode(); return *yytext; }
60
61    {SP}+                   { addSrcCode(); }
62
63    {newline}               { yylineno++; }
64
65    .                       { printf("illegal character! ... 0x%02X '%s' in line #%d\n",
66                                     *yytext, yytext, yylineno);
67                          }
68
69    %%

. . .  . . .
94    void addSrcCode(void)
95    {
96        if ( emptySrc )
97            srcBuf[0] = '\0';
98
99        if ( (strlen(srcBuf) + yyleng) < SRC_BUF_SIZE )
100       {
101           strcat(srcBuf, yytext);
102           emptySrc = 0;
103       }
104   }
105
106   char *getSrcCode(void)
107   {
108       if ( emptySrc )
109           return NULL;
110
111       emptySrc = 1;
112       return skipSP(srcBuf);
113   }
```
/p16ecc/cc_source_3.5/cc.l   Code-3.17

- #54~61: add source code matched into the buffer (`srcBuf`).
- #94~104: function `addSrcCode()` that adds the source code to the node.
- #106~113: function `getSrcCode()` that fetches the contents from the source code buffer.

The following functions are defined in common.c, will handle the insertion.

```
194   src_t *srcNode(char *s)
195   {
196       if ( s != NULL )
197       {
198           src_t *sp = (src_t*)MALLOC(sizeof(src_t));
199           sp->fileName = dupStr(currentFile);
200           sp->lineNum  = yylineno;
201           sp->srcCode  = dupStr(skipSP(s));
202           return sp;
203       }
204       return NULL;
205   }
206
207   void addSrc(node *np, char *s)
208   {
209       np->id.src = srcNode(s);
210   }
```
/p16ecc/cc_source_3.5/common.c   Code-3.18

- #194~205: generate `src_t` data and fill in all the source information.

- #207~210: add the data to the node.

3.5.3 Insert source code in the nodes

In cc.y, shown in Code-3.19, when a parsing rule catches a valid input, it uses the functions to either insert the corresponding source code into the generated node or clear the source code buffer.

```
12   #define ADD_SRC(n)   addSrc(n, getSrcCode())
13   #define CLR_SRC()    getSrcCode()
14

... ...
41   program
42       : external_definitions        { progUnit = $1; }
43       ;
44   external_definitions
45       : function_definition          { $$ = makeList($1); CLR_SRC(); }
46       | external_definitions
47         function_definition          { $$ = appendList($1, $2); CLR_SRC(); }
48       ;
49   function_definition
50       : type_specifier
51         function_declarator
52         compound_statement           { $$ = oprNode(FUNC_DEF, 2, $2, $3);
53                                          $$->opr.attr = $1;
54                                        }
55       ;
56   identifier
57       : IDENTIFIER                   { $$ = idNode($1); free($1); }
58       ;
59   primary_expr
60       : identifier                   { $$ = $1; }
61       | CONSTANT                     { $$ = conNode($1); }
62       | '(' primary_expr ')'         { $$ = $2; }
63       ;
64   assignment_statement
65       : identifier
66         '=' primary_expr ';'         { $$ = oprNode('=', 2, $1, $3); ADD_SRC($$); }
67       ;
```
/p16ecc/cc_source_3.5/cc.y   Code-3.19

- #45, #47: end of a block of code, clear the source code buffer.
- #66: insert the source code into the statement node.

3.5.4 Display the parsing tree

The display routines are defined in display.h and display.cpp, shown in Code-3.20 & Code-3.21:

```
1    #ifndef _DISPLAY_H
2    #define _DISPLAY_H
3
4    void display(node *np, int level, FILE *file = NULL);
5    void display(char const *str, node *np, int level, FILE *file = NULL);
6    void dispSrc(node *np, int level, FILE *file = NULL);
7
8    void display(attrib *ap, int level, char endln, FILE *file = NULL);
9    void display(attrib *ap, FILE *file = NULL);
10   void display(attrib *ap, char ln, FILE *file = NULL);
11
12   #endif
```
/p16ecc/cc_source_3.5/display.h   Code-3.20

```cpp
14  void display(node *np, int level, FILE *file)
15  {
16      if ( np == NULL ) return;
17      if ( file == NULL ) file = stdout;
18
19      dispSrc(np, level, file);
20      switch ( np->type )
21      {
22          case NODE_CON:
23              headMark(level, file);
24              fprintf(file, "constant node: %ld\n", np->con.value);
25              break;
26
27          case NODE_ID:
28              headMark(level, file);
29              fprintf(file, "id node: %s ", np->id.name);
30              display(np->id.attr, '\n', file);
31              break;
32
33          case NODE_STR:
34              headMark(level, file);
35              fprintf(file, "string node: %s\n", np->str.str);
36              break;
37
38          case NODE_OPR:
39              headMark(level, file);
40              if ( np->opr.oper < 127 )
41                  fprintf(file, "opr node {%d}: %c ", np->opr.nops, np->opr.oper);
42              else
43                  fprintf(file, "opr node {%d}: %s ", np->opr.nops, oprName(np->opr.oper));
44              display(np->opr.attr, '\n', file);
45
46              for (int i = 0; i < np->opr.nops; i++)
47                  display(np->opr.op[i], level+1, file);
48              break;
49
50          case NODE_LIST:
51              headMark(level, file);
52              fprintf (file, "list node {%d}:\n", np->list.nops);
53
54              for (int i = 0; i < np->list.nops; i++)
55                  display(np->list.ptr[i], level+1, file);
56              break;
57      }
58  }
```

`/p16ecc/cc_source_3.5/display.cpp` Code-3.21

The display operation is called from main.cpp, as shown in Code-3.22, upon parsing completion.

```cpp
1   #include <stdio.h>
2   #include <stdlib.h>
3   #include <string>
4   #include "common.h"
5   #include "display.h"
6
7   int main(int argc, char *argv[])
8   {
9       for (int i = 1; i < argc; i++)
10      {
11          std::string str = "cpp1 ";
12          str += argv[i];
13
14          if ( system(str.c_str()) == 0 )
15          {
16              str = argv[i]; str += "_";
17
18              _main((char*)str.c_str());
19              remove(str.c_str());
20
21              display(progUnit, 0);
22          }
23      }
24      return 0;
25  }
```

`/p16ecc/cc_source_3.5/main.cpp` Code-3.22

- #14: run preprocessor first.
- #18: start parsing.
- #21: display the result.

<u>3.5.5 Text example</u>

```
F:\p16ecc\tests>cc16e ch3_t1.c
list node {1}:
 opr node {2}: FUNC_DEF {void}
 opr node {1}: FUNC_DEC
 id node: foo
 opr node {1}: {
 list node {2}:
 opr node {1}: FUNC_DATA {int}
 list node {1}:
 id node: a
 SOURCE - ch3_t1.c #4: a = 100;
 opr node {2}: =
 id node: a
 constant node: 100
```

PART-II: PIC16F Compiler (cc16e.exe)

Chapter-4

Accomplishment of Parser Design

The last chapter introduces some basic concepts of parser design. It only outlines the skeleton of the parser so that it only covers limited functions or capabilities for parsing the input. This chapter will extend the design to meet the capabilities that a C compiler required.

4.1 Expanding the parse fundamentals

Project directory: /p16ecc/cc_source_4.1
Source files: main.cpp, cc.l, cc.y, common.h, common.c, display.cpp, ascii.h, ascii.c, display.h, makefile
Output: cc16e.exe

4.1.1 Attribution and node data structure

This section gives the all basic data types and their structures listed in common.h, which are used through the design later on.

The attrib structure or content, shown in Code-4.1, is a modifier to describe the characteristics or attributions for functions or variables. For example:

Basic form	Variant forms	
int a;	int *a;	// **a** is a pointer
	int **a;	// **a** is a pointer's pointer
	int a[10];	// **a** is an array
	int *a[10];	// **a** is an array of pointers
	int a @ 0x200;	// **a** is located at address 0x200

```
14    // attribution of a node
15    typedef struct {
16        int     type;              /* basic type */
17        int     isExtern:   1;     /* extern indication flag */
18        int     isTypedef:  1;     /* typedef indication flag */
19        int     isStatic:   1;     /* static indication flag */
20        int     isUnsigned: 1;     /* unsigned indication flag */
21        int     isVolatile: 1;     /* volatile indication flag */
22        int     isFptr:     1;     /* function pointer indication flag */
23        int     isNeg:      1;     /* negative (constant) indication flag */
24        int     dataBank;          /* data storage type, BANK0, BANK1, ... , CONST */
25        int     *ptrVect;          /* pointer descrition, depth and types */
26        int     *dimVect;          /* array description, dimensions and sizes */
27        void    *atAddr;           /* storage address indication (a node pointer) */
28        void    *newData;          /* user defined data description (a node pointer) */
29        char    *typeName;         /* user defined data name */
30        void    *parList;          /* parameter lise for functions (a node pointer) */
31        int     bitField;          /* bit-field attributes */
32    } attrib;
```
/p16ecc/cc_source_4.1/common.h   Code-4.1

- #25: pointer description. It uses an array to describe the object's pointer characteristics. For example:

```
Data form                 ptrVect value
int a;                    NULL
int *a;                   {1, 0}
int const **a;            {2, CONST, 0}
```

- #26: dimension description. It uses an array to describe the characteristic. For example:

```
Data form                 dimVect value
int a;                    NULL
int a[10];                {1, 10}
int a[10][20];            {2, 10, 20}
```

- #31: bit-field description. Lower 8-bit: field lendth, higher 8-bit: start position. For example:

```
Data form                 bitFiled value
unsigned char a:1;        0x0001
unsigned char b:2;        0x0102
unsinged char c:3;        0x0303
```

For node data types, the following code (Code-4.2) expands their structures over those in the last chapter:

```
39    /* constants */
40    typedef struct {
41        int           type;        /* type: NODE_CON */
42        src_t         *src;
43        attrib        *attr;
44        long          value;
45    } conNode_t;
46
47    /* string */
48    typedef struct {
49        int           type;        /* type: NODE_STR */
50        src_t         *src;
51        attrib        *attr;
52        char          *str;        /* string value */
53    } strNode_t;
54
55    /* identifiers */
56    typedef struct {
57        int           type;        /* type: NODE_ID */
58        src_t         *src;
59        attrib        *attr;
60        union node_t  *dim;
61        union node_t  *init;
62        char          *name;
63        char          fp_decl:1; /* function pointer */
64        union node_t  *parp;       /* parameter list      */
65    } idNode_t;
66
67    /* operators */
68    typedef struct {
69        int           type;        /* type: NODE_OPR */
70        src_t         *src;
71        attrib        *attr;
72        int           oper;        /* operator */
73        int           nops;        /* counts of operands */
74        union node_t  *op[1];      /* operands (expandable) */
75    } oprNode_t;
76
77    /* list */
78    typedef struct {
79        int           type;        /* type: NODE_LIST */
80        src_t         *src;
81        int           elipsis;     /* elipsis option */
82        int           nops;        /* length of list */
83        union node_t  *ptr[1];     /* parameters list (expandable) */
84    } listNode_t;
```

/p16ecc/cc_source_4.1/common.h

Code-4.2

- #56~65: extra modifiers for `idNode_t`:

  ```
  dim            - dimension specifier (temporary);
  init           - variable's initialization statements.
  fp_decl        - flag if it's function pointer.
  ```

- #73, #82: in `oprNode_t` and `listNode_t` structures, `nops` denotes the counts of nodes appended (pointed by `*op[1]` and `*ptr[1]`). And the appended node count is variable when generated and must be allocated at the end of the structures.

4.1.2 Expanding of lexer – cc.l

The following, Code-4.3a~Code-4.3d, provides the script that covers C language reserved words (keywords) and operators.

```
26   letter                  [A-Za-z]
27   digit                   [0-9]
28   underscore              "_"
29   following_character     ({letter}|{digit}|{underscore})
30   identifier              ({letter}|{underscore}){following_character}*
31
32   nonzero_digit           [1-9]
33   oct_digit               [0-7]
34   hex_digit               [0-9a-fA-F]
35   integer_suffix          [uU]|[1L]|([uU][1L])|([1L][uU])
36   dec_constant            ({digit}|({nonzero_digit}{digit}+))
37   oct_constant            (0{oct_digit}+)
38   hex_constant            (0[xX]{hex_digit}+)
39   bin_constant            (0[bB][01]+)
40
41   char_escape_code        [ntbrfv\\'"a?]
42   oct_escape_code         ({oct_digit}{1,3})
43   hex_escape_code         (x{hex_digit}+)
44   escape_code             ({char_escape_code}|{oct_escape_code}|{hex_escape_code})
45   escape_character        (\\{escape_code})
46
47   c_char                  ([^'\\\n]|{escape_character})
48   c_char_sequence         ({c_char}+)
49   character_constant      (\L?\'{c_char_sequence}\')
50
51   s_char                  ([^"\\\n]|{escape_character})
52   s_char_sequence         ({s_char}+)
53   string_constant         \L?\"{s_char_sequence}?\"
54
55   SP                      [ \t\v\f\015]
56   newline                 [\n]
```
/p16ecc/cc_source_4.1/cc.l   Code-4.3a

```
64   {dec_constant}{integer_suffix}? {
65                                   addSrcCode(); yylval.i = atoi(yytext);
66                                   return constantType(yytext);
67                                   }
68   {hex_constant}{integer_suffix}? {
69                                   addSrcCode(); yylval.i = aConstant(yytext, C_HEX);
70                                   return constantType(yytext);
71                                   }
72   {oct_constant}{integer_suffix}? {
73                                   addSrcCode(); yylval.i = aConstant(yytext, C_OCT);
74                                   return constantType(yytext);
75                                   }
76   {bin_constant}{integer_suffix}? {
77                                   addSrcCode(); yylval.i = aConstant(yytext, C_BIN);
78                                   return constantType(yytext);
79                                   }
80   {character_constant}     { addSrcCode(); yylval.i = aConstant(yytext, C_CHAR);
81                              return C_CONSTANT;
82                              }
83   {string_constant}        { addSrcCode();
84                              yytext[yyleng - 1] = 0;
85                              yylval.s = parseRawStr(yytext+1);
86                              return STRING;
87                              }
```
/p16ecc/cc_source_4.1/cc.l   Code-4.3b

- #64~82: rule to match an integer constant. It covers all kind of radix integers (char, binary, octal, decimal and hex-decimal).

- #83~87: rule to match a string constant. It needs to be reformed its storage format. For example, string "hello, world.\n",

    `yytext` content   {0x22, 0x68, 0x65, 0x6C, 0x6C, 0x6F, 0x2C, 0x20, 0x77, 0x6F, 0x72, 0x6C, 0x64, 0x2E, **0x5C**, **0x6E**, 0x22, 0x00}

    `yylval` content   {0x68, 0x65, 0x6C, 0x6C, 0x6F, 0x2C, 0x20, 0x77, 0x6F, 0x72, 0x6C, 0x64, 0x2E, **0x0A**, 0x00}

```
 88   "unsigned"{SP}+"char"      { addSrcCode(); return UCHAR; }
 89   "unsigned"{SP}+"int"       { addSrcCode(); return UINT; }
 90   "unsigned"{SP}+"short"     { addSrcCode(); return USHORT; }
 91   "unsigned"{SP}+"long"      { addSrcCode(); return ULONG; }
 92   ("signed"{SP}+)?"char"     { addSrcCode(); return CHAR; }
 93   ("signed"{SP}+)?"short"    { addSrcCode(); return SHORT; }
 94   ("signed"{SP}+)?"int"      { addSrcCode(); return INT; }
 95   ("signed"{SP}+)?"long"     { addSrcCode(); return LONG; }
 96   "void"                     { addSrcCode(); return VOID; }
 97   "break"                    { addSrcCode(); return BREAK; }
 98   "case"                     { addSrcCode(); return CASE; }
 99   "const"                    { addSrcCode(); return CONST; }
100   "continue"                 { addSrcCode(); return CONTINUE; }
101   "default"                  { addSrcCode(); return DEFAULT; }
102   "do"                       { addSrcCode(); return DO; }
103   "else"                     { addSrcCode(); return ELSE; }
104   "enum"                     { addSrcCode(); return ENUM; }
105   "extern"                   { addSrcCode(); return EXTERN; }
106   "for"                      { addSrcCode(); return FOR; }
107   "goto"                     { addSrcCode(); return GOTO; }
108   "if"                       { addSrcCode(); return IF; }
109   "register"                 { addSrcCode(); return REGISTER; }
110   "return"                   { addSrcCode(); return RETURN; }
111   "sizeof"                   { addSrcCode(); return SIZEOF; }
112   "static"                   { addSrcCode(); return STATIC; }
113   "switch"                   { addSrcCode(); return SWITCH; }
114   "volatile"                 { addSrcCode(); return VOLATILE; }
115   "while"                    { addSrcCode(); return WHILE; }
116   "interrupt"                { addSrcCode(); return INTERRUPT; }
117   "union"                    { addSrcCode(); return UNION; }
118   "struct"                   { addSrcCode(); return STRUCT; }
119   "typedef"                  { addSrcCode(); return TYPEDEF; }
120   "..."                      { addSrcCode(); return ELIPSIS; }
121   "_linear_"                 { addSrcCode(); yylval.i = LINEAR;  return MEM_BANK; }
122
123   {identifier}               { addSrcCode(); yylval.s = dupStr(yytext); return IDENTIFIER; }
```
/p16ecc/cc_source_4.1/cc.l   Code-4.3c

- #88~121: rules to match C language reserved words.
- #123: The rule for matching an identifier must be located after the reserved words recognition.

```
125   "+="      { addSrcCode(); yylval.i = ADD_ASSIGN; return ASSIGN_OP; }
126   "-="      { addSrcCode(); yylval.i = SUB_ASSIGN; return ASSIGN_OP; }
127   "*="      { addSrcCode(); yylval.i = MUL_ASSIGN; return ASSIGN_OP; }
128   "/="      { addSrcCode(); yylval.i = DIV_ASSIGN; return ASSIGN_OP; }
129   "%="      { addSrcCode(); yylval.i = MOD_ASSIGN; return ASSIGN_OP; }
130   "<<="     { addSrcCode(); yylval.i = LEFT_ASSIGN; return ASSIGN_OP; }
131   ">>="     { addSrcCode(); yylval.i = RIGHT_ASSIGN; return ASSIGN_OP; }
132   "&="      { addSrcCode(); yylval.i = AND_ASSIGN; return ASSIGN_OP; }
133   "^="      { addSrcCode(); yylval.i = XOR_ASSIGN; return ASSIGN_OP; }
134   "|="      { addSrcCode(); yylval.i = OR_ASSIGN;  return ASSIGN_OP; }
135   "=="      { addSrcCode(); yylval.i = EQ_OP; return EQUALITY_OP; }
136   "!="      { addSrcCode(); yylval.i = NE_OP; return EQUALITY_OP; }
137   "<="      { addSrcCode(); yylval.i = LE_OP; return RELATIONAL_OP; }
138   ">="      { addSrcCode(); yylval.i = GE_OP; return RELATIONAL_OP; }
139   [<>]      { addSrcCode(); yylval.i = *yytext; return RELATIONAL_OP; }
140   "<<"      { addSrcCode(); yylval.i = LEFT_OP; return SHIFT_OP; }
141   ">>"      { addSrcCode(); yylval.i = RIGHT_OP; return SHIFT_OP; }
142   "++"      { addSrcCode(); return INC_OP; }
143   "--"      { addSrcCode(); return DEC_OP; }
144   "||"      { addSrcCode(); return OR_OP; }
145   "&&"      { addSrcCode(); return AND_OP; }
146   "->"      { addSrcCode(); return PTR_OP; }
```
/p16ecc/cc_source_4.1/cc.l   Code-4.3d

In cc.y, shown in Code-4.4, the parser script's definition section defines all token types and their `yylval` values.

```
18    %}
19
20    %expect 1
21
22    %union {
23        char    *s;   // string value
24        int      i;   // integer value
25        node    *n;
26        attrib *a;
27        int     *p;
28    }
29
30    %token CHAR INT SHORT LONG UCHAR UINT USHORT ULONG VOID
31    %token FUNC_HEAD FUNC_DECL DATA_DECL
32    %token _IF _ELSE _ENDIF _IFDEF _IFNDEF DEFINE UNDEF PRAGMA
33    %token BREAK CASE CONST CONTINUE DEFAULT DO ELSE ENUM EXTERN
34    %token FOR GOTO IF REGISTER RETURN SIZEOF STATIC SWITCH
35    %token VOLATILE WHILE INTERRUPT UNION STRUCT TYPEDEF ELIPSIS
36    %token LINEAR
37    %token ADD_ASSIGN SUB_ASSIGN MUL_ASSIGN DIV_ASSIGN MOD_ASSIGN
38    %token LEFT_ASSIGN RIGHT_ASSIGN AND_ASSIGN XOR_ASSIGN OR_ASSIGN
39    %token EQ_OP NE_OP LE_OP GE_OP LEFT_OP RIGHT_OP INC_OP DEC_OP
40    %token OR_OP AND_OP PTR_OP FPTR CAST CALL
41    %token POST_INC POST_DEC PRE_INC PRE_DEC POS_OF ADDR_OF NEG_OF
42    %token EOL LABEL
43
44    %token <i> CONSTANT L_CONSTANT U_CONSTANT UL_CONSTANT C_CONSTANT MEM_BANK
45    %token <i> RELATIONAL_OP EQUALITY_OP SHIFT_OP ASSIGN_OP
46    %token <s> IDENTIFIER IDENTIFIER_ STRING AASM
47    %token <a> TYPEDEF_NAME
```
`/p16ecc/cc_source_4.1/cc.y`   Code-4.4

- #20: it tells the following parsing rules containing one ambiguity (it's caused by "if … else …" grammar).
- #30~47: token types (some of them have their own values denoted by `yylval`).

4.1.2 Parsing rules for function declaration/definition

Code-4.5a and Code-4.5b present the parsing rules for function declarations and definitions.

```
412   func_definition
413       : func_declarator
414         func_body                   { $$ = oprNode(FUNC_DECL, 2, $1, $2);
415                                         $$->opr.attr = newAttr(INT);
416                                       }
417       | declaration_specifiers
418         func_declarator2            { if ( $1->type == 0 )
419                                         {
420                                            if ( $2->id.attr && $2->id.attr->ptrVect )
421                                               yyerror("function pointer type not defined!");
422                                            else
423                                               $1->type = INT;
424                                         }
425                                       }
426         func_body                   { $$ = oprNode(FUNC_DECL, 2, $2, $4);
427                                         $$->opr.attr = $1;
428
429                                         if ( $2->id.attr )
430                                         {
431                                            $$->opr.attr->ptrVect = $2->id.attr->ptrVect;
432                                            $2->id.attr->ptrVect = NULL;
433                                            delAttr($2->id.attr);
434                                            $2->id.attr = NULL;
435                                         }
436                                       }
437       | INTERRUPT
438         identifier '(' ')'
439         compound_statement          { node *np = oprNode(FUNC_HEAD, 1, $2);
440                                         $$ = oprNode(FUNC_DECL, 2, np, $5);
441                                         $$->opr.attr = newAttr(INTERRUPT);
442                                       }
443       ;
```
`/p16ecc/cc_source_4.1/cc.y`   Code-4.5a

- #413~416: rule to match function declaration/definition that has no return value type. In this case, the return type is inherited as `int`.
- #417~436: rule to match function declaration/definition that specifies the return value type explicitly.
- #437~442: interrupt service function.

```
444   func_declarator
445       : identifier '('
446         parameter_type_list ')'    { int n = parListCheck($3);
447                                       if ( n > 0 )
448                                       {
449                                          $$ = oprNode(FUNC_HEAD, 2, $1, $3);
450                                          ADD_SRC($$);
451                                       }
452                                       else
453                                       {
454                                          delNode($3);
455                                          if ( n == 0 )
456                                          {
457                                             $$ = oprNode(FUNC_HEAD, 1, $1);
458                                             ADD_SRC($$);
459                                          }
460                                          else
461                                          {
462                                             yyerror("parameter error!");
463                                             CLR_SRC();
464                                          }
465                                       }
466                                      }
467       ;
468   func_declarator2    /* may include 'pointer' */
469       : func_declarator          { $$ = $1; }
470       | pointer func_declarator  { $$ = $2;
471                                    $2->id.attr = newAttr(0);
472                                    $2->id.attr->ptrVect = $1;
473                                    ADD_SRC($$);
474                                   }
475       ;
476   func_body
477       : ';'                       { $$ = NULL; }
478       | compound_statement        { $$ = $1;   }
479       ;
/p16ecc/cc_source_4.1/cc.y
```

Code-4.5b

- #476~479: `function_body` tells the difference between declaration and definition (`NULL` vs. `compound_statement` node).

4.1.3 Implementation of operation precedence

The parsing rules need to cover the operation precedence for the operation. For example, the input code

```
x = a + b * c;
```

has two arithmetic operations ('+' and '*'). According to the arithmetic rules, the multiplication '*' must be done prior to addition '+'. In C language, things are more complicated than the example above. For instance,

```
if (a == 0 || b == 1) x = 1;   →    if ((a == 0) || (b == 1)) x = 1;
```

The parsing rule part in **cc.y** has a number of rules to generate nonterminals for different precedence levels, which is listed in Table-4.1 from high to low. As the rule, operations with lower precedence always yields to the operation with higher precedence.

	nonterminal	operations/examples
Highest precedence	`primary_exp`	*number, string* or *identifier*
	`postfix_expr`	`x++, x--, x.y, x->y`
	`unary_expr`	`++x, --x, *x, &x`
	`cast_expr`	`(char)x`
	`multiplicative_expr`	`x * y, x / y, x % y`
	`additive_expr`	`x + y, x - y`
	`shift_expr`	`x >> N, x << N`
	`relational_expr`	`x > y, x >= y, x < y, x <= y`
	`equality_expr`	`x == y, x != y`
	`and_expr`	`x & y`
	`exclusive_or_expr`	`x ^ y`
	`inclusive_or_expr`	`x | y`
	`logical_and_expr`	`x && y`
	`logical_or_expr`	`x || y`
Lowest precedence	`conditional_expr`	`(a == 1)? x : y`

Table-4.1

Here, shown in Code-4.6, is the parsing rules to parse the operation

```
a + b * c
```

```
283   additive_expr
284       : multiplicative_expr              { $$ = $1; }
285       | additive_expr '+'
286         multiplicative_expr              { $$ = oprNode('+', 2, $1, $3); }
287       | additive_expr '-'
288         multiplicative_expr              { $$ = oprNode('-', 2, $1, $3); }
289       ;
290   multiplicative_expr
291       : cast_expr                        { $$ = $1; }
292       | multiplicative_expr '*'
293         cast_expr                        { $$ = oprNode('*', 2, $1, $3); }
294       | multiplicative_expr '/'
295         cast_expr                        { $$ = oprNode('/', 2, $1, $3); }
296       | multiplicative_expr '%'
297         cast_expr                        { $$ = oprNode('%', 2, $1, $3); }
298       ;
```
<table><tr><td>/p16ecc/cc_source_4.1/cc.y</td><td>Code-4.6</td></tr></table>

It can be simply illustrated as

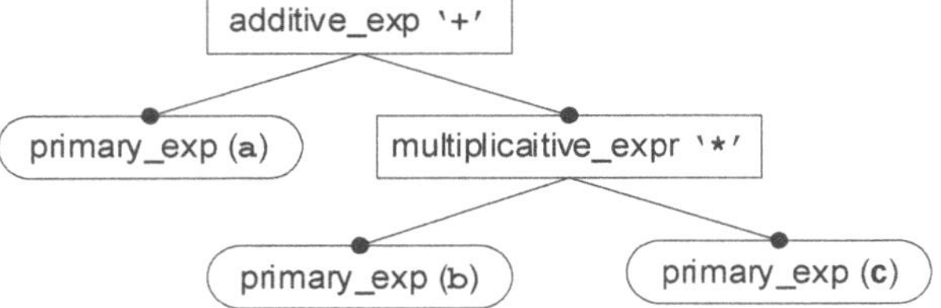

Figure-4.1

And its complete parsing procedure is

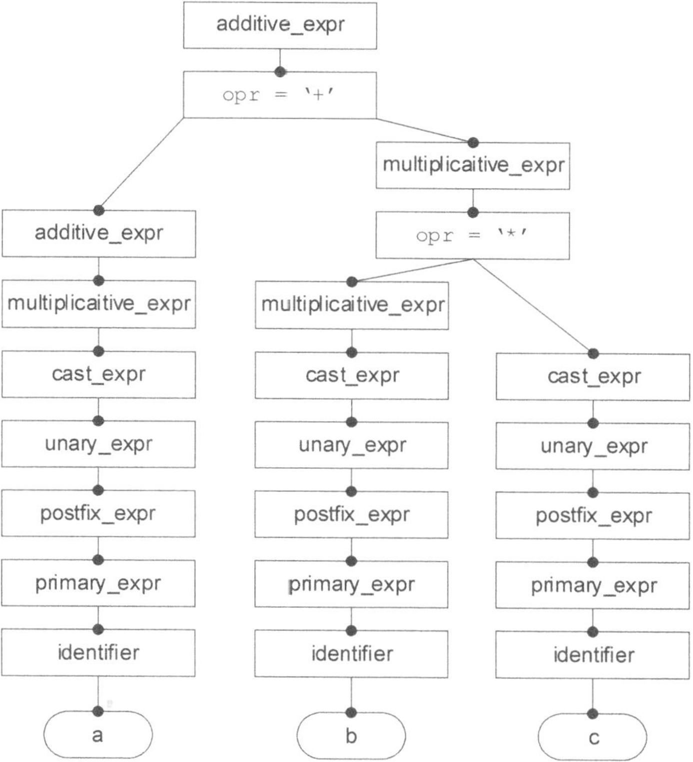

Figure-4.2

4.1.4 Accomplishment of variable declaration/definition

Variable declaration/definition is more complicated than most people expected, since there are a number of modifiers that could appear. The overall rules are shown in Code-4.7.

```
87   declaration
88     : declaration_specifiers ';' {if ( $1->type == ENUM ) {
89                                     $$ = $1->newData;
90                                     $1->newData = NULL;
91                                     delAttr($1);
92                                     ADD_SRC($$);
93                                   }
94                                   else
95                                   {
96                                     $$ = oprNode(DATA_DECL, 0);
97                                     $$->opr.attr = $1;
98                                     ADD_SRC($$);
99                                   }
100                                  }
101    | declaration_specifiers
102      init_declarator_list ';'  { $$ = oprNode(DATA_DECL, 1, $2);
103                                   $$->opr.attr = $1;
104                                   ADD_SRC($$);
105                                  }
106    ;
```
/p16ecc/cc_source_4.1/cc.y

Code-4.7

- #88~100: variable declaration/definition without variable instance list.
- #101~105: variable declaration/definition with variable instance list.

`declaration_specifiers` not only include the basic variable type (including user-defined data types), but also all specifiers (location, active region, …), as shown in Code-4.8.

```
107    declaration_specifiers
108        : storage_class_specifier      { $$ = newAttr(0);
109                                          if ( assignAcce($$, $1) )
110                                              yyerror("storage type redefined!");
111                                        }
112        | storage_class_specifier
113          declaration_specifiers        { $$ = $2;
114                                          if ( assignAcce($$, $1) )
115                                              yyerror("storage type redefined!");
116                                        }
117        | type_specifier
118          declaration_specifiers        { $$ = $1; mergeAttr($$, $2); delAttr($2); }
119        | type_specifier                { $$ = $1; }
120        | type_qualifier
121          declaration_specifiers        { $$ = $2; assignAcce($$, $1); }
122        ;
123    storage_class_specifier
124        : REGISTER                      { $$ = REGISTER; }
125        | STATIC                        { $$ = STATIC; /*}
126        | EXTERN                        { $$ = EXTERN; }
127        | TYPEDEF                       { $$ = TYPEDEF; */}
128        | MEM_BANK                      { $$ = $1; }
129        ;
130    type_qualifier
131        : CONST                         { $$ = CONST; }
132        | VOLATILE                      { $$ = VOLATILE; }
133        ;
134    type_specifier
135        : VOID                          { $$ = newAttr(VOID); }
136        | CHAR                          { $$ = newAttr(CHAR); }
137        | SHORT                         { $$ = newAttr(SHORT); }
138        | INT                           { $$ = newAttr(INT); }
139        | LONG                          { $$ = newAttr(LONG); }
140        | UCHAR                         { $$ = newAttr(CHAR);   $$->isUnsigned = 1; }
141        | USHORT                        { $$ = newAttr(SHORT);  $$->isUnsigned = 1; }
142        | UINT                          { $$ = newAttr(INT);    $$->isUnsigned = 1; }
143        | ULONG                         { $$ = newAttr(LONG);   $$->isUnsigned = 1; }
144        ;
```
`/p16ecc/cc_source_4.1/cc.y`   Code-4.8

- #134~143: basic data types.
- #123~133: all other specifiers.

`init_declarator_list` does not contain only variable names; it may also contain variable initialization statements.

```
153    init_declarator_list
154        : init_declarator               { $$ = makeList($1); }
155        | init_declarator_list ','
156          init_declarator               { $$ = appendList($1, $3); }
157        ;
158    init_declarator
159        : declarator                    { $$ = $1; }
160        | declarator '='
161          initializer                   { $$ = $1; $$->id.init = $3; }
162        ;
```
`/p16ecc/cc_source_4.1/cc.y`   Code-4.9

- #160~161: variable with initialization.

As for `declarator`, it can be a basic identifier, an array, or a function pointer, shown in Code-4.10.

```
163    declarator
164        : direct_declarator            { $$ = $1; }
165        | pointer direct_declarator    { $$ = $2;
166                                         if ( $2->id.attr == NULL )
167                                             $2->id.attr = newAttr(0);
168                                         $2->id.attr->ptrVect = $1;
169                                       }
170        ;
171    direct_declarator
172        : identifier                   { $$ = $1; }
173        | direct_declarator '[' ']'    { $$ = $1;
174                                         $$->id.dim = appendList($$->id.dim, conNode(0, INT));
175                                       }
176        | direct_declarator
177          '[' constant_expr ']'        { $$ = $1;
178                                         $$->id.dim = appendList($$->id.dim, $3);
179                                       }
180        | '(' '*' identifier ')' '('
181          parameter_type_list ')'      { int n = parListCheck($6);
182                                         $$ = $3;
183                                         $$->id.fp_decl = 1;
184                                         if ( n > 0 )
185                                             $$->id.parp = $6;
186                                         else
187                                         {
188                                             delNode($6);
189                                             if ( n < 0 )
190                                                 yyerror("parameter list error!");
191                                         }
192                                       }
193        | '(' '*' identifier
194          '[' constant_expr ']' ')' '('
195          parameter_type_list ')'      { int n = parListCheck($9);
196                                         $$ = $3;
197                                         $$->id.dim = makeList($5);
198                                         $$->id.fp_decl = 1;
199                                         if ( n > 0 )
200                                             $$->id.parp = $9;
201                                         else
202                                         {
203                                             delNode($9);
204                                             if ( n < 0 )
205                                                 yyerror("parameter list error!");
206                                         }
207                                       }
208        |  '(' '*' ')' '('
209          parameter_type_list ')'      { int n = parListCheck($5);
210                                         $$ = idNode("");
211                                         $$->id.fp_decl = 1;
212                                         if ( n > 0 )
213                                             $$->id.parp = $5;
214                                         else
215                                         {
216                                             delNode($5);
217                                             if ( n < 0 )
218                                                 yyerror("parameter list error!");
219                                         }
220                                       }
221        ;
```
/p16ecc/cc_source_4.1/cc.y   Code-4.10

- #165~169: variable with pointer specifiers.
- #173~179: variable is an array (single/multiple dimension).
- #180~220: variable is a function pointer.

For the test code:

```
int x, y[10];
```

Its parse tree is shown in Figure-4.3.

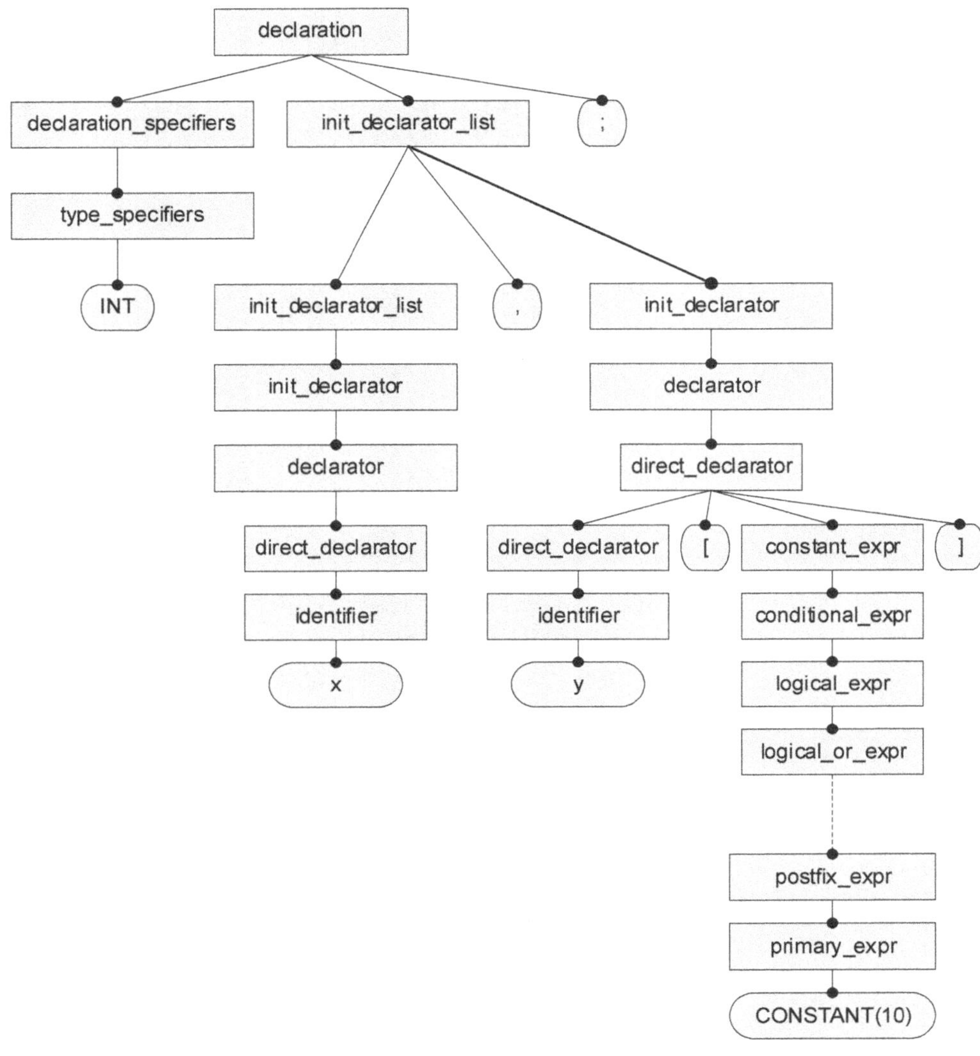

Figure-4.3

And the running result is:

```
F:\p16ecc\tests>cc16e ch4_t1.c
list node {1}:
 ch4_t1.c #1: int x, y[10];
 opr node {1}: DATA_DECL {int}
 list node {2}:
  id node: x
  id node: y
  list node {1}:
   constant node: 10 {int}
```

4.2 Parsing preprocessor statements

Project directory: /p16ecc/cc_source_4.2
Source files: main.cpp, cc.l, cc.y, common.h, common.c, display.cpp, ascii.h, ascii.c, display.h, makefile
Output: cc16e.exe

As described in Chapter 2, cpp1 only handles #include statements from input. Other preprocessor statements like #if, #else, #ifdef, …, still remain in the input. The following table shows the preprocessor statements supported in **cc16e**, in addition to #include.

Preprocessor statements supported	
#define	macro definition
#undef	remove macro definition
#if … #else … #endif	conditional compilation control
#ifdef, #ifndef	conditional compilation control
#asm … #endasm	assembly code insertion
#pragma	additional compilation control

Table-4.2

4.2.1 Expanding lexer cc.l

In the lexer cc.l, the recognition rules for those preprocessor directives are added, along with the new processing states.

```
10    extern int yyparse();
11    int errors=0;
12    char *currentFile = NULL;
13    int   preProcFlag = 0;
14    int   preProcType;

   ...

60    %x ASMCODE
61    %x PREPROC
62    %%
```
/p16ecc/cc_source_4.2/cc.l

Code-4.11

- #13~14: new processing variables.
- #60~61: new processing states.

```
68   ^{SP}*"#"{SP}*"define"   { preProcFlag = 1; BEGIN(PREPROC);
69                              addSrcCode(); return preProcType = DEFINE; }
70   ^{SP}*"#"{SP}*"undef"    { preProcFlag = 1; BEGIN(PREPROC);
71                              addSrcCode(); return preProcType = UNDEF; }
72   ^{SP}*"#"{SP}*"pragma"   { preProcFlag = 1; BEGIN(PREPROC);
73                              addSrcCode(); return preProcType = PRAGMA; }
74   ^{SP}*"#"{SP}*"ifdef"    { preProcFlag = 1; BEGIN(PREPROC);
75                              addSrcCode(); return preProcType = _IFDEF; }
76   ^{SP}*"#"{SP}*"ifndef"   { preProcFlag = 1; BEGIN(PREPROC);
77                              addSrcCode(); return preProcType = _IFNDEF; }
78   ^{SP}*"#"{SP}*"else".*   { addSrcCode(); return _ELSE; }
79   ^{SP}*"#"{SP}*"endif".*  { addSrcCode(); return _ENDIF; }
80   ^{SP}*"#"{SP}*"if"{SP}+  { preProcFlag = 1;
81                              addSrcCode(); return preProcType = _IF; }
82   <PREPROC>{SP}+           { addSrcCode(); }
83   <PREPROC>{identifier}"(" { yylval.s = dupStr(yytext);
84                              yylval.s[yyleng - 1] = '\0';
85                              addSrcCode();
86                              BEGIN(INITIAL); return IDENTIFIER_;
87                            }
88   <PREPROC>{identifier}    { yylval.s = dupStr(yytext);
89                              addSrcCode();
90                              BEGIN(INITIAL); return IDENTIFIER;
91                            }
92   ^{SP}*"#asm".*           { preProcFlag = 1; preProcType = AASM;
93                              BEGIN(ASMCODE);
94                            }
95   <ASMCODE>{newline}       { yylineno++; }
96   <ASMCODE>^{SP}*"#endasm".*  { emptySrc = 1; preProcType = 0;
97                              BEGIN(INITIAL); }
98   <ASMCODE>.*              { yylval.s = dupStr(yytext); return AASM; }
```

/p16ecc/cc_source_4.2/cc.l   Code-4.12

- #68, #70, #72, #74, #76: for #define, #undefine, #pragma, #ifdef, #ifndef preprocessor starts, it will go to PREPROC state, expecting an identifier to be followed.
- #92~98: in AASM state, output assembly code line by line.

Note: preProcFlag, when set, indicates that it's in preprocessor handling status; while proProcType indicates the type of preprocessor statement being handled.

Preprocessor statements (status) are ended by '\n', as shown in Code-4.13.

```
192  {newline}                   { yylineno++;
193                                if ( preProcFlag )
194                                {
195                                    preProcFlag = 0;
196                                    switch ( preProcType )
197                                    {
198                                        case DEFINE:
199                                        case UNDEF:
200                                        case PRAGMA:
201                                        case _IF:
202                                            return EOL;
203                                    }
204                                }
205                              }
```

/p16ecc/cc_source_4.2/cc.l   Code-4.13

- #198~202: return token type EOL for some preprocessor statements.

4.2.2 Expanding parser cc.y

On the parser side, the expansion is shown in Code-4.14.

```
85   external_definition
86       : declaration               { $$ = $1; CLR_SRC(); }
87       | func_definition           { $$ = $1; CLR_SRC(); }
88       | declaration_specifiers
89         init_declarator_list
90         '@' conditional_expr ';'  { $$ = oprNode(DATA_DECL, 1, $2);
91                                     $$->opr.attr = $1;
92                                     $1->atAddr = $4;
93                                     ADD_SRC($$);
94                                   }
95       | DEFINE identifier EOL     { $$ = oprNode(DEFINE, 1, $2);
96                                     ADD_SRC($$); }
97       | DEFINE identifier
98         expr EOL                  { $$ = oprNode(DEFINE, 2, $2, $3);
99                                     ADD_SRC($$); }
100      | DEFINE IDENTIFIER_
101        id_list ')' expr EOL      { $$ = oprNode(DEFINE, 3, idNode($2), $3, $5);
102                                    ADD_SRC($$); free($2);
103                                  }
104      | DEFINE IDENTIFIER_
105        id_list ')' EOL          { $$ = oprNode(DEFINE, 2, idNode($2), $3);
106                                   ADD_SRC($$); free($2);
107                                 }
108      | UNDEF identifier EOL     { $$ = oprNode(UNDEF, 1, $2);
109                                   ADD_SRC($$); }
110      | _IFDEF identifier
111        opt_external_definitions
112        op_else
113        _ENDIF                   { $$ = oprNode(_IFDEF, 3, $2, $3, $4);
114                                   ADD_SRC($$); }
115      | _IFNDEF identifier
116        opt_external_definitions
117        op_else
118        _ENDIF                   { $$ = oprNode(_IFNDEF, 3, $2, $3, $4);
119                                   ADD_SRC($$); }
120      | if_condition
121        opt_external_definitions
122        op_else
123        _ENDIF                   { $$ = oprNode(_IF, 3, $1, $2, $3);
124                                   ADD_SRC($$); }
125      | PRAGMA identifier EOL    { $$ = oprNode(PRAGMA, 1, $2);
126                                   ADD_SRC($$);   }
127      | PRAGMA identifier
128        conditional_expr  EOL    { $$ = oprNode(PRAGMA, 2, $2, $3);
129                                   ADD_SRC($$); }
130      ;

  ...
157  if_condition
158      : _IF conditional_expr EOL  { $$ = $2; ADD_SRC($$); }
159      ;
160  op_else
161      : _ELSE
162        opt_external_definitions  { $$ = $2; }
163      |                           { $$ = NULL; }
164      ;
```

                                    Code-4.14

- #95~107: rules for #define statements (four types of format).
- #108~109: rule for #undef statement.
- #110~114: rule for #ifdef ... #else ... #endif statement.
- #115~119: rule for #ifndef ... #else ... #endif statement.
- #120~124: rule for #if ... #else ... #endif statement.
- #125~129: rules for #pragma statement.

Note:
(1) According to the C compiler's design strategy, all the preprocessor statements should be handled in the C preprocessor (**cpp1**);
(2) The way of handling preprocessor statements described in this chapter is very weak. It can't support some situations, such as

```
#define uint8_t  unsigned char
```

4.2.3 Test example

For statement

```
#if x
#define SEED 10
#endif
```

Its parsing tree will be

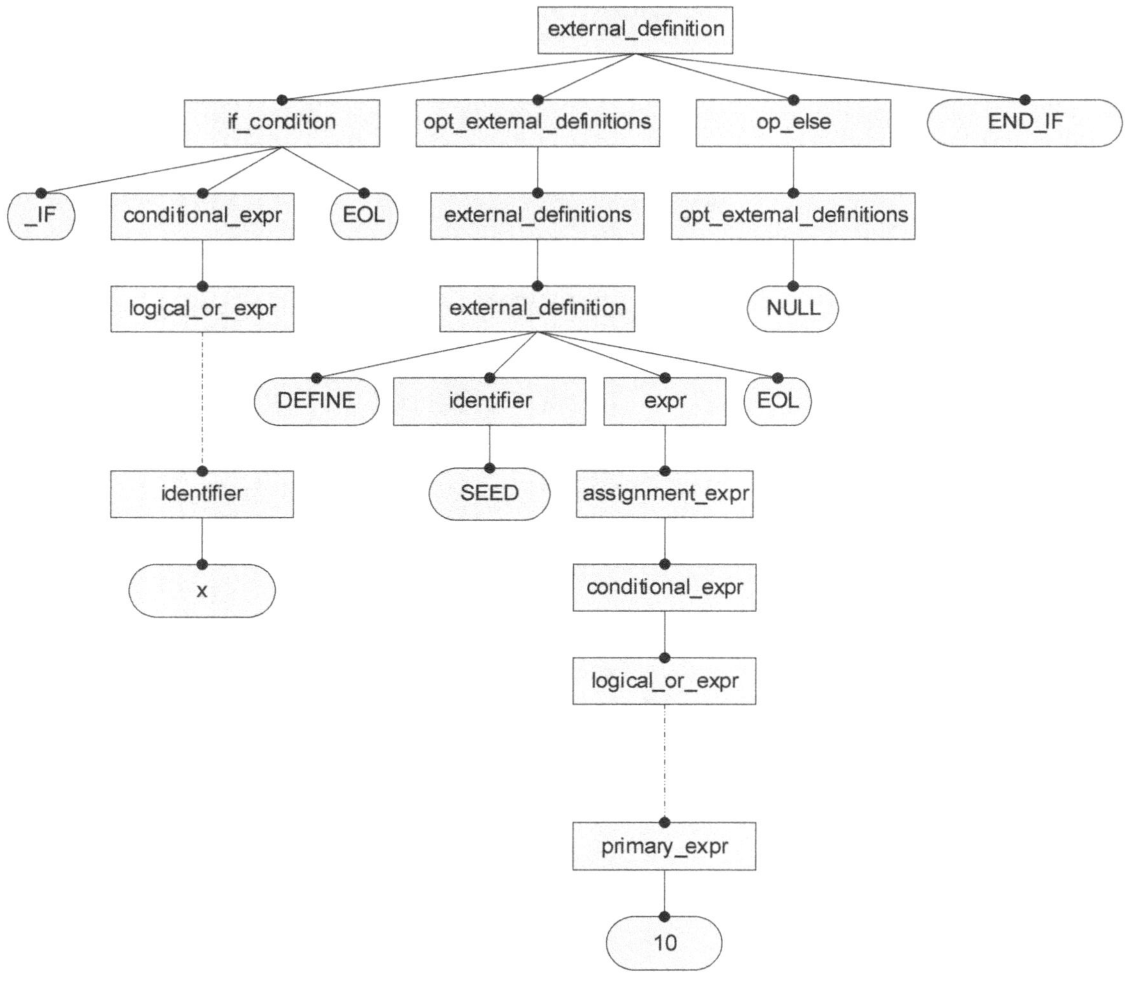

Figure-4.4

4.3 Parsing structured data

<u>Project directory</u>: /p16ecc/cc_source_4.3

In C language, `struct` and `union` statements supply a way for programmers to design or create their own type data. It allows users to define a shorter form of integers (bit-field), which is frequently used for MCU register access. This section also covers the parsing for `enum` statement.

4.3.1 Expanding parser cc.y

In cc.y definition part, it adds the nonterminals for supporting the statements, shown in Code-4.15.

```
50   %type <n> program external_definitions external_definition
51   %type <n> identifier
52   %type <n> declaration init_declarator_list init_declarator
53   %type <n> declarator direct_declarator
54   %type <a> type_specifier type_specifier2 declaration_specifiers
55   %type <i> storage_class_specifier type_qualifier pointer_acce
56   %type <p> pointer
57   %type <n> expr constant_expr conditional_expr logical_or_expr
58   %type <n> logical_and_expr inclusive_or_expr exclusive_or_expr and_expr
59   %type <n> equality_expr relational_expr shift_expr additive_expr
60   %type <n> multiplicative_expr cast_expr unary_expr primary_expr postfix_expr
61   %type <i> unary_operator assignment_operator
62   %type <n> argument_expr_list assignment_expr
63   %type <n> parameter_list parameter_type_list parameter_declaration
64   %type <n> initializer initializer_list
65   %type <n> func_definition func_declarator func_declarator2 func_body
66   %type <n> compound_statement statement_list general_statement statement
67   %type <n> expression_statement labeled_statement selection_statement
68   %type <n> else_statement case_statement_list case_statement case_action
69   %type <n> jump_statement iteration_statement opt_expr
70   %type <n> case_op_action while_expr
71   %type <n> if_condition opt_external_definitions op_else id_list
72   %type <a> enum_specifier
73   %type <n> enumerator_list opt_enumerator_expr enumerator
74   %type <a> struct_or_union_specifier
75   %type <n> struct_declaration_list
76   %type <n> struct_declaration struct_declarator_list
77   %type <i> struct_or_union
78   %type <s> opt_identifier
```
/p16ecc/cc_source_4.3/cc.y   Code-4.15

And in rules part, the following expanding will include the new types of data.

```
199   type_specifier
200     : VOID                        { $$ = newAttr(VOID); }
201     | CHAR                        { $$ = newAttr(CHAR); }
202     | SHORT                       { $$ = newAttr(SHORT); }
203     | INT                         { $$ = newAttr(INT); }
204     | LONG                        { $$ = newAttr(LONG); }
205     | UCHAR                       { $$ = newAttr(CHAR);   $$->isUnsigned = 1; }
206     | USHORT                      { $$ = newAttr(SHORT);  $$->isUnsigned = 1; }
207     | UINT                        { $$ = newAttr(INT);    $$->isUnsigned = 1; }
208     | ULONG                       { $$ = newAttr(LONG);   $$->isUnsigned = 1; }
209     | enum_specifier              { $$ = $1; }
210     | struct_or_union_specifier   { $$ = $1; }
211     ;
```
/p16ecc/cc_source_4.3/cc.y   Code-4.16

- #209: add specifier for ENUM.
- #210: add new specifier for `struct` and `union`.

```
635   enum_specifier
636       : ENUM
637         '{' enumerator_list '}'    { $$ = newAttr(ENUM);
638                                        $$->newData = oprNode(ENUM, 1, $3);
639                                      }
640       | ENUM
641         '{'
642         enumerator_list ',' '}'      { $$ = newAttr(ENUM);
643                                        $$->newData = oprNode(ENUM, 1, $3);
644                                      }
645       ;
646   enumerator_list
647       : enumerator                   { $$ = makeList($1); }
648       | enumerator_list ','
649         enumerator                   { $$ = appendList($1, $3); }
650       ;
651   enumerator
652       : identifier
653         opt_enumerator_expr          { $$ = $1;
654                                        $$->id.init = $2;
655                                      }
656       ;
657   opt_enumerator_expr
658       : '=' constant_expr            { $$ = $2; }
659       |                              { $$ = NULL; }
660       ;
```
`/p16ecc/cc_source_4.3/cc.y`   Code-4.17

For statement

```
enum {ONE=1, TWO, THREE};
```

The parsing tree is shown in Figure-4.4.

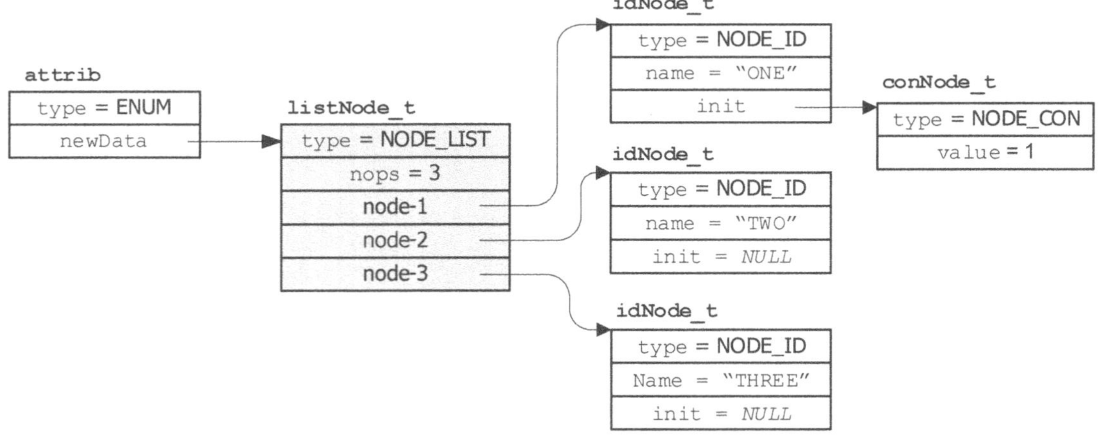

Figure-4.4

4.3.3 Parsing rule for struct/union

At the grammar level, `struct` and `union` statements are the same. The difference between them lies in how they allocate memory for data members. More precisely, the difference happens at the overlapping of data members. The Code-4.18 shows the parsing rules for struct/union statement.

```
661    struct_or_union_specifier
662        : struct_or_union
663          opt_identifier
664          '{'
665          struct_declaration_list
666          '}'                          { $$ = newAttr($1);
667                                         $$->typeName   = $2;
668                                         $$->newData    = $4;
669                                         bitFieldOffsetAssign($$->newData);
670                                       }
671        | struct_or_union
672          IDENTIFIER                   { $$ = newAttr($1);
673                                         $$->typeName = $2;
674                                       }
675        ;
676    struct_or_union
677        : STRUCT                       { $$ = STRUCT; }
678        | UNION                        { $$ = UNION; }
679        ;
680    opt_identifier
681        :                             { $$ = NULL; }
682        | IDENTIFIER                   { $$ = $1; }
683        ;
684    struct_declaration_list
685        : struct_declaration           { $$ = makeList($1); }
686        | struct_declaration_list
687          struct_declaration           { $$ = appendList($1, $2); }
688        ;
689    struct_declaration
690        : type_specifier
691          struct_declarator_list ';' { $$ = oprNode(DATA_DECL, 1, $2);
692                                         $$->opr.attr = $1;
693                                       }
694        | type_specifier
695          identifier ':' CONSTANT ';'
696                                       { $$ = NULL;
697                                         if ( $1->type != CHAR )
698                                            yyerror("bit field type error!");
699                                         else if ( $4 > 8 || $4 <= 0 )
700                                            yyerror("bit field size error!");
701                                         else
702                                         {
703                                            $$ = oprNode(DATA_DECL, 1, makeList($2));
704                                            $1->bitField    = ($4 < 8)? $4: 0;
705                                            $1->isUnsigned = 1;
706                                            $$->opr.attr    = $1;
707                                         }
708                                       }
709        ;
/p16ecc/cc_source_4.3/cc.y                                    Code-4.18
```

- #662~670: `struct/union` definition.
- #671~675: `struct/union` usage.
- #689~708: `struct/union` internal data members, including bit-field definition (#694~708).
  Note, (1) only allow `char` type being the carrier for bit-field; (2) the offset of bit-field member cannot be determined this moment.
- #669: assign offsets for bit-field variables after all declaration lines in `struct/union`.

As a practice, readers can try making the parsing tree for `struct/union` statement.

4.4 Parsing typedef statement

<u>Project directory</u>: /p16ecc/cc_source_4.4

Overall, `typedef` statement is not necessary for C/C++ programming. It just creates an alias for another data type. It does not create a new data type; rather, it gives an existing one a new name. On the other side, `typedef` is frequently used by C/C++ software programmers. The reason that makes it popular could be (1) it brings a method to make data types more 'meaningful', improving the readability in short; (2) it usually helps code porting from project to project. For example, the following statements can appear often in daily programming:

```
typedef unsigned char uint8_t;
typedef unsigned int  uint16_t;
typedef unsigned long uint32_t;
```

The `typedef` statement has the format in C as

```
typedef exist_data_type new_type_name ;
```

where *new_type_name* is simply an identifier, beyond the reserved (key) words in C.

It brings a question to the lexer (cc.l). When 'catching' an identifier in the lexer, what and how does it return the token type to the parser (a regular identifier vs. a type specifier)?

4.4.1 Expanding lexer for typedef

In cc.l, it adds a variable, `ignoreTypedef`, to the C code section for status transition. It also needs to set up a table, `newNameList`, to record the new type names.

```
10    extern int yyparse();
11    int fatalError=0;
12    char *currentFile = NULL;
13    int  preProcFlag = 0;
14    int  preProcType;
15    int  ignoreTypedef = 0;
16    NameList *newNameList = NULL;
17    NameList *sysIncludeList = NULL;
```
/p16ecc/cc_source_4.4/cc.l   Code-4.19

* #15~16: new items added.

In the rule section, when catching an identifier, it checks both `ignoreTypede` (bit0) and `newNameList` to determine what to return.

```
168   {identifier}            { NameList *dp = searchName(newNameList, yytext);
169                             addSrcCode();
170                             if ( dp == NULL || (ignoreTypedef & 1) )
171                             {
172                                yylval.s = dupStr(yytext);
173                                return IDENTIFIER;
174                             }
175                             yylval.a = newAttr(TYPEDEF_NAME);
176                             yylval.a->typeName = dupStr(dp->name);
177                             return TYPEDEF_NAME;
178                             }
```
/p16ecc/cc_source_4.4/cc.l   Code-4.20

- #168: search `newNameList`, to see if the identifier has been defined as a data type.
- #170~174: if the identifier not in the list or `ignoreTypedef` is set, then return it as a regular identifier.
- #175~177: otherwise, it's considered as a new type specifier (name). The new name will be added to the table in the parser. The following, Figure-4.5, is the flowchart for the algorithm.

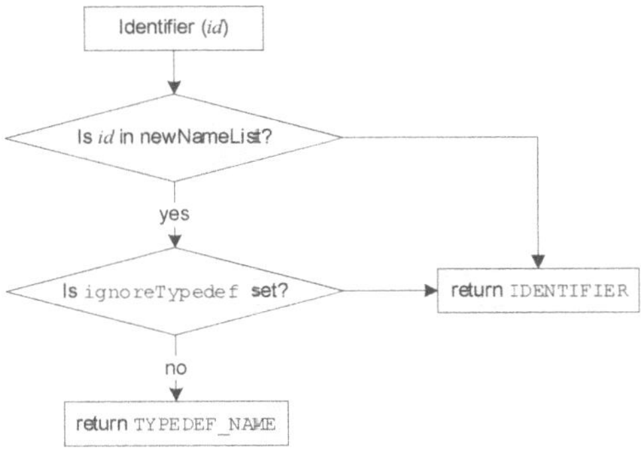

Figure-4.5

4.4.2 Expanding paser for typedef

In cc.y, it manipulates `ignoreTypedef` to control the parsing `typedef` statement, as shown in Code-4.21.

```
136  declaration
137      : declaration_specifiers ';' {if ( $1->type == ENUM ) {
138                                        $$ = $1->newData;
139                                        $1->newData = NULL;
140                                        delAttr($1);
141                                        ADD_SRC($$);
142                                    }
143                                    else
144                                    {
145                                        $$ = oprNode(DATA_DECL, 0);
146                                        $$->opr.attr = $1;
147                                        ADD_SRC($$);
148                                    }
149                                    ignoreTypedef &= ~1;
150                                    }
151      | declaration_specifiers
152        init_declarator_list ';'  { $$ = oprNode(DATA_DECL, 1, $2);
153                                    $$->opr.attr = $1;
154                                    ADD_SRC($$);
155                                    ignoreTypedef &= ~1;
156                                    }
157      | AASM                      { $$ = oprNode(AASM, 1, strNode($1));
158                                    ADD_SRC($$); }
159      | TYPEDEF                   { ignoreTypedef &= ~1; }
160        type_specifier            { ignoreTypedef |=  1; }
161        init_declarator_list ';'  { $3->isTypedef = 1;
162                                    if ( $5 && $5->list.nops == 1 )
163                                    {
164                                        char *id = $5->list.ptr[0]->id.name;
165                                        $$ = oprNode(TYPEDEF, 1, idNode(id));
166                                        $$->opr.attr = $3;
167                                        addName(&newNameList, id);
168                                        ADD_SRC($$);
169                                    }
170                                    else
171                                    {
172                                        $$ = NULL;
173                                        delAttr($3);
174                                        yyerror("incorrect 'typedef' format!");
175                                        CLR_SRC();
176                                    }
177                                    ignoreTypedef &= ~1;
178                                    delNode($5);
179                                    }
180      ;
```

- #159~179: the parsing rule breaks into three stages, and `ignoreTypedef` flips along the parsing stages, as shown in Figure4.6.

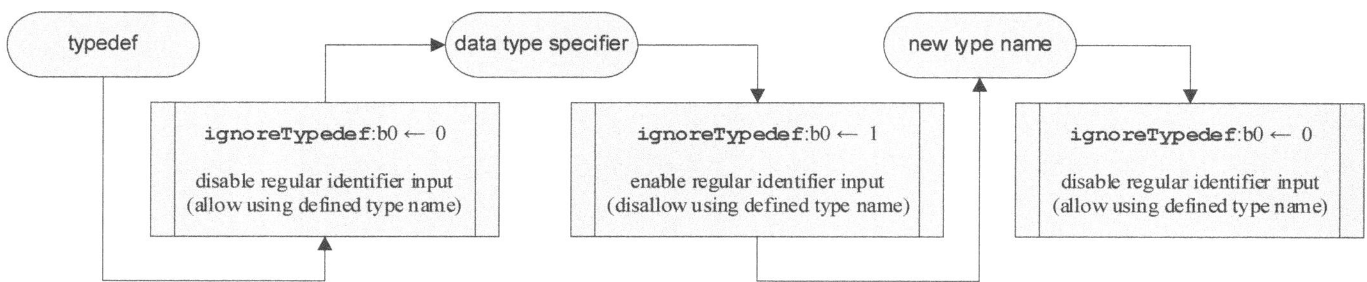

Figure4.6

When lexer returns `TYPEDEF_NAME`, it will be identified as a data type specifier, as shown in Code-4.22.

```
224    type_specifier
225      : VOID                        { $$ = newAttr(VOID); }
226      | CHAR                        { $$ = newAttr(CHAR); }
227      | SHORT                       { $$ = newAttr(SHORT); }
228      | INT                         { $$ = newAttr(INT); }
229      | LONG                        { $$ = newAttr(LONG); }
230      | UCHAR                       { $$ = newAttr(CHAR);  $$->isUnsigned = 1; }
231      | USHORT                      { $$ = newAttr(SHORT); $$->isUnsigned = 1; }
232      | UINT                        { $$ = newAttr(INT);   $$->isUnsigned = 1; }
233      | ULONG                       { $$ = newAttr(LONG);  $$->isUnsigned = 1; }
234      | enum_specifier              { $$ = $1; }
235      | struct_or_union_specifier   { $$ = $1; }
236      | TYPEDEF_NAME                { $$ = $1; }
237      ;
```


- #236: `TYPEDEF_NAME` is added to the specifier rule list.

The following is a test example:

ch4_t2.c	`typedef unsigned char uint8_t;`
	
	`typedef struct {`
	`  char *name;`
	`  uint8_t age;`
	`} Student_t;`
	
	`Student_t Lisa;`

	`F:\p16ecc\tests>cc16e ch4_t2.c`
	`list node {3}:`
	` ch4_t2.c #1: typedef unsigned char uint8_t;`
	` opr node {1}: (oper. token: 300) {typedef unsigned char}`
	` id node: uint8_t`
	` ch4_t2.c #6: typedef struct   char *name;  uint8_t age; Student_t;`
	` opr node {1}: (oper. token: 300) {typedef STRUCT}`
	` id node: Student_t`
	` ch4_t2.c #8: Student_t Lisa;`
	` opr node {1}: DATA_DECL {(Student_t) TYPE_NAME}`
	` list node {1}:`
	`  id node: Lisa`

So far, the parser design is finished. The parsing result of the input, the parsing tree, will be stored in `progUnit`, waiting for further processing. The parsing tree is another form of the input (C source file) and presents the syntactic structure of the input.

As the conclusion of the chapter, the whole parser source code **cc.y** is listed below.

```
%{
#include <stdio.h>
#include <stdarg.h>
#include <string.h>
#include "common.h"

extern int  yyerror(char *);
extern int  yylex(), yylineno;
extern FILE *yyin;
extern int  ignoreTypedef;

int yyparse(void);

#define ADD_SRC(n)   addSrc(n, getSrcCode())
#define CLR_SRC()    getSrcCode()
static int parListCheck(node *np);
static void bitFieldOffsetAssign(node *p);

#define YYDEBUG 1
%}

%expect 1

%union {
    char    *s;  // string value
    int      i;  // integer value
    node    *n;
    attrib *a;
    int     *p;
}

%token CHAR INT SHORT LONG UCHAR UINT USHORT ULONG VOID
%token FUNC_HEAD FUNC_DECL DATA_DECL
%token _IF _ELSE _ENDIF _IFDEF _IFNDEF DEFINE UNDEF PRAGMA
%token BREAK CASE CONST CONTINUE DEFAULT DO ELSE ENUM EXTERN
%token FOR GOTO IF REGISTER RETURN SIZEOF STATIC SWITCH
%token VOLATILE WHILE INTERRUPT UNION STRUCT TYPEDEF ELIPSIS
%token LINEAR
%token ADD_ASSIGN SUB_ASSIGN MUL_ASSIGN DIV_ASSIGN MOD_ASSIGN
%token LEFT_ASSIGN RIGHT_ASSIGN AND_ASSIGN XOR_ASSIGN OR_ASSIGN
%token EQ_OP NE_OP LE_OP GE_OP LEFT_OP RIGHT_OP INC_OP DEC_OP
%token OR_OP AND_OP PTR_OP FPTR CAST CALL
%token POST_INC POST_DEC PRE_INC PRE_DEC POS_OF ADDR_OF NEG_OF
%token EOL LABEL

%token <i> CONSTANT L_CONSTANT U_CONSTANT UL_CONSTANT C_CONSTANT MEM_BANK
%token <i> RELATIONAL_OP EQUALITY_OP SHIFT_OP ASSIGN_OP
%token <s> IDENTIFIER IDENTIFIER_ STRING AASM
%token <a> TYPEDEF_NAME

%type <n> program external_definitions external_definition
%type <n> identifier
%type <n> declaration init_declarator_list init_declarator
%type <n> declarator direct_declarator
%type <a> type_specifier type_specifier2 declaration_specifiers
%type <i> storage_class_specifier type_qualifier pointer_acce
%type <p> pointer
%type <n> expr constant_expr conditional_expr logical_or_expr
%type <n> logical_and_expr inclusive_or_expr exclusive_or_expr and_expr
%type <n> equality_expr relational_expr shift_expr additive_expr
%type <n> multiplicative_expr cast_expr unary_expr primary_expr postfix_expr
%type <i> unary_operator assignment_operator
%type <n> argument_expr_list assignment_expr
%type <n> parameter_list parameter_type_list parameter_declaration
%type <n> initializer initializer_list
%type <n> func_definition func_declarator func_declarator2 func_body
```

```
%type <n> compound_statement statement_list general_statement statement
%type <n> expression_statement labeled_statement selection_statement
%type <n> else_statement case_statement_list case_statement case_action
%type <n> jump_statement iteration_statement opt_expr
%type <n> case_op_action while_expr
%type <n> if_condition opt_external_definitions op_else id_list
%type <n> op_if_statement op_else_statement
%type <a> enum_specifier
%type <n> enumerator_list opt_enumerator_expr enumerator
%type <a> struct_or_union_specifier
%type <n> struct_declaration_list
%type <n> struct_declaration struct_declarator_list
%type <i> struct_or_union
%type <s> opt_identifier

%start program

%%

program
    : external_definitions          { progUnit = $1; }
    ;
external_definitions
    : external_definition           { $$ = makeList($1); }
    | external_definitions
      external_definition           { $$ = appendList($1, $2); }
    ;
external_definition
    : declaration                   { $$ = $1; CLR_SRC(); }
    | func_definition               { $$ = $1; CLR_SRC(); }
    | declaration_specifiers
      init_declarator_list
      '@' conditional_expr ';'      { $$ = oprNode(DATA_DECL, 1, $2);
                                      $$->opr.attr = $1;
                                      $1->atAddr = $4;
                                      ADD_SRC($$);
                                    }
    | UNDEF identifier EOL          { $$ = oprNode(UNDEF, 1, $2);
                                      ADD_SRC($$);}
    | DEFINE identifier EOL         { $$ = oprNode(DEFINE, 1, $2);
                                      ADD_SRC($$); }
    | DEFINE identifier
      expr EOL                      { $$ = oprNode(DEFINE, 2, $2, $3);
                                      ADD_SRC($$); }
    | DEFINE IDENTIFIER_
      id_list ')' expr EOL          { $$ = oprNode(DEFINE, 3, idNode($2), $3, $5);
                                      ADD_SRC($$); free($2);
                                    }
    | _IFDEF identifier
      opt_external_definitions
      op_else
      _ENDIF                        { $$ = oprNode(_IFDEF, 3, $2, $3, $4);
                                      ADD_SRC($$); }
    | _IFNDEF identifier
      opt_external_definitions
      op_else
      _ENDIF                        { $$ = oprNode(_IFNDEF, 3, $2, $3, $4);
                                      ADD_SRC($$); }
    | if_condition
      opt_external_definitions
      op_else
      _ENDIF                        { $$ = oprNode(_IF, 3, $1, $2, $3);
                                      ADD_SRC($$); }
    | PRAGMA identifier EOL         { $$ = oprNode(PRAGMA, 1, $2);
                                      ADD_SRC($$);   }
    | PRAGMA identifier
      conditional_expr  EOL         { $$ = oprNode(PRAGMA, 2, $2, $3);
                                      ADD_SRC($$); }
    ;
declaration
    : declaration_specifiers ';' {if ( $1->type == ENUM ) {
                                      $$ = $1->newData;
                                      $1->newData = NULL;
                                      delAttr($1);
                                      ADD_SRC($$);
                                    }
```

```
                                             else
                                             {
                                               $$ = oprNode(DATA_DECL, 0);
                                               $$->opr.attr = $1;
                                               ADD_SRC($$);
                                             }
                                             ignoreTypedef &= ~1;
                                           }
    | declaration_specifiers
      init_declarator_list ';'         { $$ = oprNode(DATA_DECL, 1, $2);
                                         $$->opr.attr = $1;
                                         ADD_SRC($$);
                                         ignoreTypedef &= ~1;
                                       }
    | AASM                             { $$ = oprNode(AASM, 1, strNode($1));
                                         ADD_SRC($$); }
    | TYPEDEF                          { ignoreTypedef &= ~1; }
      type_specifier                   { ignoreTypedef |=  1; }
      init_declarator_list ';'         { $3->isTypedef = 1;
                                         if ( $5 && $5->list.nops == 1 )
                                         {
                                           char *id = $5->list.ptr[0]->id.name;
                                           $$ = oprNode(TYPEDEF, 1, idNode(id));
                                           $$->opr.attr = $3;
                                           addName(&newNameList, id);
                                           ADD_SRC($$);
                                         }
                                         else
                                         {
                                           $$ = NULL;
                                           delAttr($3);
                                           yyerror("incorrect 'typedef' format!");
                                           CLR_SRC();
                                         }
                                         ignoreTypedef &= ~1;
                                         delNode($5);
                                       }
    ;
opt_external_definitions
    : external_definitions             { $$ = $1; }
    |                                  { $$ = NULL; }
    ;
if_condition
    : _IF conditional_expr EOL         { $$ = $2; ADD_SRC($$); }
    ;
op_else
    : _ELSE
      opt_external_definitions         { $$ = $2; }
    |                                  { $$ = NULL; }
    ;
id_list
    :                                  { $$ = NULL; }
    | identifier                       { $$ = makeList ($1); }
    | id_list ',' identifier           { $$ = appendList($1, $3); }
    ;
declaration_specifiers
    : storage_class_specifier          { $$ = newAttr(0);
                                         if ( assignAcce($$, $1) )
                                            yyerror("storage type redefined!");
                                       }
    | storage_class_specifier
      declaration_specifiers           { $$ = $2;
                                         if ( assignAcce($$, $1) )
                                            yyerror("storage type redefined!");
                                       }
    | type_specifier
      declaration_specifiers           { $$ = $1; mergeAttr($$, $2); delAttr($2); }
    | type_specifier                   { $$ = $1; }
    | type_qualifier
      declaration_specifiers           { $$ = $2; assignAcce($$, $1); }
    ;
storage_class_specifier
    : REGISTER                         { $$ = REGISTER; }
    | STATIC                           { $$ = STATIC; }
    | EXTERN                           { $$ = EXTERN; }
    | MEM_BANK                         { $$ = $1; }
```

```
     ;
type_qualifier
    : CONST                       { $$ = CONST; }
    | VOLATILE                    { $$ = VOLATILE; }
    ;
type_specifier
    : VOID                        { $$ = newAttr(VOID); }
    | CHAR                        { $$ = newAttr(CHAR); }
    | SHORT                       { $$ = newAttr(SHORT); }
    | INT                         { $$ = newAttr(INT); }
    | LONG                        { $$ = newAttr(LONG); }
    | UCHAR                       { $$ = newAttr(CHAR);  $$->isUnsigned = 1; }
    | USHORT                      { $$ = newAttr(SHORT); $$->isUnsigned = 1; }
    | UINT                        { $$ = newAttr(INT);   $$->isUnsigned = 1; }
    | ULONG                       { $$ = newAttr(LONG);  $$->isUnsigned = 1; }
    | enum_specifier              { $$ = $1; }
    | struct_or_union_specifier   { $$ = $1; }
    | TYPEDEF_NAME                { $$ = $1; }
    ;
type_specifier2
    : declaration_specifiers      { $$ = $1; }
    | declaration_specifiers
      pointer                     { $$ = $1; appendPtr($$, $2);
                                    if ( $1->type == 0 )
                                       yyerror ("pointer type not defined!");
                                  }
    ;
init_declarator_list
    : init_declarator             { $$ = makeList($1); }
    | init_declarator_list ','
      init_declarator             { $$ = appendList($1, $3); }
    ;
init_declarator
    : declarator                  { $$ = $1; }
    | declarator '='
      initializer                 { $$ = $1; $$->id.init = $3; }
    ;
declarator
    : direct_declarator           { $$ = $1; }
    | pointer direct_declarator   { $$ = $2;
                                    if ( $2->id.attr == NULL )
                                       $2->id.attr = newAttr(0);
                                    $2->id.attr->ptrVect = $1;
                                  }
    ;
direct_declarator
    : identifier                  { $$ = $1; }
    | direct_declarator '[' ']'   { $$ = $1;
                                    $$->id.dim = appendList($$->id.dim, conNode(0, INT));
                                  }
    | direct_declarator
      '[' constant_expr ']'       { $$ = $1;
                                    $$->id.dim = appendList($$->id.dim, $3);
                                  }
    | '(' '*' identifier ')' '('
      parameter_type_list ')'     { int n = parListCheck($6);
                                    $$ = $3;
                                    $$->id.fp_decl = 1;
                                    if ( n > 0 )
                                       $$->id.parp = $6;
                                    else
                                    {
                                       delNode($6);
                                       if ( n < 0 )
                                          yyerror("parameter list error!");
                                    }
                                  }
    | '(' '*' identifier
      '[' constant_expr ']' ')' '('
      parameter_type_list ')'     { int n = parListCheck($9);
                                    $$ = $3;
                                    $$->id.dim = makeList($5);
                                    $$->id.fp_decl = 1;
                                    if ( n > 0 )
                                       $$->id.parp = $9;
                                    else
```

```
                                      {
                                        delNode($9);
                                        if ( n < 0 )
                                          yyerror("parameter list error!");
                                      }
                                    }
    |   '(' '*' ')' '('
        parameter_type_list ')'    { int n = parListCheck($5);
                                      $$ = idNode(""):
                                      $$->id.fp_decl = 1;
                                      if ( n > 0 )
                                        $$->id.parp = $5;
                                      else
                                      {
                                        delNode($5);
                                        if ( n < 0 )
                                          yyerror("parameter list error!");
                                      }
                                    }
    ;
identifier
    : IDENTIFIER                   { $$ = idNode($1): free($1); }
    ;
pointer_acce
    : CONST                        { $$ = CONST; }
    | MEM_BANK                     { $$ = $1; }
    ;
pointer
    : '*'                          { $$ = newPtr(0); }
    | '*' pointer                  { $$ = includePtr(0, $2); }
    | '*' pointer_acce             { $$ = newPtr($2): }
    | '*' pointer_acce pointer     { $$ = includePtr($2, $3); }
    ;
constant_expr
    : conditional_expr             { $$ = $1; }
    ;
conditional_expr
    : logical_or_expr              { $$ = $1; }
    | logical_or_expr '?' expr
      ':' conditional_expr         { $$ = oprNode('?', 3, $1, $3, $5); }
    ;
logical_or_expr
    : logical_and_expr             { $$ = $1; }
    | logical_or_expr OR_OP
      logical_and_expr             { $$ = oprNode(OR_OP, 2, $1, $3); }
    ;
logical_and_expr
    : inclusive_or_expr            { $$ = $1; }
    | logical_and_expr AND_OP
      inclusive_or_expr            { $$ = oprNode(AND_OP, 2, $1, $3); }
    ;
inclusive_or_expr
    : exclusive_or_expr            { $$ = $1; }
    | inclusive_or_expr '|'
      exclusive_or_expr            { $$ = oprNode('|', 2, $1, $3); }
    ;
exclusive_or_expr
    : and_expr                     { $$ = $1; }
    | exclusive_or_expr '^'
      and_expr                     { $$ = oprNode('^', 2, $1, $3); }
    ;
and_expr
    : equality_expr                { $$ = $1; }
    | and_expr '&'
      equality_expr                { $$ = oprNode('&', 2, $1, $3); }
    ;
equality_expr
    : relational_expr              { $$ = $1; }
    | equality_expr EQUALITY_OP
      relational_expr              { $$ = oprNode($2, 2, $1, $3); }
    ;
relational_expr
    : shift_expr                   { $$ = $1; }
    | relational_expr RELATIONAL_OP
      shift_expr                   { $$ = oprNode($2, 2, $1, $3); }
    ;
```

```
shift_expr
    : additive_expr
    | shift_expr SHIFT_OP
      additive_expr                 { $$ = oprNode($2, 2, $1, $3); }
    ;
additive_expr
    : multiplicative_expr           { $$ = $1; }
    | additive_expr '+'
      multiplicative_expr           { $$ = oprNode('+', 2, $1, $3); }
    | additive_expr '-'
      multiplicative_expr           { $$ = oprNode('-', 2, $1, $3); }
    ;
multiplicative_expr
    : cast_expr                     { $$ = $1; }
    | multiplicative_expr '*'
      cast_expr                     { $$ = oprNode('*', 2, $1, $3); }
    | multiplicative_expr '/'
      cast_expr                     { $$ = oprNode('/', 2, $1, $3); }
    | multiplicative_expr '%'
      cast_expr                     { $$ = oprNode('%', 2, $1, $3); }
    ;
cast_expr
    : unary_expr                    { $$ = $1; }
    | '(' type_specifier2 ')'
      cast_expr                     { if ( $2->type == VOID &&
                                           $2->ptrVect == NULL )
                                        yyerror("invalid cast type - 'void'!");
                                      $$ = oprNode(CAST, 1, $4);
                                      $$->opr.attr = $2; }
    ;
unary_expr
    : postfix_expr                  { $$ = $1; }
    | INC_OP unary_expr             { $$ = oprNode(PRE_INC, 1, $2); }
    | DEC_OP unary_expr             { $$ = oprNode(PRE_DEC, 1, $2); }
    | unary_operator cast_expr      { $$ = $1? oprNode($1, 1, $2): $2; }
    | SIZEOF '(' unary_expr ')'     { $$ = oprNode(SIZEOF, 1, $3); }
    | SIZEOF
      '(' type_specifier2 ')'       { $$ = oprNode(SIZEOF, 0); $$->opr.attr = $3; }
    ;
unary_operator
    : '&'                           { $$ = ADDR_OF; }
    | '*'                           { $$ = POS_OF; }
    | '+'                           { $$ = 0; }
    | '-'                           { $$ = NEG_OF; }
    | '~'                           { $$ = '~'; }
    | '!'                           { $$ = '!'; }
    ;
postfix_expr
    : primary_expr                  { $$ = $1; }
    | postfix_expr
      '[' expr ']'                  { $$ = oprNode('[', 2, $1, $3); }
    | postfix_expr '('
      argument_expr_list ')'        { $$ = oprNode(CALL, 2, $1, $3); }
    | postfix_expr '(' ')'          { $$ = oprNode(CALL, 2, $1, NULL); }
    | postfix_expr '.'
      identifier                    { $$ = oprNode('.', 2, $1, $3); }
    | postfix_expr PTR_OP
      identifier                    { $$ = oprNode(PTR_OP, 2, $1, $3); }
    | postfix_expr INC_OP           { $$ = oprNode(POST_INC, 1, $1); }
    | postfix_expr DEC_OP           { $$ = oprNode(POST_DEC, 1, $1); }
    ;
primary_expr
    : identifier                    { $$ = $1; }
    | CONSTANT                      { $$ = conNode($1, INT); }
    | C_CONSTANT                    { $$ = conNode($1, CHAR); }
    | U_CONSTANT                    { $$ = conNode($1, INT); $$->con.attr->isUnsigned = 1; }
    | L_CONSTANT                    { $$ = conNode($1, LONG); }
    | UL_CONSTANT                   { $$ = conNode($1, LONG); $$->con.attr->isUnsigned = 1; }
    | STRING                        { $$ = strNode($1); free($1); }
    | '(' expr ')'                  { $$ = $2; }
    ;
expr
    : expr ',' assignment_expr      { $$ = oprNode(',', 2, $1, $3); }
    | assignment_expr               { $$ = $1; }
    ;
assignment_operator
```

```
	    : '='                           { $$ = '='; }
	    | ASSIGN_OP                     { $$ = $1; }
	    ;
assignment_expr
	    : conditional_expr              { $$ = $1; }
	    | unary_expr assignment_operator
	      assignment_expr               { $$ = oprNode($2, 2, $1, $3); }
	    ;
argument_expr_list
	    : assignment_expr               { $$ = makeList($1); }
	    | argument_expr_list ','
	      assignment_expr               { $$ = appendList($1, $3); }
	    ;
parameter_type_list
	    :                               { $$ = NULL; }
	    | parameter_list                { $$ = $1; }
	    | parameter_list ',' ELIPSIS{ $$ = $1; $$->list.elipsis = 1; }
	    ;
parameter_list
	    : parameter_declaration         { $$ = makeList($1); }
	    | parameter_list ','
	      parameter_declaration         { $$ = appendList($1, $3); }
	    ;
parameter_declaration
	    : type_specifier
	      direct_declarator             { $1->isFptr  = $2->id.fp_decl;
	                                      $1->parList = $2->id.parp;
	                                      $2->id.parp = NULL;
	                                      $$ = $2;
	                                      $$->id.attr = $1;
	                                    }
	    | type_specifier
	      pointer direct_declarator { $1->isFptr  = $3->id.fp_decl;
	                                      $1->parList = $3->id.parp;
	                                      $3->id.parp = NULL;
	                                      $1->ptrVect = $2;
	                                      $$ = $3;
	                                      $$->id.attr = $1;
	                                    }
	    | type_specifier                { $$ = idNode("");
	                                      $$->id.attr = $1;
	                                    }
	    | type_specifier pointer        { $$ = idNode("");
	                                      $1->ptrVect = $2;
	                                      $$->id.attr = $1;
	                                    }
	    ;
initializer
	    : assignment_expr               { $$ = $1; }
	    | '{' initializer_list '}'      { $$ = $2; }
	    | '{'
	      initializer_list ',' '}'      { $$ = $2; }
	    ;
initializer_list
	    : initializer                   { $$ = makeList($1); }
	    | initializer_list
	      ',' initializer               { $$ = appendList($1, $3); }
	    ;
func_definition
	    : func_declarator
	      func_body                     { $$ = oprNode(FUNC_DECL, 2, $1, $2);
	                                      $$->opr.attr = newAttr(INT);
	                                    }
	    | declaration_specifiers
	      func_declarator2              { if ( $1->type == 0 )
	                                      {
	                                        if ( $2->id.attr && $2->id.attr->ptrVect )
	                                            yyerror("function pointer type not defined!");
	                                        else
	                                            $1->type = INT;
	                                      }
	                                    }
	      func_body                     { $$ = oprNode(FUNC_DECL, 2, $2, $4);
	                                      $$->opr.attr = $1;

	                                      if ( $2->id.attr )
```

```
                                        {
                                            $$->opr.attr->ptrVect = $2->id.attr->ptrVect;
                                            $2->id.attr->ptrVect = NULL;
                                            delAttr($2->id.attr);
                                            $2->id.attr = NULL;
                                        }
                                    }
    | INTERRUPT
      identifier '(' ')'
      compound_statement            { node *np = oprNode(FUNC_HEAD, 1, $2);
                                        $$ = oprNode(FUNC_DECL, 2, np, $5);
                                        $$->opr.attr = newAttr(INTERRUPT);
                                    }
    ;
func_declarator
    : identifier '('
      parameter_type_list ')'       { int n = parListCheck($3);
                                        if ( n > 0 )
                                        {
                                            $$ = oprNode(FUNC_HEAD, 2, $1, $3);
                                            ADD_SRC($$);
                                        }
                                        else
                                        {
                                            delNode($3);
                                            if ( n == 0 )
                                            {
                                                $$ = oprNode(FUNC_HEAD, 1, $1);
                                                ADD_SRC($$);
                                            }
                                            else
                                            {
                                                yyerror("parameter error!");
                                                CLR_SRC();
                                            }
                                        }
                                    }
    ;
func_declarator2    /* may include 'pointer' */
    : func_declarator               { $$ = $1; }
    | pointer func_declarator       { $$ = $2;
                                        $2->id.attr = newAttr(0);
                                        $2->id.attr->ptrVect = $1;
                                        ADD_SRC($$);
                                    }
    ;
func_body
    : ';'                           { $$ = NULL; }
    | compound_statement            { $$ = $1;    }
    ;
compound_statement
    : '{' '}'                       { $$ = oprNode('{', 0);      CLR_SRC(); }
    | '{' statement_list '}'        { $$ = oprNode('{', 1, $2); CLR_SRC(); }
    ;
statement_list
    : general_statement             { $$ = makeList($1);
                                        $$ = appendList($$, oprNode(';', 0)); }
    | statement_list
      general_statement             { $$ = appendList($1, $2);
                                        $$ = appendList($$, oprNode(';', 0)); }
    ;
general_statement
    : statement                     { $$ = $1; CLR_SRC(); }
    | declaration                   { $$ = $1; CLR_SRC(); }
    ;
statement
    : compound_statement            { $$ = $1; }
    | expression_statement          { $$ = $1; }
    | labeled_statement             { $$ = $1; }
    | selection_statement           { $$ = $1; }
    | jump_statement                { $$ = $1; }
    | iteration_statement           { $$ = $1; }
    | _IFDEF identifier
      op_if_statement
      op_else_statement
      _ENDIF                        { $$ = oprNode(_IFDEF, 3, $2, $3, $4); }
```

```
    | _IFNDEF identifier
      op_if_statement
      op_else_statement
      _ENDIF                     { $$ = oprNode(_IFNDEF, 3, $2, $3, $4); }
    | if_condition
      op_if_statement
      op_else_statement
      _ENDIF                     { $$ = oprNode(_IF, 3, $1, $2, $3); }
    | DEFINE identifier EOL      { $$ = oprNode(DEFINE, 1, $2); }
    | DEFINE identifier
      conditional_expr EOL       { $$ = oprNode(DEFINE, 2, $2, $3); }
    | DEFINE IDENTIFIER_
      id_list ')' expr EOL       { $$ = oprNode(DEFINE, 3, idNode($2), $3, $5);
                                   free($2); ADD_SRC($$);
                                 }
    ;
op_if_statement
    :                            { $$ = NULL; }
    | statement_list             { $$ = $1;    }
    ;
op_else_statement
    :                            { $$ = NULL; }
    | _ELSE statement_list       { $$ = $2;    }
    ;
expression_statement
    : ';'                        { $$ = oprNode(';', 0); ADD_SRC($$); }
    | expr ';'                   { $$ = $1; ADD_SRC($$); }
    ;
labeled_statement
    : identifier ':'             { $$ = oprNode(LABEL, 1, $1); ADD_SRC($$); }
    ;
selection_statement
    : IF '(' expr ')'            { ADD_SRC($3); }
      statement
      else_statement            { $$ = oprNode(IF. 3, $3, $6, $7); }
    | SWITCH '(' expr ')' '{'    { ADD_SRC($3); }
      case_statement_list '}'    { $$ = oprNode (SWITCH, 2, $3, $7); }
    ;
else_statement
    :                            { $$ = NULL; }
    | ELSE statement             { $$ = $2; }
    ;
case_statement_list
    : case_statement             { $$ = makeList($1); }
    | case_statement_list
      case_statement             { $$ = appendList($1, $2); }
    ;
case_statement
    : CASE constant_expr ':'     { ADD_SRC($2); }
      case_op_action             { $$ = oprNode(CASE, 2, $2, $5); }
    | DEFAULT ':'                { CLR_SRC(); }
      case_action                { $$ = oprNode(DEFAULT, 1, $4); }
    ;
case_action
    : statement                  { $$ = makeList($1); }
    | case_action
      statement                  { $$ = appendList($1, $2); }
    ;
case_op_action
    :                            { $$ = NULL; }
    | case_action                { $$ = $1; }
    ;
jump_statement
    : GOTO identifier ';'        { $$ = oprNode(GOTO, 1, $2); ADD_SRC($$); }
    | CONTINUE ';'               { $$ = oprNode(CONTINUE, 0); ADD_SRC($$); }
    | BREAK ';'                  { $$ = oprNode(BREAK, 0);    ADD_SRC($$); }
    | RETURN ';'                 { $$ = oprNode(RETURN, 0);   ADD_SRC($$); }
    | RETURN expr ';'            { $$ = oprNode(RETURN, 1, $2); ADD_SRC($$); }
    ;
while_expr
    : WHILE '(' expr ')'         { $$ = $3; ADD_SRC($$); }
    ;
iteration_statement
    : while_expr statement       { $$ = oprNode(WHILE, 2, $1, $2); }
    | DO                         { CLR_SRC(); }
      statement while_expr ';'   { $$ = oprNode(DO, 2, $3, $4); }
```

```
      | FOR '(' opt_expr ';'
              opt_expr ';'
              opt_expr ')'        { CLR_SRC(); }
        statement                 { $$ = oprNode(FOR, 4, $3, $5, $7, $10); }
      ;
opt_expr
      :                           { $$ = NULL; }
      | expr                      { $$ = $1; ADD_SRC($$); }
      ;
enum_specifier
      : ENUM
        '{' enumerator_list '}'   { $$ = newAttr(ENUM);
                                    $$->newData = oprNode(ENUM, 1, $3);
                                  }
      | ENUM
        '{'
        enumerator_list ',' '}'   { $$ = newAttr(ENUM);
                                    $$->newData = oprNode(ENUM, 1, $3);
                                  }

      ;
enumerator_list
      : enumerator                { $$ = makeList($1); }
      | enumerator_list ','
        enumerator                { $$ = appendList($1, $3); }
      ;
enumerator
      : identifier
        opt_enumerator_expr       { $$ = $1;
                                    $$->id.init = $2;
                                  }

      ;
opt_enumerator_expr
      : '=' constant_expr         { $$ = $2; }
      |                           { $$ = NULL; }
      ;
struct_or_union_specifier
      : struct_or_union
        opt_identifier
        '{'
        struct_declaration_list
        '}'                       { $$ = newAttr($1);
                                    $$->typeName = $2;
                                    $$->newData  = $4;
                                    bitFieldOffsetAssign($$->newData);
                                  }
      | struct_or_union
        IDENTIFIER                { $$ = newAttr($1);
                                    $$->typeName = $2;
                                  }

      ;
struct_or_union
      : STRUCT                    { $$ = STRUCT; }
      | UNION                     { $$ = UNION; }
      ;
opt_identifier
      :                           { $$ = NULL; }
      | IDENTIFIER                { $$ = $1; }
      ;
struct_declaration_list
      : struct_declaration        { $$ = makeList($1); }
      | struct_declaration_list
        struct_declaration        { $$ = appendList($1, $2); }
      ;
struct_declaration
      : type_specifier
        struct_declarator_list ';' { $$ = oprNode(DATA_DECL, 1, $2);
                                     $$->opr.attr = $1;
                                   }
      | type_specifier
        identifier ':' CONSTANT ';'
                                   { $$ = NULL;
                                     if ( $1->type != CHAR )
                                       yyerror("bit field type error!");
                                     else if ( $4 > 8 || $4 <= 0 )
                                       yyerror("bit field size error!");
                                     else
```

```
                                        {
                                $$ = oprNode(DATA_DECL, 1, makeList($2));
                                $1->bitField   = ($4 < 8)? $4: 0;
                                $1->isUnsigned = 1;
                                $$->opr.attr   = $1;
                                        }
                                }
        ;
struct_declarator_list
    : declarator                    { $$ = makeList($1); }
    | struct_declarator_list ','
      declarator                    { $$ = appendList($1, $3); }
    ;
%%

static int parListCheck(node *np)
{
    int i, n = LIST_LENGTH(np);
    for (i = 0; i < n; i++)
    {
        node *nnp = LIST_NODE(np, i);
        attrib *attr = nnp->id.attr;

        if ( attr->type == VOID && !attr->ptrVect && !attr->isFptr )
            return (n == 1 && !strlen(nnp->id.name))? 0: -1;
    }
    return n;
}

static void bitFieldOffsetAssign(node *p)
{
    if ( p && p->type == NODE_LIST )
    {
        int i, offset = 0;
        for (i = 0; i < p->list.nops; i++)
        {
            node *np1 = LIST_NODE(p, i);
            if ( np1->type == NODE_OPR && np1->opr.oper == DATA_DECL && np1->opr.nops == 1 )
            {
                attrib *ap = np1->opr.attr;
                if ( ap && ap->bitField )
                {
                    if ( (offset + ap->bitField) > 8 ) offset = 0;

                    ap->bitField |= (offset << 8);           // set offset for the field
                    offset       += (ap->bitField & 0xff);  // adjust offset for next...
                    continue;
                }
            }
            offset = 0;
        }
    }
}
```

Chapter-5

Pre-scan for Parsing Tree

As described in the last chapter, the parsing tree generated from `yyparse()` contains pretty much all the information from the input (C source code). Actually, the parsing tree is still in 'customized' format, which usually becomes obstacles for compiling later on. To make compilation easy and straightforward, the following must be done during the pre-scan.

1. Eliminate preprocessor statement lines (`#define`, `#undef`, `#if` ... `#else` ... `#endif`), besides `#pragma`.
2. Remove those unused codes within conditional compilation control, as well as those repeated inclusions of header files.
3. Merge the operations on constants.
4. Replace macro definitions (`#define` and `enum`) at the places where they are referenced.
5. Restore the type specifiers to their original forms.
6. Syntax validations (variable names shall be defined before being used).

After pre-scan, the parsing tree will become cleaner and slimmer. Here are some examples that illustrate the changes before and after pre-scan:

Example 5.1 - ch5_t1.c

<u>Before</u>

```
typedef unsigned char BYTE;
BYTE a;
```

<u>After</u>

```
unsigned char a;
```

Example 5.2 - ch5_t2.c

<u>Before</u>

```
int foo(int n)
{
#define SEED 10
  if ( n ) {
    enum{V1=1,V2,V3};
    n = V3 + SEED;
  }
  else {
    enum{V1=101,V2,V3};
    n = V3 + SEED;
  }
  return n;
}
```

<u>After</u>

```
int foo(int n)
{
  if ( n ) {
    n = 13;
  }
  else {
    n = 113;
  }
  return n;
}
```

5.1   Preparations for pre-scan

<u>Project directory</u>: /p16ecc/cc_source_5.1
<u>New source files</u>: nlist.h, nlist.cpp, sizer.h, sizer.cpp

In this chapter, some new data structures or classes are added to the design, before going to pre-scan.

5.1.1   <u>Name nodes and list</u>

Naturally, shown as Code-5.1, it's a good idea to create a search list or table that records the symbols, along with their definitions. Each symbol defined by a #define, enum, or user-defined type specifier name, along with its definition (type, name, etc.), will be wrapped in a node (Nnode) and added to the list. The nodes in the list are organized into layers corresponding to the nested braces in which they are defined.

As shown in Example 5.2 above, symbols V1 … V3 have been defined twice (in both 'if' and 'else' braced blocks) by enum statements; and symbol SEED is defined outside of the "if … else…" statement. In C, the search for a symbol definition always starts from the current braced layer and then its parent layer, but never from a sibling braced layer. The search can be limited to the same layer or go through the entire list. For this reason, the Nnode data class has a double-linked structure.

```
1    #ifndef _NLIST_H
2    #define _NLIST_H
3    #include "common.h"
4
5    class Nnode {
6        public:
7            Nnode(int _type, char *_name=NULL, node *np1=NULL, node *np2=NULL);
8            ~Nnode();
9
10           int      type;            // Nnode type - DEFINE/ENUM
11           char     *name;           // Nnode name
12           node     *np[2];          // parameters used in definition
13           attrib   *attr;           // attributes
14           Nnode    *next;           // link pointer to the same block
15           Nnode    *parent;         // link pointer to the parent
16
17           int      nops(void)       { return (np[0] == NULL)? 0:
18                                              (np[1] == NULL)? 1: 2; }
19    };
20
21    class Nlist {
22        private:
23            Nnode *list;
24            Nnode *find(char *name, int type);
25
26        public:
27            Nlist ();
28            ~Nlist();
29            void     addLayer(void);
30            void     delLayer(void);
31            Nnode    *add(char *name, int type, node *np1=NULL, node *np2=NULL);
32            void     del(char *name);
33            Nnode    *search(char *name, int type=0);
34            attrib   *search(attrib *attr);
35    };
36
37    #endif
```
/p16ecc/cc_source_5.1/nlist.h   Code-5.1

- #5~19: node class that records a single name for #define or enum definition. Note, *next is for the symbols that appear in the same braced block; and *parent is for its right outside block.
- #21~36: class for search list, which is constructed based on Nnode.

Figure-5.1 shows the architecture of the search list/table.

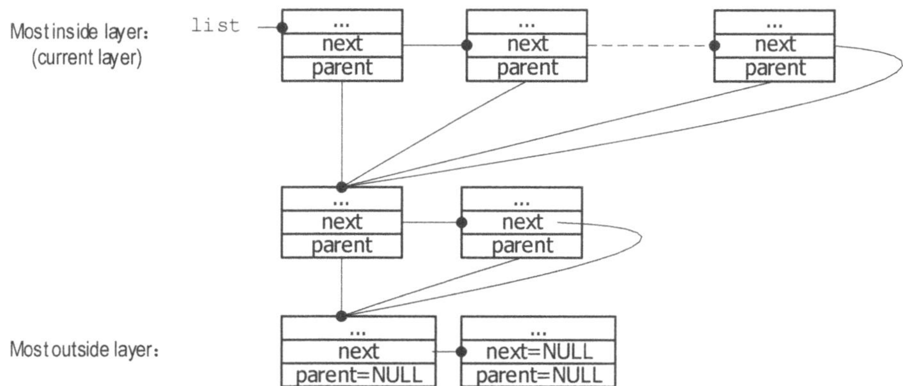

Figure-5.1

As shown in Firgure-5.1,
- All the nodes in the same layer (braced block) have the same value for parent.
- The last node in a layer has the same value for both next and parent.
- A NULL value for parent in a node indicates the most outside layer.

A blank Nnode (all fields zeroed) is generated when a new layer is created. The most outside layer will be created when Nlist is instantiated (see Code-5.2).

```
28   ///////////////////////////////////////////////////////////////////////
29   Nlist :: Nlist()
30   {
31       list = NULL;
32       addLayer();
33   }
34
35   Nlist :: ~Nlist()
36   {
37       while ( list ) {
38           Nnode *next = list->next;
39           delete list;
40           list = next;
41       }
42   }
43
44   void Nlist :: addLayer(void)
45   {
46       Nnode *p   = new Nnode(0);
47       p->next    = list;
48       p->parent  = list;
49       list       = p;
50   }
```
/p16ecc/cc_source_5.1/nlist.cpp

Code-5.2

- #29~33: Nlist constructor. It creates the most outside layer.
- #44~50: create a new layer with a blank Nnode.

The `Nlist` is dynamically changing as the pre-scan proceeds. When pre-scan reaches a compound statement in the parsing tree, it adds a new layer by calling `addLayer()` into the list, which becomes the current layer, and then removes the layer from the list right after the statement.

Code-5.3 shows adding a symbol to the list. For a new symbol defined by #define, it will be added to the most outside layer; otherwise, it's just added to the current layer.

```
81   Nnode *Nlist :: add(char *name, int type, node *np1, node *np2)
82   {
83       if ( find(name, type) ) return NULL;
84
85       Nnode *p = new Nnode(type, name, np1, np2);
86       if ( np1 ) delSrc(np1), np1->id.src = NULL;
87       if ( np2 ) delSrc(np2), np2->id.src = NULL;
88
89       if ( type == DEFINE )
90       {
91           Nnode *p1 = list;
92           while ( p1->parent ) p1 = p1->next;
93
94           p->next = p1;                    // p1 -> the most outside layer.
95
96           for (Nnode *p2 = list; p2->parent; p2 = p2->next)
97           {
98               if ( p2->parent == p1 ) p2->parent = p;
99               if ( p2->next   == p1 ) p2->next   = p;
100          }
101
102          if ( list == p1 ) list = p; // don't forget this!
103      }
104      else
105      {
106          p->next   = list;
107          p->parent = list->parent;
108          list = p;
109      }
110      return p;
111  }
```
/p16ecc/cc_source_5.1/nlist.cpp Code-5.3

- #89~103: for symbols from #define,
    - searching for the most outside layer (#92).
    - insert the new node at the front of the layer (#94).
    - adjust the pointers in the next layer (#96~100).
- #104~109: for other types of symbols, simply insert it in the front of the current layer.

5.1.2 Sizer class tool

The sizer class provides a set of functions for measuring object sizes in bytes, such as the total length and offsets of data members in a struct. Also, the functions in the class will help trace an alias defined by a `typedef` back to its original type. The class definition is shown in Code-5.4.

```
 1   #ifndef _SIZER_H
 2   #define _SIZER_H
 3   #include "common.h"
 4
 5   typedef enum {
 6       TOTAL_SIZE = 1, ATTR_SIZE, INDIR_SIZE, SUBDIM_SIZE
 7   } SIZER_OPTION;
 8
 9   class Sizer {
10       public:
11           int size(attrib *ap, SIZER_OPTION opt);
12           int size(bool neg_val, int val);
13
14           int     memberCount(attrib *ap);
15           attrib *memberAttr(attrib *ap, int index);
16           attrib *memberAttr(attrib *ap, char *id_name);
17           attrib *memberAttrClone(attrib *ap, int index);
18           attrib *memberAttrClone(attrib *ap, char *id_name);
19           int     memberOffset(attrib *ap, char *id_name);
20
21       private:
22           int size(node *np, int type, bool total_size = false);
23   };
24
25   extern Sizer sizer;
26
27   #endif
```

/p16ecc/cc_source_5.1/sizer.hCode-5.4

- #11: a variable or function return value size is always obtained from its attributes.
- #12: it gives an instant's size based on the value.
- #15~18: routines that obtain object's original data type.
- #19: member in a `struct/union` offset.

In **cc16e**, there are four basic types of data, and their lengths are defined as in Table-5.1:

Data type	Length in bytes
char	1
int	2
short	3
long	4

Table-5.1

All types of pointers are 2 bytes in length.

Code-5.5 is a function in the class that returns the size of an object.

```cpp
12     int Sizer :: size(attrib *ap, SIZER_OPTION opt)
13     {
14         int n = 0;
15
16         if ( opt == INDIR_SIZE )
17         {
18             if ( ptrWeight(ap) > 0 )
19             {
20                 attrib *attr = cloneAttr(ap);
21                 reducePtr(attr);
22                 n = size(attr, TOTAL_SIZE);
23                 delAttr(attr);
24                 return n;
25             }
26         }
27         if ( opt == SUBDIM_SIZE )
28         {
29             attrib *attr = cloneAttr(ap);
30             decDim(attr);
31             n = size(attr, TOTAL_SIZE);
32             delAttr(attr);
33             return n;
34         }
35
36         int dim_size = (opt == ATTR_SIZE)? 1: dimScale(ap);
37
38         if ( ap->ptrVect || ap->isFptr )
39             n = 2;
40         else
41             switch ( ap->type )
42             {
43                 case CHAR:  n = (BF_OFFSET(ap) && opt == TOTAL_SIZE)? 0: 1; break;
44                 case INT:
45                 case SHORT:
46                 case LONG:  n = ap->type - CHAR + 1;     break;
47
48                 case STRUCT:
49                 case UNION:
50                     if ( ap->newData )
51                     {
52                         n = size((node*)ap->newData, ap->type, opt == TOTAL_SIZE);
53                         break;
54                     }
55                 default:
56                     n = -1;
57                     if ( ap->type )
58                         printf("something wrong! - %d\n", ap->type); // it should never happen!
59             }
60
61         return n * dim_size;
62     }
```

/p16ecc/cc_source_5.1/sizer.cpp   Code-5.5

To measure an integer, shown in Code-5.6, it needs to consider its absolute value, if it is a negative value. It just gives the type that belongs (char, int, short, long).

```cpp
64     int Sizer :: size(bool neg_val, int val)
65     {
66         unsigned int value = (unsigned int)val;
67
68         if ( neg_val )
69         {
70             if ( val < 0 )
71                 return (val >= -128    )? CHAR:
72                        (val >= -32768   )? INT:
73                        (val >= -8388608)? SHORT: LONG;
74
75             return (val < 128    )? CHAR:
76                    (val < 32768   )? INT:
77                    (val < 8388608)? SHORT: LONG;
78         }
79
80         return (value <= 0xffU    )?  CHAR:
81                (value <= 0xffffU   )?  INT:
82                (value <= 0xffffffU)?  SHORT: LONG;
83     }
```

/p16ecc/cc_source_5.1/sizer.cpp   Code-5.6

- #68~78: for a negative integer.
- #80~82: for a positive integer.

The following routine, shown in Code-5.7, helps finding the offset of a member in struct/union.

```
211    int Sizer :: memberOffset(attrib *ap, char *id_name)
212    {
213        if ( ap->type == STRUCT )
214        {
215            int offset = 0;
216            node *np = (node*)ap->newData;
217            for (int i = 0; i < LIST_LENGTH(np); i++)     // lines of declair
218            {
219                node *pp = LIST_NODE(np, i);
220                for (int j = 0; j < LIST_LENGTH(pp); j++)
221                {
222                    node *ppp = LIST_NODE(pp, j);        // variable id list
223                    node *inp = OPR_NODE(ppp, 0);
224
225                    if ( strcmp(inp->id.name, id_name) == 0 ) // found member
226                        return (BF_OFFSET(ppp->opr.attr) > 0)? offset - 1: offset;
227
228                    offset += size(ppp->opr.attr, TOTAL_SIZE);
229                }
230            }
231        }
232        return 0;
233    }
```
/p16ecc/cc_source_5.1/sizer.cpp

Code-5.7

- #225~226: when found the member, adjust the offset by -1 if it's not the first bit-field member.
- #232: offset is always 0 for a union member.

5.2   Pre-scan process

<u>Project directory</u>: /p16ecc/cc_source_5.2
<u>New source files</u>: prescan.h, prescan.cpp

The pre-scan operation is also encapsulated in a class, as shown in Code-5.8. The scanning will traverse the entire parsing tree, node by node.

```
1    #ifndef _PRESCAN_H
2    #define _PRESCAN_H
3    #include "common.h"
4
5    class PreScan
6    {
7        public:
8            PreScan(Nlist *_nlist);
9            node *scan(node *np);
10
11       private:
12           Nlist *nlist;
13           int    errCount;
14           int    tagSeed;
15           src_t *src;
16           int    enumNeg;  // min item size in an enum set
17           int    enumPos;  // max item size in an enum set
18
19           node *scan(idNode_t  *ip);
20           node *scan(oprNode_t *op);
21           int    mergeCon(int n1, int n2, int oper);
22           node *replacePar (node *np, node *name_par, node *real_par);
23           void   errPrint(const char *msg, char *extra=NULL);
24           char *tagLabel(void);
25           void   attrRestore(attrib **ap, bool update_flags=false);
26           bool   StUnNameCheck(node *np, char *name);
27    };
28
29    #endif
```
/p16ecc/cc_source_5.2/prescan.h   Code-5.8

- #8: the constructor of the class. It brings in the initialized `Nlist`.
- #9: function called from `main()` to start pre-scan operation.
- #19 and #20: these two functions are used to scan specific nodes (`idNode_t` and `oprNode_t`) in parsing tree.

5.2.1 <u>Start pre-scan</u>

Pre-scan happens right after the parsing tree generated, in `main()`, shown as Code-5.9.

```
 1    #include <stdio.h>
 2    #include <stdlib.h>
 3    #include <string>
 4    #include "common.h"
 5    #include "display.h"
 6    #include "nlist.h"
 7    #include "prescan.h"
 8
 9    int main(int argc, char *argv[])
10    {
11        for (int i = 1; i < argc; i++)
12        {
13            std::string buffer = "cpp1 ";
14            buffer += argv[i];
15
16            if ( system(buffer.c_str()) == 0 )
17            {
18                buffer = argv[i]; buffer += "_";
19
20                int rtcode = _main((char*)buffer.c_str());
21                remove(buffer.c_str());
22
23                display(progUnit, 0);
24                if ( rtcode == 0 )
25                {
26                    Nlist nlist;
27                    PreScan preScan(&nlist);
28                    progUnit = preScan.scan(progUnit);
29                }
30
31                //display(progUnit, 0);
32            }
33        }
34        return 0;
35    }
```
/p16ecc/cc_source_5.2/main.cpp   Code-5.9

- #26: create/instantiate the `Nlist`. It's now empty. (it could be initialized with contents defined in the command line)
- #27: create/instantiate `PreScan` class.
- #28: start pre-scan, and get the result that has been scanned. (after pre-scan the result is still a parsing tree)

Note, as a test, we can print the parsing tree, `progUnit`, before and after pre-scan to compare the differences.

As defined in **common.h**, there are five different nodes that are unified as `node`. And only three of them need to be scanned, as shown in Code-5.10.

```
15   PreScan :: PreScan(Nlist *list) // constructor
16   {
17       nlist = list;
18       errCount = 0;
19       tagSeed  = 0;
20       src      = NULL;
21   }
22
23   node *PreScan :: scan(node *np)
24   {
25       if ( np != NULL )
26       {
27           GET_COMMENT(np);
28
29           switch ( np->type )
30           {
31               case NODE_ID:    // scan an Id node
32                   return scan((idNode_t*)np);
33
34               case NODE_STR:   // scan a String node
35               case NODE_CON:   // scan a Constant node
36                   break;
37
38               case NODE_OPR:   // scan an Operation node
39                   return scan((oprNode_t*)np);
40
41               case NODE_LIST: // scan a List node
42                   for (int i = 0; i < np->list.nops; i++)
43                       np->list.ptr[i] = scan(np->list.ptr[i]);
44                   break;
45           }
46       }
47       return np;
48   }
```

/p16ecc/cc_source_5.2/prescan.cpp   Code-5.10

- #15~21: constructor of class `PreScan`.
- #31~32: scan an ID node.
- #34~36: bypass scanning for String and Constant nodes (return them back directly).
- #38~39: scan an Oper node.
- #41~44: scan a List node (scan the nodes in the list).

<u>5.2.3 Scan ID Node</u>

When reaching an ID node in the parsing tree, it will check the Nlist to see if the ID name has been defined by `#define` or `enum` statement. Copy and replacement will happen if found. Plus, additional scanning for other fields in the node are needed as well. The code is shown in Code-5.11.

```
50   node *PreScan :: scan(idNode_t *ip)
51   {
52       Nnode *dp = nlist->search(ip->name);
53       if ( dp && (dp->type == ENUM || dp->type == DEFINE) && dp->nops() == 1 )
54       {
55           node *new_np = scan(cloneNode(dp->np[0]));
56           new_np->con.src = ip->src, ip->src = NULL;
57
58           if ( new_np->type == NODE_ID )
59           {
60               if ( ip->init ) moveNode(&new_np->id.init, &ip->init);
61               if ( ip->dim  ) moveNode(&new_np->id.dim , &ip->dim);
62               if ( ip->parp ) moveNode(&new_np->id.parp, &ip->parp);
63           }
64           else if ( ip->init || ip->dim || ip->parp )
65               errPrint("error in #define replacement");
66
67           CLR_COMMENT((node*)ip), delNode((node*)ip);
68           ip = (idNode_t*)new_np;
69       }
70
71       if ( ip->type == NODE_ID )
72       {
73           ip->init = scan(ip->init);
74           ip->parp = scan(ip->parp);
75           ip->dim  = scan(ip->dim);
76           if ( ip->dim )  // check if dimensions are defined...
77           {
78               node *lp = ip->dim; // get dimension list
79               for (int i = 0; i < lp->list.nops; i++)
80                   if ( lp->list.ptr[i]->type != NODE_CON )
81                   {
82                       errPrint("id dimension undefined!", ip->name);
83                       break;
84                   }
85           }
86       }
87       return (node*)ip;
88   }
```

/p16ecc/cc_source_5.2/prescan.cpp   Code-5.11

- #52: search ID name in `Nlist`.
- #53~69: if found, replace the node with the defined value (#55). And also copy all the extra operations (#60~62).
- #73~75: scan the fields for other operations (function call parameters, initialization statements and array dimensions).

5.2.4 Scan Oper Node – Compound Statement

As mentioned earlier, when reaching a compound statement, it needs to create a new layer in `Nlist`; and remove the layer after.

```
90    node *PreScan :: scan(oprNode_t *op)
91    {
92        Nnode *ndp = NULL;
93        attrib *ap = NULL, *aap;
94        node   *np = NULL, *p1 = NULL, *p2 = NULL;
95        bool  ins_aap = false;
96
97        switch ( op->oper )
98        {
99            case '{':
100               if ( op->nops > 0 )
101               {
102                   nlist->addLayer();
103                   op->op[0] = scan(op->op[0]);
104                   nlist->delLayer();
105               }
106               return (node*)op;
```
/p16ecc/cc_source_5.2/prescan.cpp
Code-5.12

5.2.5 Scan Oper Node – #define and #undef

`#define` statement can be classified as three forms (described in cc.y):

#define statement form	Parsing tree generation in cc.y	
#define *id*	DEFINE identifier EOL	$$ = oprNode(DEFINE, 1, $2);
#define *id expr*	DEFINE identifier expr EOL	$$ = oprNode(DEFINE, 2, $2, $3);
#define *id* (*par_list*) *expr*	DEFINE IDENTIFIER_ id_list ')' expr EOL	$$ = oprNode(DEFINE, 3, idNode($2), $3, $5);

```
108           case DEFINE:    // add a macro definition
109               if ( op->nops > 1 ) p1 = op->op[1], op->op[1] = NULL;
110               if ( op->nops > 2 ) p2 = op->op[2], op->op[2] = NULL;
111
112               ndp = nlist->add(op->op[0]->id.name, DEFINE, p1, p2);
113               if ( ndp == NULL )
114               {
115                   errPrint("redefined name", op->op[0]->id.name);
116                   delNode(p1), delNode(p2);
117               }
118
119               CLR_COMMENT(op); delNode((node*)op);
120               return NULL;
121
122           case UNDEF:     // remove a macro definition
123               nlist->del(op->op[0]->id.name);
124               CLR_COMMENT(op); delNode((node*)op);
125               return NULL;
```
/p16ecc/cc_source_5.2/prescan.cpp
Code-5.13

- #109~110: fetch *expr* and *par_list*, if there are.
- #112: add the *id* symbol to `Nlist`, together with its *expr* part.
- #119~120: remove the node.
- #122~125: for `#undef` statement/node, simply remove the symbol from `Nlist`.

Note, both types of the nodes will be removed by pre-scan, replaced by **NULL**.

Compiling control preprocessors (`#ifdef`, `#ifndef` and `#if`) have the similar syntax structure. They are treated pretty much the same way in pre-scan.

	Parsing tree generation in cc.y	

`#ifdef` *id*
...
`#else`
...
`#endif`
	`_IFDEF identifier`
	`opt_external_definitions`
	`op_else`
	`_ENDIF`
		`$$ = oprNode(_IFDEF, 3, $2, $3, $4);`

`#ifndef` *id*
...
`#else`
...
`#endif`
	`_IFNDEF identifier`
	`opt_external_definitions`
	`op_else`
	`_ENDIF`
		`$$ = oprNode(_IFNDEF, 3, $2, $3, $4);`

`#if` *expr*
...
`#else`
...
`#endif`
	`if_condition`
	`opt_external_definitions`
	`op_else`
	`_ENDIF`
		`$$ = oprNode(_IF, 3, $1, $2, $3);`

As shown in Code-5.14, the return is one of the two branches (code blocks) based on the evaluation in the statement. For `#ifdef` and `#ifndef`, the evaluation is to check if the symbol has been defined or not, but for `#if`, *expr* could be an expression (math or logic). A zero or non-zero result value determines the evaluation.

```
127     case _IFDEF:
128         if ( nlist->search(op->op[0]->id.name, DEFINE) )
129             np = scan(op->op[1]), op->op[1] = NULL;
130         else if ( op->nops > 2 )
131             np = scan(op->op[2]), op->op[2] = NULL;
132
133         CLR_COMMENT(op); delNode((node*)op);
134         return np;
135
136     case _IFNDEF:
137         if ( !nlist->search(op->op[0]->id.name, DEFINE) )
138             np = scan(op->op[1]), op->op[1] = NULL;
139         else if ( op->nops > 2 )
140             np = scan(op->op[2]), op->op[2] = NULL;
141
142         CLR_COMMENT(op); delNode((node*)op);
143         return np;
144
145     case _IF:
146         op->op[0] = scan(op->op[0]);
147         if ( op->op[0] && op->op[0]->type == NODE_CON )
148         {
149             if ( op->op[0]->con.value )
150             {
151                 np = scan(op->op[1]);
152                 op->op[1] = NULL;
153             }
154             else
155             {
156                 np = scan(op->op[2]);
157                 op->op[2] = NULL;
158             }
159         }
160         else
161             errPrint("unsolved expr in '#if' ");
162
163         CLR_COMMENT(op); delNode((node*)op);
164         return np;
```

- #127~134: scan *#ifdef … #endif* statement.
- #136~143: scan *#ifndef … #endif* statement.
- #145~164: scan *#if … #endif* statement.

5.2.7 Scan Oper Node – CALL

The CALL node in the parsing tree is not necessarily a function call. Rather, it can be a macro definition. In this case, replacement will take place, as shown in Code-5.15.

```
166        case CALL:
167            ndp = nlist->search(op->op[0]->id.name, DEFINE);
168            op->op[1] = scan(op->op[1]);    // parameters...
169            if ( ndp && ndp->np[1] && equListLength(op->op[1], ndp->np[0]) )
170            {
171                node *new_np = cloneNode(ndp->np[1]);
172                for (int i = 0; op->op[1] && i < op->op[1]->list.nops; i++)
173                {
174                    node *name_par = ndp->np[0]->list.ptr[i];
175                    node *real_par =  op->op[1]->list.ptr[i];
176                    new_np = replacePar(new_np, name_par, real_par);
177                }
178
179                new_np = scan(new_np);
180                new_np->id.src = op->src;
181                op->src = NULL; delNode((node*)op);
182                GET_COMMENT(new_np);
183                return new_np;
184            }
```
/p16ecc/cc_source_5.2/prescan.cpp   Code-5.15

- #167: search Nlist.
- #171~177: replacement takes place. The replacement does not only happen to the id name but also applies to parameters if any (#176).

5.2.8 Scan Oper Node – FUNC_DECL

This covers both function declaration and definition. In **cc16e**, it doesn't allow function parameters to contain any type of array, like

```
int func(int a[10])
{
. . .
}
```

Instead, it allows a null array like

```
int func(int a[])
{
. . .
}
```

which is then converted to

```
int func(int *a)
{
. . .
}
```

Besides, variable types need to be restored to their original or native form if it's an alias, by calling `attrRestore()`. Code-5.16 shows the pre-scan for FUNC_DECL.

```
192            case FUNC_DECL:
193                // scan function returning value type
194                attrRestore(&op->attr);
195
196                np = op->op[0]; // FUNC_HEAD
197                if ( np && np->type == NODE_OPR && np->opr.nops > 1 )
198                {
199                    np = OPR_NODE(np, 1);          // parameter list
200                    for (int i = 0; i < np->list.nops; i++)
201                    {   // scan function parameter list
202                        node *parp = np->list.ptr[i];
203                        attrRestore(&parp->id.attr);
204
205                        int list_length = LIST_LENGTH(parp->id.dim);
206                        if ( list_length > 0 && parp->id.attr )
207                        {
208                            while ( list_length-- )
209                            {
210                                node *np = LIST_NODE(parp->id.dim, list_length);
211                                if ( np->con.value != 0 ) errPrint("null dim only!");
212                                insertPtr(parp->id.attr, parp->id.attr->dataBank);
213                            }
214                            delNode((node*)parp->id.dim);
215                            parp->id.dim = NULL;
216                        }
217                    }
218                }
219
220                op->op[1] = scan(op->op[1]);    // scan func body
221                return (node*)op;
```
/p16ecc/cc_source_5.2/prescan.cpp   Code-5.16

- #194: function return type restoration.
- #203: function parameter type restoration.
- #200~217: null array conversion.
- #220: scan function body.

5.2.9 Scan Oper Node – ENUM

An `enum` statement defines a set of symbols or creates a set of symbolized constants. Each symbol in the set has either its assigned value or a default value that starts at 0 and increases by 1 for each subsequent symbol.

Example 5.3

enum statement	Result
enum {SEED1, SEED2, SEED3};	SEED1 = 0, SEED2 = 1, SEED3 = 2

Example 5.4

enum statement	Result
enum {SEED1=100, SEED2, SEED3=200};	SEED1 = 100, SEED2 = 101, SEED3 = 200

Further, sometimes it is important to know what the max length a symbol in the set can be.

Example 5.5

enum statement	Result
typedef enum { SEED1=254, SEED2, SEED3} SEED_t; . . .	SEED1 = 254, SEED2 = 255, SEED3 = **256**

<table>
<tr><td>

```c
int func(SEED_t seed)
{
    switch ( seed )
    {
      case SEED1: return 1;
      case SEED2: return 2;
      case SEED3: return 3;
    }
}
```

</td><td></td></tr>
</table>

where the max length of a `seed` is 2 bytes.

It shows that it must evaluate all the symbols in the set to determine the maximum length required to hold the value for each symbol. And it's done during the pre-scan, as shown in Code-5.17.

```
223    case ENUM:
224        enumNeg = 0;
225        enumPos = 0;
226        np = op->op[0]; // a list of exprs
227        if ( np && np->type == NODE_LIST )
228        {
229            int  enum_v = 0;     // start default value
230            char enum_s = 0;     // sign of enum value
231            for (int i = 0; i < np->list.nops; i++, enum_v++)
232            {
233                node *p  = np->list.ptr[i];
234                node *ep = p->id.init = scan(p->id.init);
235                if ( ep )
236                {   // enum item with init. value
237                    if ( ep->type == NODE_CON )
238                    {
239                        enum_v = ep->con.value; // item value
240                        enum_s = ep->con.attr->isNeg;
241                    }
242                    else
243                        errPrint("unsolved value in enum");
244                }

246                if ( enum_v >= 0 ) enum_s = 0;

248                if ( enum_s ) enumNeg = MIN(enumNeg, enum_v);
249                else          enumPos = MAX((unsigned int)enumPos,
250                                            (unsigned int)enum_v);

252                int type = sizer.size(enum_s != 0, enum_v);
253                node *cp = conNode(enum_v, type);
254                if ( nlist->add(p->id.name, ENUM, cp, NULL) == NULL )
255                {
256                    errPrint("id redefined in enum", p->id.name);
257                    delNode(cp);
258                }
259            }
260        }
261        CLR_COMMENT(op), delNode((node*)op);
262        return NULL;
```

`/p16ecc/cc_source_5.2/prescan.cpp`   Code-5.17

- #229: default start value for symbols = 0, which increases by 1 for the next symbol's value (#230).
- #233: fetch an item from the symbol list.
- #234: obtain its assigned value if it exists.
- #235~244: confirmation of the value.
- #248~250: obtain the min and max values for the symbol list, and save them in `enumNeg` and `enumPos`. They will be used later.
- #252~254: create a constant node and save it into `Nlist`, with its symbol.
- #260: remove `ENUM` node.

5.2.10 Scan Oper Node – DATA_DECL

This node is for data variable declaration/definition in C. The most common instance could be like

Example 5.6
```
    int x, *y, z[10];
```

Pre-scan also covers `struct/union` definition and usages. The definition will be added to `Nlist` since they are user-defined data types.

Plus, during the pre-scan, the variable list will be split into a series of individual `DATA_DECL` nodes, as illustrated below.

Before	After
`int x, *y, z[10];`	`int x; int *y; int z[10];`

Code-5.18 shows how pre-scan scans `struct/union` node.

```
264        case DATA_DECL:
265            aap = ap = op->attr;
266            ap->atAddr = scan((node*)ap->atAddr);
267            if ( ap->type == STRUCT || ap->type == UNION )
268            {
269                if ( ap->newData )  // defined a struct/union here
270                {
271                    if ( ap->typeName == NULL )
272                        ap->typeName = dupStr(tagLabel());
273
274                    ndp = nlist->add(ap->typeName, ap->type);
275
276                    if ( ndp == NULL )
277                        errPrint("struc/union redefined!", ap->typeName);
278                    else
279                    {   // scan the format of struct/union
280                        ap->newData = scan((node*)ap->newData);
281
282                        // check struct/union member names.
283                        if ( StUnNameCheck((node*)ap->newData, NULL) == false )
284                            errPrint("struct/union member id duplicated!");
285
286                        ndp->attr = cloneAttr(ap);
287                    }
288                }
289                else                   // use a struct/union here
290                {
291                    if ( (ndp = nlist->search(ap->typeName, ap->type)) == NULL )
292                        errPrint("struct/union not defined!");
293
294                    ins_aap = (ndp == NULL || ndp->attr == NULL);
295                    aap     = (ndp == NULL || ndp->attr == NULL)? newAttr(ap->type): ndp->attr;
296                    if ( ins_aap && ap->typeName ) aap->typeName = dupStr(ap->typeName);
297                }
298            }
299            else if ( ap->type == TYPEDEF_NAME )
300            {
301                attrRestore(&op->attr, true);
302                aap = ap = op->attr;
303            }
```
`/p16ecc/cc_source_5.2/prescan.cpp`   Code-5.18

- #265: scan the address field, which allocates the variable to a fixed address.
- #269~288: `struct/union` definition. Add the definition into `Nlist` (#274).
- #291: `struct/union` usage, check if it's been defined.
- #294~296: create an `attrib` for `aap`, which will be referenced for instances of the struct/union, when a struct/union member references its own struct/union type (self-reference). See in Example 5.7.

Example 5.7
```
    struct Data {
        int x, *y;
        struct Data *p; // self-reference
            ...
    };
```

Code-5.19 shows how to split the variable list into a series of single DATA_DECL nodes. Also, it will move the type modifiers (pointer, dimensions) and some other modifiers to the target place as shown in Figure-5.1.

```
int *x[3];
```

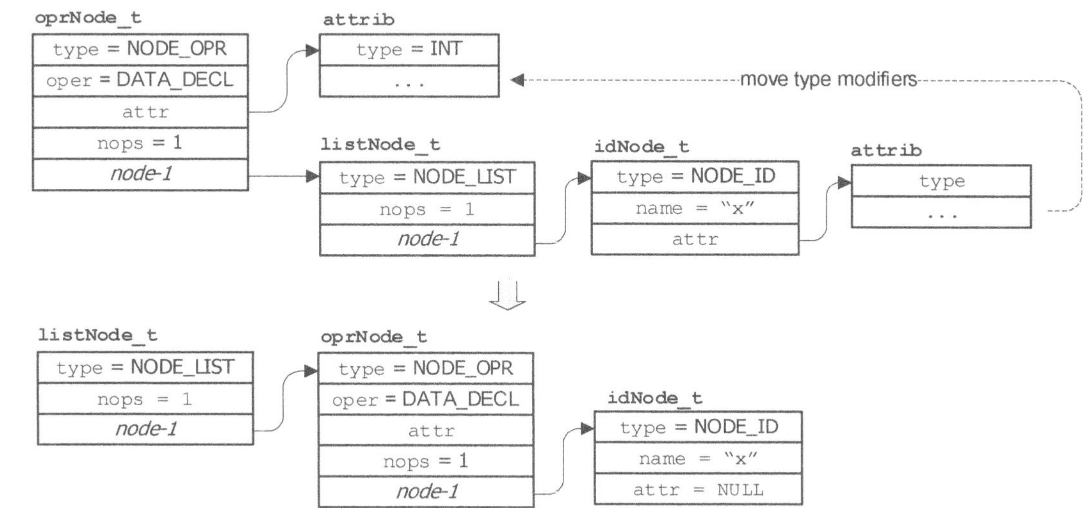

Figure-5.1

```
311             np = op->op[0]; // list of variables
312             for (int i = 0; i < np->list.nops; i++)
313             {
314                 node *inp = scan(np->list.ptr[i]);
315                 node *dnp = oprNode(DATA_DECL, 1);   // create a data node.
316
317                 dnp->opr.attr = cloneAttr(aap);
318
319                 if ( inp->id.attr )
320                 {
321                     dnp->opr.attr->ptrVect = inp->id.attr->ptrVect;
322                     inp->id.attr->ptrVect  = NULL;
323                     delAttr(inp->id.attr); inp->id.attr = NULL;
324                 }
325
326                 inp->id.parp = scan(inp->id.parp);
327                 inp->id.dim  = scan(inp->id.dim);
328
329                 dnp->opr.attr->dimVect = makeDim(inp->id.dim);
330                 dnp->opr.attr->isFptr  = inp->id.fp_decl;
331                 dnp->opr.attr->parList = inp->id.parp;
332                 inp->id.parp = NULL;
333
334                 dnp->opr.op[0]  = inp;
335                 np->list.ptr[i] = dnp;
336
337                 if ( mem_addr >= 0 )
338                 {
339                     delNode((node*)dnp->opr.attr->atAddr);
340                     dnp->opr.attr->atAddr = conNode(mem_addr, INT);
341                     mem_addr += sizer.size(dnp->opr.attr, TOTAL_SIZE);
342                 }
343
344                 /* for data error:  'void a;' */
345                 if ( !(dnp->opr.attr->isFptr || dnp->opr.attr->ptrVect) &&
346                       dnp->opr.attr->type == VOID                       )
347                     errPrint("invalid data type: ", inp->id.name);
348
349                 if ( ins_aap && !dnp->opr.attr->ptrVect )
350                     errPrint("self contained member: ", inp->id.name);
351             }
352
353             op->op[0] = NULL;
354             delNode((node*)op);
355             op = (oprNode_t *)np;
356         }
```

/p16ecc/cc_source_5.2/prescan.cpp

Code-5.19

- #314: get a variable node from list.
- #315: create a `DATA_DECL` node for it.
- #321~332: move the modifiers.
- #349~350: self-reference member must be in pointer type.
- #354: remove old `DATA_DECL` node.
- #355: return reformed `DATA_DECL` node list.

The following is a test for `DATA_DECL` pre-scan.

```
int x, *y, z[10];
```

Before pre-scan

```
list node {1}:
 ch5_t3.c #1: int x, *y, z[10];
 opr node {1}: DATA_DECL {int}
 list node {3}:
  id node: x
  id node: y { *}
  id node: z
  list node {1}:
   constant node: 10 {int}
```

After pre-scan

```
list node {1}:
 list node {3}:
 opr node {1}: DATA_DECL {int}
  id node: x
 opr node {1}: DATA_DECL {int *}
  id node: y
 opr node {1}: DATA_DECL {int[10]}
  id node: z
```

5.2.11 Scan Oper Node – TYPEDEF

`typedef` statement can be classified into two forms:

1. Create an alias for an existing data type.

2. Create an alias for a `struct`/`union` or `enum`.

Examples

```
typedef unsigned char uint8_t;

typedef struct {
    char name[32];
    int age;
} Student_t;

typedef enum {
  SEED1, SEED2, SEED3
} Seed_t;
```

All these typedef statements shall add a alias name to `Nlist`, with their alias names in pre-scan. The Code-5.20 shows the operation.

```
361    case TYPEDEF:
362        ap = op->attr;
363        ap->newData = scan((node*)ap->newData);
364
365        if ( ap->type == STRUCT || ap->type == UNION )
366        {
367            if ( !StUnNameCheck((node*)ap->newData, NULL) )
368                errPrint("struct/union member id duplicated!");
369        }
370
371        np  = op->op[0];
372        ndp = nlist->add(np->id.name, TYPEDEF);
373
```

```
374             if ( ndp )
375             {
376                 if ( ap->type == ENUM )
377                 {
378                     int neg_size = sizer.size(enumNeg < 0, enumNeg);
379                     int pos_size = sizer.size(enumNeg < 0, enumPos);
380                     ndp->attr = newAttr(MAX(neg_size, pos_size));
381                     ndp->attr->isUnsigned = enumNeg? 0: 1;
382                 }
383                 else
384                 {
385                     attrib *aap = nlist->search(ap);
386                     if ( aap )
387                         ndp->attr = cloneAttr(aap);
388                     else
389                         errPrint("typedef name not defined!", ap->typeName);
390                 }
391             }
392             else
393                 errPrint("redefined type_name!", np->id.name);
394
395             delNode((node*)op);
396             return NULL;
```

                                    Code-5.20

- #363: scan for struct/union or enum definition.
- #367: check if data members in struct/union have duplication.
- #372: add the alias name in Nlist.
- #377~382: for enum statement, figure out the max length (in bytes) that symbol values take.

5.2.12 Scan Oper Node – SIZEOF

In C/C++, `sizeof()` is an operator, not a function, used to determine the size in bytes for a variable or data type. In syntax, `sizeof()` is treated as an integer constant, and its value is computed at compile time. `sizeof()` can be applied to either a data type or a data object. As shown in Code-5.21, the pre-scan only sizes the data type and a string constant. Data object sizing will be left for a later compilation step, and it may be delayed until the linking process.

```
398         case SIZEOF:
399             op->op[0] = scan(op->op[0]);
400             if ( op->nops == 0 )
401             {
402                 int n = 0;
403                 ap = op->attr;
404                 switch ( ap->type )
405                 {
406                     case TYPEDEF_NAME:
407                         ndp = nlist->search(ap->typeName, TYPEDEF);
408                         if ( ndp ) ap = ndp->attr;
409                         else errPrint("sizeof type error!", ap->typeName);
410                         break;
411
412                     case STRUCT:
413                     case UNION:
414                         if ( ap->newData == NULL )
415                         {
416                             ndp = nlist->search(ap->typeName, ap->type);
417                             if ( ndp ) ap = ndp->attr;
418                             else errPrint("sizeof type error!", ap->typeName);
419                         }
420                         break;
421                 }
422
423                 if ( ap == NULL )
424                     errPrint("data type not defined!");
425                 else {
426                     n = sizer.size(ap, TOTAL_SIZE);
427                     if ( n < 0 )
428                         errPrint("sizeof type error!");
429                 }
430
431                 CLR_COMMENT(op), delNode((node*)op);
432                 return conNode(n, INT);
433             }
434             else if ( op->op[0]->type == NODE_STR )
435             {
436                 int n = strlen(op->op[0]->str.str);
437                 CLR_COMMENT(op), delNode((node*)op);
438                 return conNode(n+1, INT);
439             }
440             break;
```
`/p16ecc/cc_source_5.2/prescan.cpp`   Code-5.21

- #401~433: `sizeof()` is for a data type.
- #435~439: `sizeof()` is for a string constant.

5.2.13 Scan Oper Node – Math & Logic Operations

Operations on constants can be merged into a single constant node. This is not only an optimization but also sometimes necessary for compilation. For instance (ch5_t4.c),

```
#define SEED1 5
#define SEED2 (SEED1 + 1)
. . .
#if (SEED2 > SEED1)
. . .
#endif
```

To determine the compilation procedure, the exact values for `SEED1` and `SEED2` must be provided in the #if statements. Code-5.22 shows how to merge constants in binary and unary math & logic operations.

```
452        for (int i = 0; i < op->nops; i++)
453            op->op[i] = scan(op->op[i]);
454
455        switch ( op->oper )
456        {
457            case '+': case '-': case '*': case '/': case '%':
458            case '&': case '|': case '^':
459            case LEFT_OP: case RIGHT_OP:
460            case EQ_OP: case NE_OP: case AND_OP: case OR_OP:
461            case '>': case '<': case LE_OP: case GE_OP:
462                if ( op->op[0]->type == NODE_CON &&
463                     op->op[1]->type == NODE_CON )
464                {
465                    int n = mergeCon(op->op[0]->con.value,
466                                     op->op[1]->con.value, op->oper);
467                    CLR_COMMENT(op); delNode((node*)op);
468                    return conNode(n, INT);
469                }
470                break;
471
472            case '~':
473            case '!':
474            case NEG_OF:
475                if ( op->op[0]->type == NODE_CON )
476                {
477                    int n = op->op[0]->con.value;
478                    n = (op->oper == '~')? ~n:
479                        (op->oper == '!')? (n? 0:1):
480                                            -n;
481
482                    np = op->op[0]; op->op[0] = NULL;
483                    np->con.value = n;
484                    if ( op->oper == NEG_OF ) np->con.attr->isNeg ^= 1;
485                    CLR_COMMENT(op); delNode((node*)op);
486                    return np;
487                }
488                break;
489        }
490        return (node*)op;
```
/p16ecc/cc_source_5.2/prescan.cpp   Code-5.22

- #457~470: merge constants in binary math & logic operations.
- #472~488: constant conversion for unary operations.

The following is the test result of **ch5_t5.c**.

```
enum {ONE=10, TWO, THREE};
int n = ONE + TWO*THREE;
```

<table>
<tr><td>

<u>Before pre-scan</u>
```
list node {2}:
 ch5_t5.c #1: enum ONE=10, TWO, THREE;
 opr node {1}: (oper. token: 285)
 list node {3}:
  id node: ONE
  init...:
  constant node: 10 {int}
  id node: TWO
  id node: THREE
 ch5_t5.c #2: int n = ONE + TWO * THREE;
 opr node {1}: DATA_DECL {int}
 list node {1}:
  id node: n
  init...:
  opr node {2}: +
   id node: ONE
   opr node {2}: *
   id node: TWO
   id node: THREE
```

</td><td>

<u>After pre-scan</u>
```
list node {2}:
 list node {1}:
 opr node {1}: DATA_DECL {int}
  id node: n
  init...:
  constant node: 142 {int}
```

</td></tr>
</table>

Chapter-6

P-code Generation

After parsing, the input C source code was converted into a parse tree. This chapter will discuss how to translate the contents of the parsing tree into an intermediate code called P-code (Pseudo Code) or microcode. P-code is similar to the assembly code; every line of P-code describes a simple operation that is still independent of the target MCU.

The P-code is a 3-tuple structure, and its format can be described as

$$\{op\text{-}code, [t_1, t_2, t_3]\}$$

where *op-code* determines the operation type, while t_1, t_2 and t_3 are the operation items. After processing, the entire parsing tree will be converted into a series of P-codes as output for further processing. For example,

C code operation	P-code (before optimization)	P-code (after optimization)
`z = x + y`	$\{`+`, [t, x, y]\}$ $\{`=`, [z, t]\}$	$\{`+`, [z, x, y]\}$

6.1 Basic classes for constructing P-code

<u>Project directory</u>: /p16ecc/cc_source_6.1
<u>New source files</u>: pnode.h, pnode.cpp, item.h, item.cpp

This section presents the data classes that build up the P-code.

6.1.1 Operation item class – Item

As described above, the operation items are fundamental data elements. They are represented by a class – Item. Class Item not only covers different types of operand data but also is the element of the compiling stack for the 'shift-reduce' operation. Code-6.1a and Code-6.1b show the class definition.

```
 7   typedef enum {
 8       STR_ITEM = 1,    // string Item
 9       CON_ITEM,        // constant Item
10       ID_ITEM,         // id Item
11       TEMP_ITEM,       // temp Item
12       ACC_ITEM,        // ACC Item
13       LBL_ITEM,        // label Item
14       IMMD_ITEM,       // immediate Item
15       INDIR_ITEM,      // indirect address Item
16       DIR_ITEM,        // direct address Item
17       PTR_ITEM,        // pointer Item
18       PID_ITEM,        // pointer variable Item
19   } ITEM_TYPE;
```
/p16ecc/cc_source_6.1/item.h   Code-6.1a

- #7~19: there are 11 types of items.

```
21   typedef union {
22       int      i;
23       char     *s;
24       void     *p;
25   } value;
26
27   #define SIGNED(ip)        ip->acceSign()
28
29   ////////////////////////////////////////////////////////////////////////////
30   class Item {
31
32       public:
33           ITEM_TYPE    type;     // item type
34           value        val;
35           int          bias;     // bias for indirect access
36           attrib       *attr;    // attributes
37           void         *home;    // owner of id (ID_ITEM only)
38           Item         *next;    // linking chain
39
40       public:
41           Item(ITEM_TYPE t);
42           ~Item();
43
44           Item *clone (void);
45           void updateAttr (attrib *ap, int data_bank = -1);
46           void updatePtr  (int *p);
47           void updateName (char *new_name);
48           bool isWritable (void);
49           bool isOperable (void);
50           bool isMonoVal  (void);
51           bool isAccePtr  (void);
52           bool isBitVal   (void);
53           attrib *acceAttr(void);
54           int   stepSize  (void);
55           int   acceSize  (void);
56           int   acceSign  (void);
57           int   storSize  (void);
58           bool  isBF      (void);
59   };
```

/p16ecc/cc_source_6.1/item.h   Code-6.1b

- #21~25: the `union` that holds item value.
- #30~59: item class definition. Note, `*next` (#38) is for stack operations.
    - #33~34: The `type` field indicates the type of the item and determines how the item value is expressed in the `val` union; see the table below.

Item type	Usage	Expressed in union `value`
STR_ITEM	string item	val.s
CON_ITEM	constant item	val.i
ID_ITEM	variable/identifier item	val.s
TEMP_ITEM	temporary item	val.i
ACC_ITEM	accumulator item	val.i
LBL_ITEM	label item	val.s
IMMD_ITEM	address of an identifier item	val.s
INDIR_ITEM	indirect access item (through a temporary item)	val.i
DIR_ITEM	direct access item	val.i
PTR_ITEM	pointer item	val.p
PID_ITEM	indirect access item (through an identifier item)	val.s

Table-6.1

The following (Code-6.2) outlines the functions used for Item class usages.

92

```
61    Item *intItem(int val, attrib *ap=NULL);
62    Item *strItem(char *str);
63    Item *lblItem(int lbl);
64    Item *lblItem(char *str);
65    Item *idItem (char *name, attrib *ap=NULL);
66    Item *ptrItem(void *p);
67    Item *immdItem(char *str, attrib *ap);
68    Item *tmpItem(int index, attrib *ap=NULL);
69    Item *accItem(attrib *attr);
70
71    bool moveMatch(Item *ip0, Item *ip1);
72    attrib *maxMonoAttr(Item *ip0, Item *ip1, int op);
73    attrib *maxSizeAttr(Item *ip0, Item *ip1);
74    bool comparable(int code, Item *ip0, Item *ip1);
75    bool same(Item *ip0, Item *ip1, bool fsr_check = false);
76    bool sameTemp(Item *ip0, Item *ip1);
77    bool overlap(Item *ip0, Item *ip1);
78    bool related(Item *ip0, Item *ip1);
79
80    #define IS_TEMP(ip,idx)    (ip->type == TEMP_ITEM && ip->val.i == idx)
```
/p16ecc/cc_source_6.1/item.h   Code-6.2

- #61~69: create the desired type of Item class.

6.1.2 P-code class – Pnode

Every instance of the Pnode class is a line P-code. Its structure is straightforward, as shown in Code-6.3.

```
4     #include "item.h"
5
6     enum {
7         P_FUNC_BEG = 1024, P_FUNC_END, P_SRC_CODE,
8         P_JZ, P_JNZ, P_JEQ, P_JNE, P_JGT, P_JGE, P_JLT, P_JLE,
9         P_CAST, P_ARG_PASS, P_ARG_CLEAR, P_CALL, P_JBZ, P_JBNZ,
10        P_JZ_INC, P_JZ_DEC, P_JNZ_INC, P_JNZ_DEC,
11        P_MOV, P_DJNZ, P_IJNZ, P_BRANCH,
12        P_SEGMENT, P_FILL, P_COPY, P_ASMFUNC,
13    };
14
15    ////////////////////////////////////////////////////////////////////////
16    class Pnode {
17
18        public:
19            int     type;
20            Item    *items[3];
21            Pnode   *last;
22            Pnode   *next;
23
24        public:
25            Pnode(int type);
26            Pnode(int type, Item *ip0);
27            ~Pnode();
28
29            void updateItem(int index, Item *ip);
30            void updateName(int index, char *str);
31            void insert(Pnode *pp);
32            Pnode *end(void);
33    };
34
35    void addPnode(Pnode **list, Pnode *pnp);
36    void delPnodes(Pnode **list);
```
/p16ecc/cc_source_6.1/pnode.h   Code-6.3

- #6~13: expanded Pnode types. (the Pnode types were originally the token types defined in cc.h)
- #20: operand items.
- #21~22: pointers (*last and *next) for constructing a double-link list. It will make insert and delete Pnode operations easier.

6.1.3 Parsing tree traversal and compiling stack

A parsing tree is constructed with the nodes that represent operations in the order of precedence. P-code generation procedure uses so-called "post-order traversal" to visit all the nodes in the tree. In post-order traversal, it always visits the child nodes first, from left to right, and then the parent node. For example,

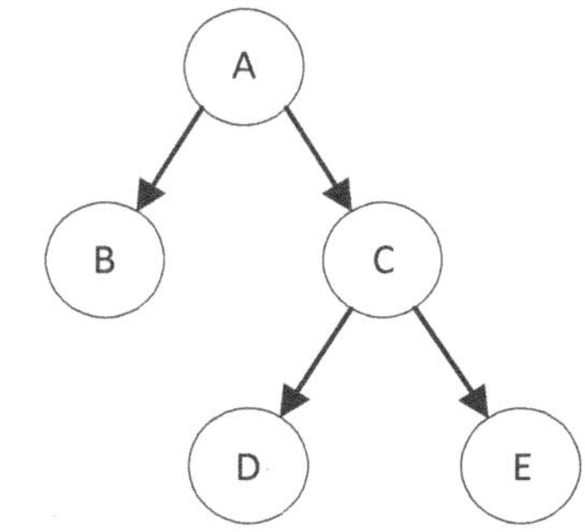

Figure-6.1

In Figure-6.1, the post-order traversal sequence of visits is: **B D E C A**.

When a child node is visited, an item (`Item` class instance) is created and pushed onto the compiling stack. At the parent node, after all child nodes are visited, it pops all items from the stack to generate a P-code, and then the result is converted into a result item to be pushed onto the stack for further operations. This procedure is similar to the 'shift-reduce' operation. The following, Table-6.2, illustrates the operations of "shift-reduce" and P-code generation in detail.

For C code:

```
x = a + b * c;
```

Stack status	Parsing tree traversal	Operation status	P-code generation
$(null)		Start from the root (the stack is empty).	
$(null)		Enter '=' P-code generation.	

$x	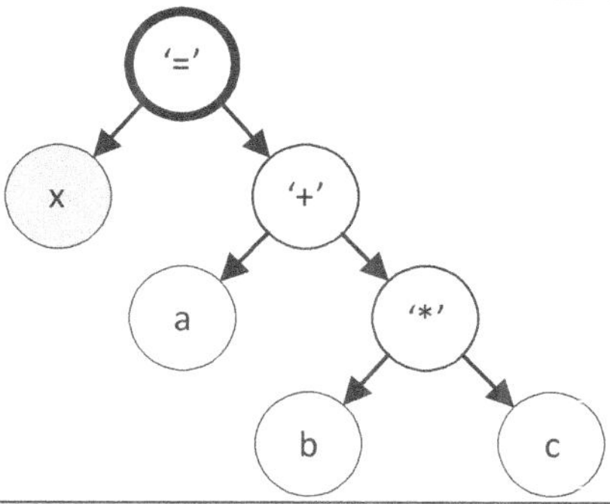 '=' → x, '+'; '+' → a, '*'; '*' → b, c	Create an item for **x**, and push it onto the stack ('shift').	
$x	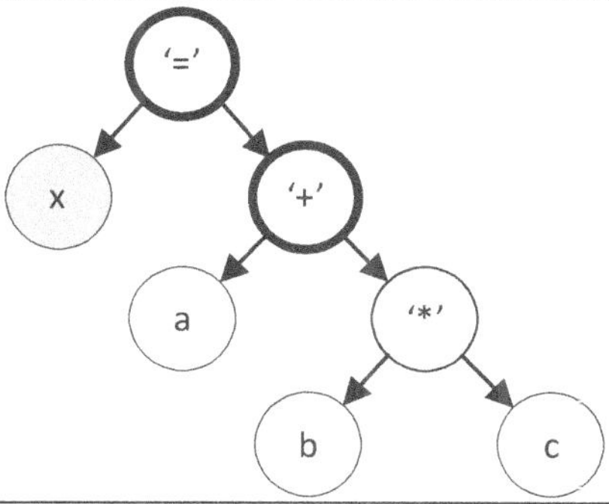 '=' → x, '+'; '+' → a, '*'; '*' → b, c	Enter '+' P-code generation.	
$x a	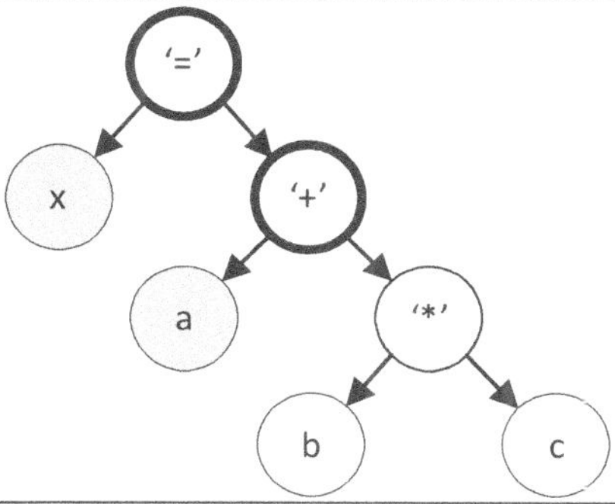 '=' → x, '+'; '+' → a, '*'; '*' → b, c	Create an item for **a**, and push it onto the stack ('shift').	
$x a	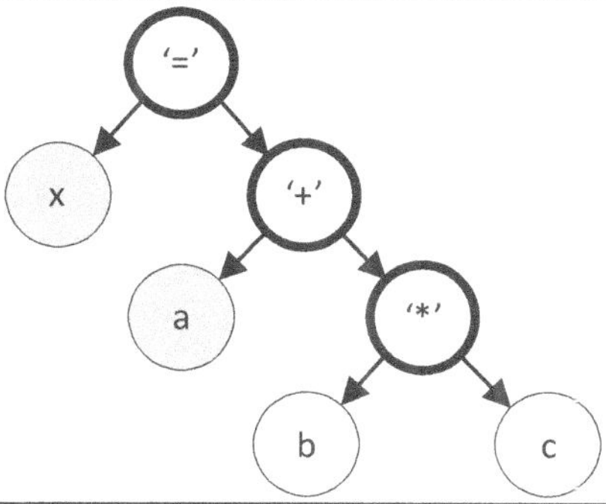 '=' → x, '+'; '+' → a, '*'; '*' → b, c	Enter '*' P-code generation.	
$x a b	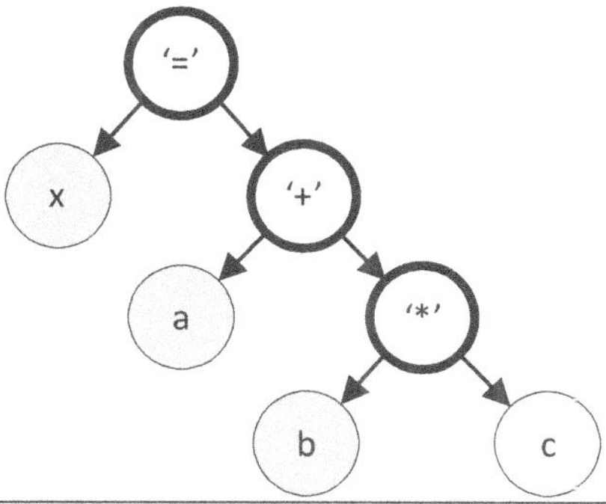 '=' → x, '+'; '+' → a, '*'; '*' → b, c	Create an item for **b**, and push it onto the stack ('shift').	

Stack	Parse tree	Action	Output (P-code)
$x a b c	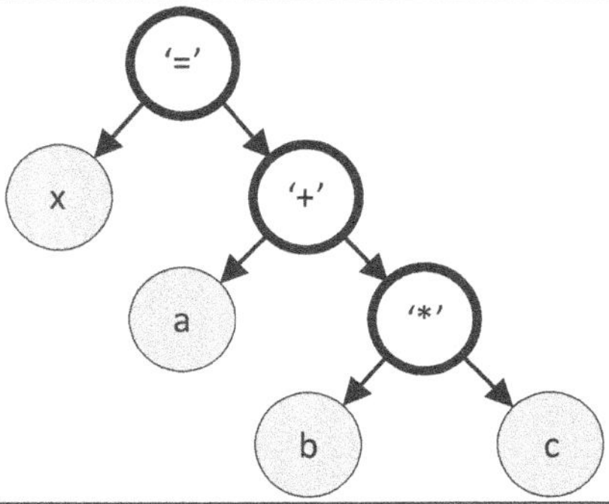	Create an item for c, and push it onto the stack ('shift').	
$x a t_1	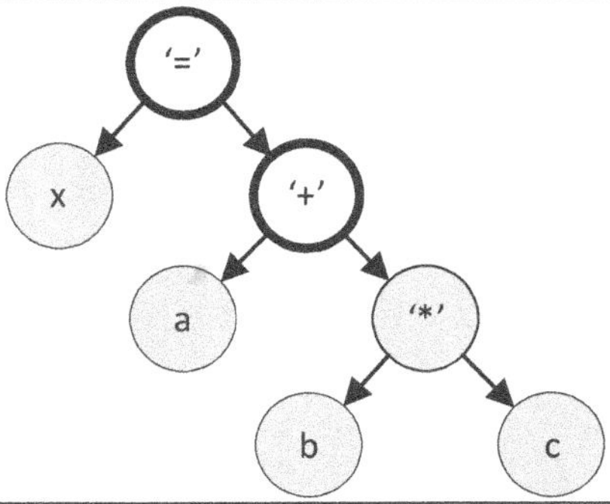	Pop items from the stack ('reduce'), and generate P-code for '*' operation. duplicate and push t_1 to the stack.	{ '*', [t_1, b, c]}
$x t_2	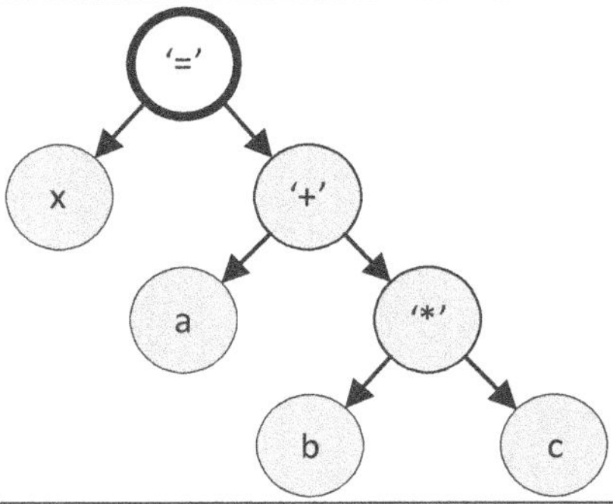	Pop items from the stack ('reduce'), and generate P-code for '+' operation. Then duplicate and push t_2 to the stack.	{ '*', [t_1, b, c]} { '+', [t_2, a, t_1]}
$x	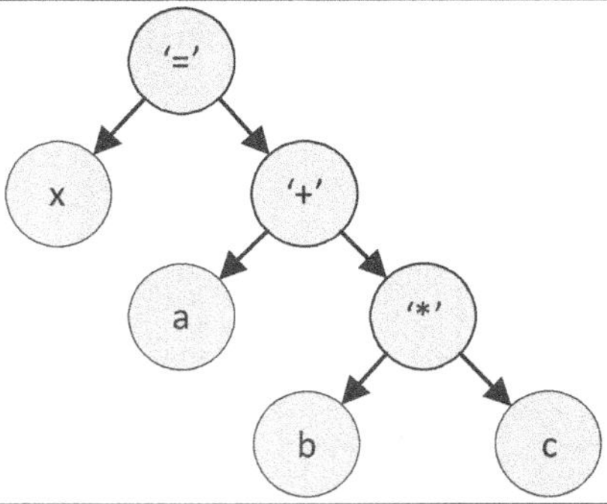	Pop items from the stack ('reduce'), and generate P-code for '=' operation. Then duplicate and push x back to the stack.	{ '*', [t_1, b, c]} { '+', [t_2, a, t_1]} { '=', [x, t_2]}

Table-6.2

Note, (1) where t_1 and t_2 are temporary variables (items); (2) item x is remained as a shadow in the stack.

6.2 Classes for managing data and functions

<u>Project directory</u>: `/p16ecc/cc_source_6.2`
<u>New source files</u>: `dlink.h`, dlink.cpp, flink.h, flink.cpp

During P-code generation, it also performs syntax checks and validations on the input. According to C language syntax rules, any variable or identifier must be defined or clarified before being referenced. Every variable/identifier has its effective region within which it can be referenced. Plus, braces are used not only to form compound statements but also to define the effective scope of their local variables.

To accomplish this, it needs to create two links that manage the data and function names for searching and validation. The following is a typical case that illustrates how variables and identifiers are defined in different regions.

Example-6.1

<table>
<tr><td>

```
    int a, b;
    void func(int x)
    {
      int n;
      if ( x ) {
        int u;
      }
      else {
        long u;
      }
    }
```

</td><td>

Region-1: external variables
function and parameter
Region-2:
  function internal variable
Region-3: it can have its own local variables

Region-4: it's in parallel with *Region-3*

</td></tr>
</table>

- Function name "`func`" belongs to *Region-1*, and function parameter "`x`" belongs to *Region-2*. Within the same region, names can't be duplicated.
- Inner region can access the names defined in the outside region, but not vice versa.
- Regions that are in parallel can have the same names for local variables (their memory allocation will overlap).

6.2.1 Data link and its node classes

Compared to `Nlist`, the data link class is more complex, as it must support additional operations. Code-6.4a and Code-6.4b show the class definitions. Among them, Code-6.4a shows the data node class `Dnode`, which holds a variable defined with all the attributes.

```
 4   class Dnode {
 5       public:
 6           char    *name;          // data name
 7           attrib *attr;           // data attributes
 8           int     index;          // id sequence index
 9           char    *func;          // owner(funtion) pointer
10           int     atAddr;         // memory location
11           int     elipsis:1;      // function with uncertain count of parameters
12           node    *parp;          // function pointer's parameters
13           Dnode   *next;
14
15       public:
16           Dnode(char *_name, attrib *_attr, int _index);
17           ~Dnode();
18
19           char *nameStr(void);
20           void  nameUpdate(char *_name);
21           int   size(void);
22
23           void dimUpdate(node *np);   // dim demension update
24           bool dimCheck(node *np);    // dim dimension check
25           bool fptrCheck(node *np);   // function pointer check
26   };
```
`/p16ecc/cc_source_6.2/dlink.h`   Code-6.4a

- #6~12: data variable information. `index` (#8) is a unique number assigned to each data `variable` because names can be duplicated.
- #13: the link for the variables in the same region.

The `D-Link` class has two types, `PAR_LINK` and `VAR_LINK`, that indicate the link's usage. Its definition and structure are illustrated in Code-6.4b. With the links, the entire data link in Example 6.1 is illustrated in Figure 6.1.

```
28   enum {LOCAL_SEARCH, WHOLE_SEARCH};   // how to search Dlink
29   enum {PAR_DLINK, VAR_DLINK};         // type of Dlink
30
31   class Dlink {
32       public:
33           int    dtype;
34           Dnode *dlist;
35           Dlink *parent, *next, *child;
36           Dnode *search(char *name);
37
38       public:
39           Dlink(int type);
40           ~Dlink();
41
42           void   add(Dlink *lk);       // add a sibling link
43           void   addChild(Dlink *lk);  // add a child link
44
45           int    dataCount(void);
46           void   add(Dnode *dp);
47           Dnode *add(node *np, attrib *attr, int index);
48           Dnode *search(char *name, int search_mode);
49           Dnode *get(int index);
50           bool   nameCheck(char *name, int end_index);
51   };
```
`/p16ecc/cc_source_6.2/dlink.h`   Code-6.4b

- #28: search modes – `LOCAL_SEARCH` and `WHOLE_SEARCH`.
- #29: `Dlink` types – `PAR_DLINK` and `VAR_DLINK`.
- #35: links used for constructing data links (see below).

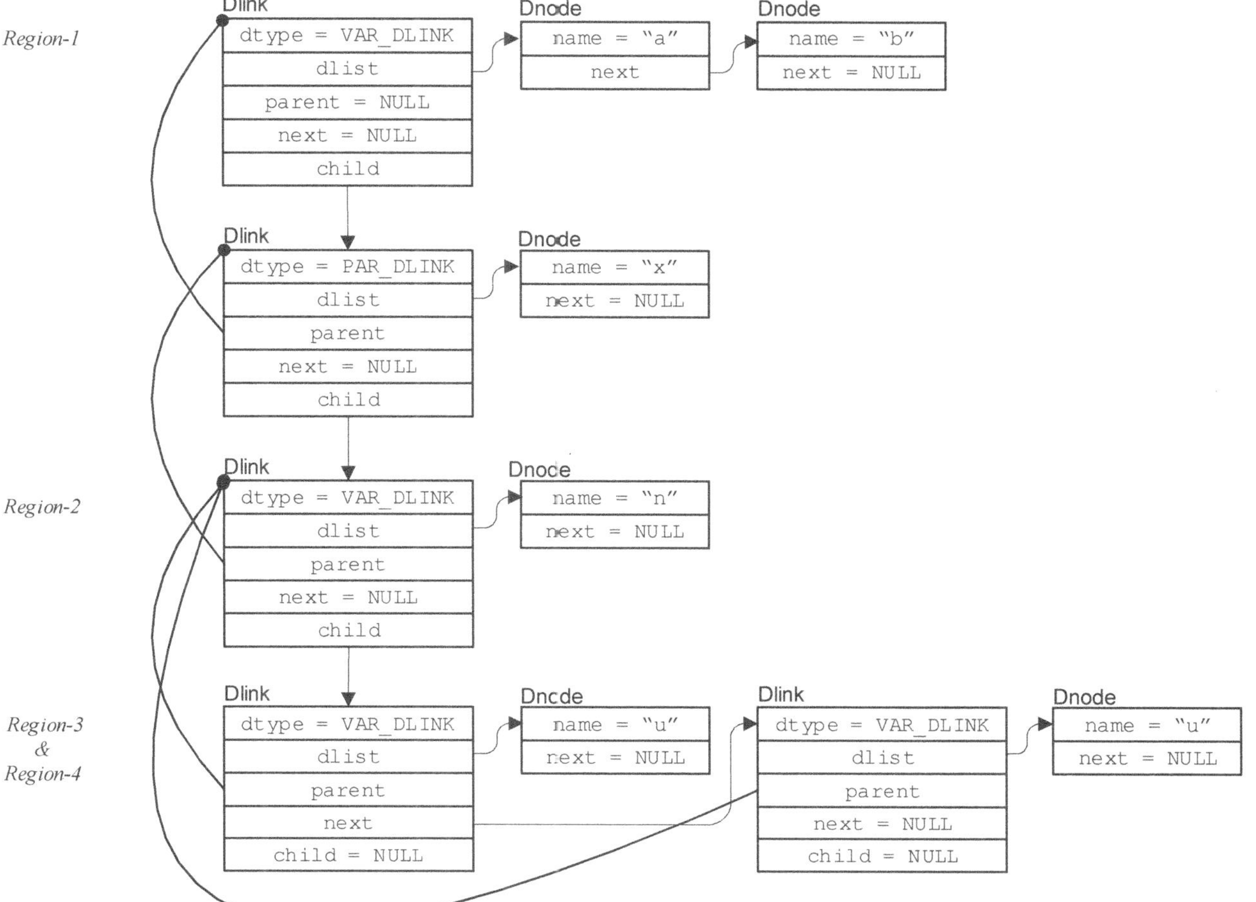

Figure-6.1

<u>6.2.2 Operations of data link class</u>

dlink.cpp lists the basic operations for both Dnode and Dlink classes. Most of them are straightforward to understand.

There are two overloaded functions in Dlink class that add a Dnode, as shown in Code-6.5.

```
116    void Dlink :: add(Dnode *dp)
117    {
118        if ( dp )
119        {
120            Dnode *dnp = dlist;
121            if ( dnp == NULL )
122                dlist = dp;
123            else
124            {
125                while ( dnp->next ) dnp = dnp->next;
126                dnp->next = dp;
127            }
128        }
129    }
130
131    Dnode *Dlink :: add(node *np, attrib *attr, int index)
132    {
133        Dnode *dp    = new Dnode(np->id.name, attr, index);
134        dp->parp     = cloneNode(np->id.parp);
135        dp->elipsis = np->id.parp? np->id.parp->list.elipsis: 0;
136        add(dp);
137        return dp;
138    }
```
/p16ecc/cc_source_6.2/dlink.cpp   Code-6.5

Code-6.6 shows how to add a Dlink as a child. It manages the multiple Dlinks in parallel.

```
140    void Dlink :: addChild(Dlink *lk)
141    {
142        lk->parent = this;
143
144        if ( child == NULL )
145            child = lk;
146        else
147        {
148            Dlink *dlk = child;
149            while ( dlk->next ) dlk = dlk->next;
150            dlk->next = lk;
151        }
152    }
```
/p16ecc/cc_source_6.2/dlink.cpp   Code-6.6

As shown in Code-6.7, searching for a variable name in a data link is based on the usage. When defining a variable, it should use LOCAL_SEARCH mode to check if name duplication occurs in the same region. Otherwise, WHOLE_SEARCH mode should be applied to check if the name has been defined in any effective region.

```
154   Dnode *Dlink :: search(char *name, int search_mode)
155   {
156       for (Dnode *dp = dlist; dp; dp = dp->next)  // search in local dlist.
157       {
158           if ( strcmp(name, dp->name) == 0 )
159               return dp;
160       }
161
162       if ( parent == NULL )   return NULL;
163
164       if ( search_mode == LOCAL_SEARCH )
165       {
166           if ( dtype == PAR_DLINK || parent->dtype != PAR_DLINK )
167               return NULL;
168       }
169
170       return parent->search(name, search_mode);
171   }
```
/p16ecc/cc_source_6.2/dlink.cpp   Code-6.7

- #156~160: search the local `dlist` for the name.
- #164~168: if it's a `LOCAL_SEARCH` mode, the search will stop at `PAR_DLINK`.
- #170: otherwise, keep searching in the parent's `dlist`.

6.2.3 Function link classes

Function link classes are simpler than data link classes. It's defined in flink.h, as shown in Code-6.8.

```
 7   typedef enum {STATIC_DATA, GENERIC_DATA} FDATA_t;
 8
 9   class Fnode {
10       public:
11           char      *name;
12           int        ftype;
13           attrib    *attr;
14           Dlink     *dlink;
15           bool       elipsis;
16           int        dIndex;
17           int        endLbl;
18           NameList *fcall;      // collect func calls.
19           Fnode     *next;
20
21       public:
22           Fnode (char *_name);
23           ~Fnode();
24
25           int      parCount(void) { return dlink->dataCount(); }
26           int      parSize(void);        // parameter total size
27           attrib *parAttr(int index); // parameter attrib.
28
29           Dnode    *getData(FDATA_t type, int *index, int *offset);
30           Dnode    *getData(Dnode *dnp, FDATA_t type, int *index, int *offset);
31           Dnode    *getData(Dlink *dlp, FDATA_t type, int *index, int *offset);
32   };
33
34   class Flink {
35       public:
36           Fnode *flist;
37
38       public:
39           Flink() { flist = NULL; }
40           ~Flink();
41
42           Fnode *search(char *name);
43           void  add(Fnode *fp);
44   };
```
/p16ecc/cc_source_6.2/flink.h   Code-6.8

- #9~32: `Fnode` class. Each instance of `Fnode` holds one function definition.

- #14: it has the Dlink link that holds and manages the internal function variables.
  - #19: it uses a single link to make a function list.
  - #18: `fcall` is a name list that keeps all the functions called.
- #34~44: `Flink` class.

6.2.4 Operations of function link class

Code-6.9a ~ Code-6.9c shows a group of overloaded functions in the `Fnode` class, which allows for retrieving a local variable `Dnode`, as well as its offset.

Code in Code-6.9a shows how to get the *i*th variable `Dnode` in a `Dnode` list (*i* is indicated by `depth`).

```
47    Dnode *Fnode :: getData(Dnode *dnp, FDATA_t type, int *depth, int *offset)
48    {
49        for (; dnp; dnp = dnp->next)
50        {
51            attrib *attr = dnp->attr;
52            if ( (attr->isStatic && type == STATIC_DATA) ||
53                 (!attr->isStatic && type == GENERIC_DATA) )
54            {
55                if ( *depth == 0 )
56                    return dnp;
57
58                (*depth)--;
59                *offset += dnp->size();
60            }
61        }
62        return NULL;
63    }
```
/p16ecc/cc_source_6.2/flink.cpp   Code-6.9a

- #47: `offset` is the sequential index of the variable in the list to retrieve.
- #52~53: two types of local variables, static or generic.
- #59: adjust the offset.

Code in Code-6.9b shows how to get the *i*th variable `Dnode` in groups of `Dlink` that are in parallel (*i* is indicated by `depth`).

```
65    Dnode *Fnode :: getData(Dlink *dlp, FDATA_t type, int *depth, int *offset)
66    {
67        Dnode *d;
68        int offs = *offset;
69        for(; dlp; dlp = dlp->next)
70        {
71            *offset = offs;
72            d = getData(dlp->dlist, type, depth, offset);
73            if ( d ) return d;
74
75            d = getData(dlp->child, type, depth, offset);
76            if ( d ) return d;
77        }
78        return NULL;
79    }
```
/p16ecc/cc_source_6.2/flink.cpp   Code-6.9b

- #69: go through all `Dlink`s that are in parallel.
- #71: start with the same offset.
- #72~73: go through `dlist` first.
- #75~76: then go through its child `Dlink`s.

Code in Code-6.9c shows how to get the *i*th variable `Dnode` in the child `Dlink` (*i* is indicated by `depth`).

```
81    Dnode *Fnode :: getData(FDATA_t type, int *depth, int *offset)
82    {
83        *offset = 0;
84
85        // skip dlist
86        return getData(dlink->child, type, depth, offset);
87    }
```
/p16ecc/cc_source_6.2/flink.cpp   Code-6.9c

6.3 Start P-code Generation

<u>Project directory</u>: /p16ecc/cc_source_6.3
<u>New source files</u>: pcoder.h, pcoder.cpp, const.h, const.cpp

P-code generation is done in the `Pcoder` class.

6.3.1 Pcoder class definition

```
14    extern Dlink *dataLink; // data link
15    extern Flink *funcLink; // func link;
16
17    class Pcoder {
18        public:
19            Pcoder();
20            ~Pcoder();
21            void run(node *np);
22
23            void  takeSrc(node *np);
24            Pnode *mainPcode;
25            Pnode *initPcode;
26            Pnode *constPcode;
27            src_t *src;
28            int   errorCount;
29            int   warningCount;
30
31            Dlink *curDlink;      // current data link
32            Fnode *curFnode;      // current func node
33
34            int   labelSeed;
35            int   getLbl(void) { return ++labelSeed; }
36            Item  *iStack;
37            Const *constGroup;
38
39            #define MAX_TEMP_INDEX  32
40            unsigned int tempIndexMask;
```
/p16ecc/cc_source_6.3/pcoder.h   Code-6.10a

- #14: `dataLink` - global data link.
- #15: `funcLink` - global function link.
- #24: `mainPcode` - running code P-code stream.
- #25: `initPcode` - initialization code P-code stream.
- #26: `constPcode` - constant string P-code stream.
- #31: `curDlink` - the current data link.
- #32: `curFnode` - the current function link.
- #34~35: `labelSeed` and `getLbl()` - the seed and function for label index generation.
- #36: `iStack` - the pointer for compiling stack.
- #37: `constGroup` - class for collecting constants that will be located in ROM.
- #39~40: maximum temporary variables/items = 32 (note, it's usually enough).

Three P-code streams will be generated, each for a different purpose. The following example shows the reason.

Example 6.1:

```
const char array[3] = {'1', '2', '3'};  // ROMed constants, they are put in constPcode.
int seed = 100;                 // initialization code; put it in initPcode.
void delay(int n)                    // running code, they are put in mainPcode.
{  while ( n-- ); }
```

```
42          private:
43              Item *makeTemp(void);
44              Item *makeTemp(attrib *ap);
45              Item *makeTemp(int size);
46              void releaseTemp(Item *ip);
47
48              void warningPrint(const char *msg, char *opt_msg);
49              void errPrint(const char *msg, char *opt_msg);
50              void errPrint(const char *msg) { errPrint(msg, NULL); }
51              void setLbl(int lbl) {
52                  addPnode(&mainPcode, new Pnode(LABEL, lblItem(lbl)));
53              }
54              void PUSH(Item *ip) {    // push in an Item
55                  ip->next = iStack;
56                  iStack = ip;
57              }
58              void CLEAR(void) {       // clear the stack
59                  while ( iStack ) {
60                      Item *inext = iStack->next;
61                      delete iStack;
62                      iStack = inext;
63                  }
64              }
65              Item *POP(void) {        // pop out an Item
66                  Item *ip = iStack;
67                  if ( ip ) iStack = ip->next;
68                  return ip;
69              }
70              Item *TOP(void) {        // get the top Item
71                  return iStack;
72              }
73              int DEPTH(void) {        // get the depth
74                  int n = 0;
75                  for(Item *ip = iStack; ip; ip = ip->next)
76                      n++;
77                  return n;
78              }
79              void DEL(void) {         // delete top Item
80                  if ( iStack ) delete POP();
81              }
82              void DEL(int n) {        // delete n Items
83                  while ( n-- ) DEL();
84              }
85  };
```

`/p16ecc/cc_source_6.3/pcoder.h`   Code-6.10a

- #54~84: functions for compiling stack operations.

6.3.2 Start Pcoder class

Code-6.11 shows the constructor and destructor of the `Pcoder` class.

```
23  Pcoder :: Pcoder()
24  {
25      memset(this, 0, sizeof(Pcoder));
26
27      dataLink = new Dlink(VAR_DLINK);     // global variable list
28      funcLink = new Flink();
29
30      curDlink = dataLink;                 // current variable list
31      curFnode = NULL;
32
33      constGroup = new Const();
34  }
35
36  Pcoder :: ~Pcoder()
37  {
38      CLEAR();
39      delete constGroup;
40      delPnodes(&mainPcode);
41      delPnodes(&initPcode);
42      delPnodes(&constPcode);
43  }
```

`/p16ecc/cc_source_6.3/pcoder.cpp`   Code-6.11

- #25: Zero out the whole class, all the contents are initialized.
- #27~31: initialize both data and function links.
- #33: create a string manager.

P-code generation got started in main(), as shown in Code-6.12.

```
1    #include <stdio.h>
2    #include <stdlib.h>
3    #include <string>
4    #include "common.h"
5    #include "display.h"
6    #include "nlist.h"
7    #include "prescan.h"
8    #include "pcoder.h"
9
10   int main(int argc, char *argv[])
11   {
12       for (int i = 1; i < argc; i++)
13       {
14           std::string buffer = "cpp1 ";
15           buffer += argv[i];
16
17           if ( system(buffer.c_str()) == 0 )
18           {
19               buffer = argv[i]; buffer += "_";
20
21               int rtcode = _main((char*)buffer.c_str());
22               remove(buffer.c_str());
23
24               display(progUnit, 0);
25               if ( rtcode == 0 )
26               {
27                   Nlist nlist;
28                   PreScan preScan(&nlist);
29                   progUnit = preScan.scan(progUnit);
30
31                   Pcoder pcoder;          // P-code generation
32                   pcoder.run(progUnit);   // run it now.
33               }
34
35               //display(progUnit, 0);
36           }
37       }
38       return 0;
39   }
```
/p16ecc/cc_source_6.3/main.cpp   Code-6.12

- #31~32: instantiate a Pcoder class and start P-code generation, after pre-scan.

Now, it's running for P-code generation, as shown in Code-6.13, in the Pcoder class. The parsing tree has five types of nodes, and the `Pcoder` class treats them separately.

```cpp
77   void Pcoder :: run(node *np)
78   {
79       if ( np == NULL ) return;
80       Dnode *dp;
81       Fnode *fp;
82       Item  *iptr;
83
84       takeSrc(np);
85       switch ( np->type )
86       {
87           case NODE_CON:  // const Node
88               PUSH(intItem(np->con.value, cloneAttr(np->con.attr)));
89               break;
90
91           case NODE_STR:  // string Node
92               iptr = idItem(constGroup->add(np->str.str));
93               iptr->attr = newAttr(CHAR);
94               iptr->type = IMMD_ITEM;
95               iptr->attr->dataBank = CONST;
96               PUSH(iptr);
97               break;
98
99           case NODE_ID:   // ID node
100              dp = curDlink->search(np->id.name, WHOLE_SEARCH);
101              fp = funcLink->search(np->id.name);
102              if ( dp )           // found in data link
103              {
104                  Item *ip = (dp->attr->dimVect)? immdItem(dp->nameStr(), dp->attr):
105                                                  idItem  (dp->nameStr(), dp->attr);
106                  ip->home = (void*)dp;
107                  PUSH(ip);
108              }
109              else if ( fp )  // found in function link
110              {
111                  Item *ip = lblItem(np->id.name);
112                  ip->home = (void*)fp;
113                  PUSH(ip);
114              }
115              else
116                  errPrint("name not found!", np->id.name);
117              break;
118
119          case NODE_LIST: // list Node
120              for (int i = 0; i < np->list.nops; i++)
121                  run(np->list.ptr[i]);
122              break;
123
124          case NODE_OPR:  // operator Node
125  //           run1((oprNode_t *)np);
126              break;
127      }
128  }
```
/p16ecc/cc_source_6.3/pcoder.cpp   Code-6.13

- #87~89: for a constant type node (NODE_CON), create a CON_ITEM item and push it into the stack.
- #91~97: for a string type node (NODE_STR), send it to the string manager constGroup to store. The manager will create a name for the string, then create an IMMD_ITEM item with its name and push it into the stack.
- #99~117: for an identifier type node (NODE_ID), search for it in both data and function links. If it's found in the data link, its alias name will be used to generate either an IMMD_ITEM or ID_ITEM item; and then be pushed into the stack (#102~108). Otherwise, an ID_ITEM is generated and pushed onto the stack (#109~114).
- #119~122: for a list-type node (NODE_LIST), it simply deals with the nodes in the list one by one for P-code generation.
- #124~126: for an operational node (NODE_OPR), it needs to be handled based on its operation type individually, which is discussed in the following sections of this chapter.

6.4 P-code Generation For Oper Nodes (1)

<u>Project directory</u>: /p16ecc/cc_source_6.4
<u>New source files</u>: pcoder0.cpp, pcoder1.cpp

<u>6.4.1 P-code for statement termination node</u>

When finishing a statement of input C source code, some cleanup is needed.

```
17    void Pcoder :: run1(oprNode_t *op)
18    {
19        attrib *ap = op->attr;
20        char *fname;
21        node *np;
22        Dlink *dlk;
23
24        switch ( op->oper )
25        {
26            case ';':
27                addPnode(&mainPcode, new Pnode(';'));
28                CLEAR();
29                // clear Temp Index mask
30                memset(&tempIndexMask, 0, sizeof(tempIndexMask));
31                break;
```
/p16ecc/cc_source_6.4/pcoder1.cpp   Code-6.14

- #27: generate a P-code that marks the termination of a statement.
- #28: clear the compiling stack (see the example shown in Table-6.2).
- #30: clear the usage of temporary variables.

<u>6.4.2 P-code for start of a compound statement</u>

As mentioned in Section 6.2, some data link operations must be performed before and after a compound statement.

```
33            case '{':
34                dlk = new Dlink(VAR_DLINK);
35                curDlink->addChild(dlk);
36                curDlink = dlk;
37                run(op->op[0]) ;
38                curDlink = dlk->parent;
39                break;
```
/p16ecc/cc_source_6.4/pcoder1.cpp   Code-6.15

- #34~35: add a new child layer to the current data link.
- #36: update the current data link.
- #37: P-code generation for the statements within a compound statement.
- #38: go back to the parent layer of the data link.

<u>6.4.3 P-code for DATA_DECL</u>

In C, the declaration and definition of a data variable can differ in appearance.

Example 6.2:
```
    extern int x;    // data variable declaration, can appear multiple times.
    int x = 100;     // data variable definition, can only appear once.
```

Example 6.3:
```
    extern int x[]; // virtual array specifier
    int x[5];          // actual array specifier
```

Code-6.16a~Code-6.16d shows the procedure of P-code generation.

```
41              case DATA_DECL:
42                  if ( op->nops > 0 && op->op[0] )
43                  {
44                      node *np = op->op[0];    // variables id node
45                      attrib *ap = op->attr;
46                      bool data_def = ap->isExtern? false: true;
```
/p16ecc/cc_source_6.4/pcoder1.cpp   Code-6.16a

- #44: get the variable id node (there is only one node after pre-scan).
- #45: get the variable's attributes.
- #46: determine if it's a declaration or definition.

```
56                      char *dname = np->id.name;   // data name
57                      Dnode *dp = curDlink->search(dname, LOCAL_SEARCH);
58
59                      if ( dp ) //
60                      {
61                          if ( cmpAttr(dp->attr, op->attr) )
62                          {
63                              errPrint("data type conflict!", dname);
64                              break;
65                          }
66
67                          if ( !dp->dimCheck(np->id.dim) )
68                          {
69                              errPrint("data dimension error!", dname);
70                              break;
71                          }
72
73                          if ( !dp->fptrCheck(np) )
74                          {
75                              errPrint("func pointer conflicts!", dname);
76                              break;
77                          }
78
79                          if ( data_def ) // data definition?
80                          {
81                              if ( dp->attr->isExtern )
82                                  dp->attr->isExtern = 0;
83                              else
84                              {
85                                  errPrint("data redefined!", dname);
86                                  break;
87                              }
88                          }
89                      }
90                      else if ( funcLink && funcLink->search(dname) )
91                      {
92                          errPrint("name redefined!", dname);
93                          break;
94                      }
```
/p16ecc/cc_source_6.4/pcoder1.cpp   Code-6.16b

- #56~57: get the variable name, and check if it's been declared or defined.
- #59~77: error check.
- #81~82: mark the flag in the data node if it's been defined.

```
 96                                    if ( dp == NULL )
 97   ─                                {
 98                                        if ( curFnode )
 99                                            dp = curDlink->add(np, ap, ++curFnode->dIndex);
100                                        else
101                                            dp = curDlink->add(np, ap, 0);
102
103                                        dp->atAddr = mem_addr;
104                                    }
105
106                                    dp->func = curFnode? curFnode->name: NULL;
```
/p16ecc/cc_source_6.4/pcoder1.cpp   Code-6.16c

- #96~104: if it's the first time of declaration/definition, add the variable into the data link.
- #106: add the owner to the data node.

There are more steps if it's a data definition.

```
108                                    if ( data_def )
109   ─                                {
110                                        dp->dimUpdate((node*)np->id.dim);
111
112                                        if ( np->id.init )  // data initialization ...
113   ─                                    {
114                                            if ( dp->func && !ap->isStatic && ap->dataBank != CONST )
115   ─                                        {
116                                                int depth = DEPTH();
117                                                run(np->id.init);
118                                                if ( DEPTH() == (depth+1) )
119   ─                                            {
120                                                    Item *ip = idItem(dp->nameStr(), dp->attr);
121                                                    ip->home = dp;
122                                                    Pnode *pnp = new Pnode('=', ip);
123                                                    pnp->items[1] = POP();
124                                                    addPnode(&mainPcode, pnp);
125                                                }
126                                            }
127                                            else if ( ap->dataBank == CONST )
128                                                romDataInit(dname, dp->attr, np->id.init, dp, mem_addr);
129                                            else
130                                                ramDataInit(dname, dp->attr, np->id.init, dp, mem_addr);
131                                        }
132                                        else if ( ap->dataBank == CONST )
133   ─                                    {
134                                            errPrint("uninitialized constant item!");
135                                        }
136                                        else if ( dp->attr->dimVect )    // array defined
137   ─                                    {
138                                            int *v = dp->attr->dimVect;
139   ─                                        for (int i = 0; i < v[0]; i++) {
140   ─                                            if ( v[i+1] <= 0 ) {
141                                                    errPrint("unknown array dimension!");
142                                                    break;
143                                                }
144                                            }
145                                        }
146                                    }
```
/p16ecc/cc_source_6.4/pcoder1.cpp   Code-6.16d

- #110: update the dimension attributes (as shown in Example 6.3).
- #112~131: initialization for the variable.
    - #114~126: a function internal non-static variable, it generates a generic assignment P-code.
    - #127~128: a ROMed variable, calling `romDataInit()` for initialization, which is in pcoder0.cpp.
    - #129~130: an external or static RAMed variable, calling `ramDataInit()` for initialization, which is in pcoder0.cpp.

<u>6.4.4 P-code for FUNC_DECL</u>

Similar to the case of DATA_DELC, function declaration and definition have the difference in the following example,

Example 6.4

```
void func(int a, int *b);
```

Function declaration - function body is NULL. Additionally, the function parameter names are virtual and optional.

```
void func(int x, int *y) {
    . . .
}
```

Function definition - the names of the function parameters must be present.

Code-6.17a~Code-6.17c shows the procedure of P-code generation.

```
150    case FUNC_DECL:
151        np = op->op[0]; // get function head
152        fname = np->opr.op[0]->id.name;
153
154        if ( curDlink->search(fname, LOCAL_SEARCH) )
155            errPrint("name redefined!", fname);      // name used as data!
156        else
157        {
158            bool func_def = (op->op[1] != NULL);    // func body exitst?
159            bool err = false;
160
161            Fnode *fp = funcLink->search(fname);     // function existed already?
162            if ( fp == NULL )   // No, it's first time ...
163            {
164                fp = new Fnode(fname);               // create Fnode
165                fp->attr = cloneAttr(ap);            // copy it's (return) type.
166                funcLink->add(fp);                   // add Fnode to Flink
167
168                if ( np->opr.nops > 1 ) {            // with parameter?
169                    node *pp = np->opr.op[1];        // parameter list
170                    fp->elipsis = pp->list.elipsis;
171                    for (int i = 0; i < pp->list.nops; i++)
172                    {
173                        node *parp = pp->list.ptr[i];
174
175                        if ( func_def && fp->dlink->search(parp->id.name, LOCAL_SEARCH) )
176                            errPrint("parameter name duplicatated!", NULL), err = true;
177
178                        Dnode *dp = fp->dlink->add(parp, parp->id.attr, 0);
179                        dp->func = fp->name;
180                    }
181                }
182            }
183            else if ( fp->endLbl > 0 && func_def )
184            {
185                errPrint("function redefined!", fname), err = true;
186            }
```

<u>/p16ecc/cc_source_6.4/pcoder1.cpp</u>   Code-6.17a

- #154~155: confirm if the function name has not been used for any data variable.
- #158: check if it's a function definition.
- #161~182: if it's the first time for the function declaration or definition, add it to the function link, together with its parameters (#171~180).

```
187                         else    // function has been declaired/defined - need confirming
188   ▄                     {
189   ▄                         if ( (np->opr.nops <= 1 && fp->dlink->dataCount() != 0) ||
190                                  (np->opr.nops >  1 && fp->dlink->dataCount() == 0) ||
191                                  (np->opr.nops >  1 && np->opr.op[1]->list.nops != fp->dlink->dataCount()) )
192   ▄                         {
193                                 errPrint("function parameter count conflict!", fname), err = true;
194                             }
195                             else if ( np->opr.nops > 1 )
196   ▄                         {
197                                 np = np->opr.op[1]; // get parameter list
198                                 for (int i = 0; !err && i < np->list.nops; i++)
199   ▄                             {
200                                     Dnode *dp = fp->dlink->get(i);
201                                     node  *pp = np->list.ptr[i];
202
203                                     if ( cmpAttr(dp->attr, pp->id.attr) )
204                                         errPrint("parameter type confilict!", fname), err = true;
205                                     else if ( dp->fptrCheck(pp) != true )
206                                         errPrint("parameter type confilict!", fname), err = true;
207                                     else if ( func_def )
208   ▄                                 {
209                                         if ( !fp->dlink->nameCheck(pp->id.name, i) )
210                                             errPrint("par name duplicated!", fname), err = true;
211                                         else
212                                             dp->nameUpdate(pp->id.name);
213                                     }
214                                 }
215                             }
216                         }
```

/p16ecc/cc_source_6.4/pcoder1.cpp   Code-6.17b

- #207~213: if it's a function definition, (1) confirm no duplication parameter name existing, (2) update the names for parameters.

```
218                         if ( func_def && !err )
219   ▄                     {
220                             curDlink->addChild(fp->dlink);  // add par dlink to parent
221                             curDlink = fp->dlink;           // update current Dlink
222                             curFnode = fp;

224                             addPnode(&mainPcode, new Pnode(P_FUNC_BEG, ptrItem(fp)));
225                             fp->endLbl = getLbl();          // set end label for function

227                             run(op->op[1]);                 // scan function body

229                             addPnode(&mainPcode, new Pnode(LABEL, lblItem(fp->endLbl)));
230                             addPnode(&mainPcode, new Pnode(P_FUNC_END, ptrItem(fp)));
231                             curFnode = NULL;
232                             curDlink = fp->dlink->parent;// recover current Dlink
233                         }
234                     }
```

/p16ecc/cc_source_6.4/pcoder1.cpp   Code-6.17c

- #218~233: for function definition, need to generate the P-codes for the function body.
    - #220: create the data link layer for the function body (*Region-2* as indicated in Example 6.1).
    - #221: update the current data link.
    - #222: set the current function link.
    - #224: generate P-code for function start.
    - #225: generate a label for the function end (termination place of the function) which might be used in the function.
    - #227: go for P-code generation of the function body.
    - #229: place the function end label here.
    - #230: generate P-code for function end.
    - #231: clear the current function link.
    - #232: recover the current data link.

111

6.5 P-code Generation For Oper Nodes (2)

Project directory: /p16ecc/cc_source_6.5
New source files: pcoder2.cpp
Modified source files: pcoder.h

6.5.1 P-code for the assignment statement '='

This is the most common operation in C programming. It has the generic form as

        x = y

According to C syntax, **x** should be treated first, and then **y**. After both **x** and **y** are treated, their result items
are pushed into the compiling stack.

```
16   void Pcoder :: run2(oprNode_t *op)
17   {
18       int depth = DEPTH();
19       Pnode *pnp;
20       Item *ip0, *ip1;
21       int code = OPR_TYPE(op);
22
23       switch ( code )
24       {
25           case '=':    // x := y
26               run(OPR_NODE(op, 0));    // scan 'x'
27               run(OPR_NODE(op, 1));    // scan 'y'
28
29               if ( DEPTH() == (depth+2) )
30               {
31                   ip1 = POP();        // get 'y'
32                   ip0 = POP();        // get 'x'
33
34                   if ( ip0->isWritable() && moveMatch(ip0, ip1) )
35                   {
36                       pnp = new Pnode('=');
37                       pnp->items[0] = ip0;
38                       pnp->items[1] = ip1;
39                       addPnode(&mainPcode, pnp);
40
41                       PUSH((ip1->type == CON_ITEM)? ip1->clone():
42                                                     ip0->clone());
43                   }
44                   else
45                   {
46                       delete ip0, delete ip1;
47                       errPrint("assignment error!");
48                   }
49               }
```
/p16ecc/cc_source_6.5/pcoder2.cpp   Code-6.18

- #26~27: scan both **x** and **y**.
- #31~32: pop the result item of **x** and **y**.
- #34: syntax validation (**x** must be writable).
- #36~39: generate the P-code for ' =' operation.
- #41~42: duplicate the result item and push it back to the stack. This is necessary. Look at the examples below:

Example 6.5: a = b = c + d;
Example 6.6: if ((x = y) != 0) {...}

In the C language, the operator '&' supports two operations. One of them is to get the address of a data variable.

C language grammar:	&X
Parsing rules:	`'&' cast_expr  { $$ = oprNode(ADDR_OF, 1, $2); }`

The operation is to obtain the address of X, which is equivalent to a constant. Where X could have extra operations that make things complicated, as in the example below.

Example 6.7:
```
int list[10];
. . .
int *getAddr(int i) {
    return &list[i];
}
```

Besides, operators '&' and '*' counteract each other (see Example 6.8), which shall be considered during the P-code generation, as shown in Code-6.19.

Example 6.8:

<u>Original C code</u>

```
int list[10];
. . .
int getValue(int i) {
    return list[i];
}
```

<u>Equivalent code</u>

```
int list[10];
. . .
int getValue(int i) {
    return *&list[i];
}
```

Example 6.9:

<u>Original C code</u>

```
int list[10];
. . .
int *getAddr() {
    return &list;
}
```

<u>Equivalent code</u>

```
int list[10];
. . .
int *getAddr() {
    return list;
}
```

```
54              case ADDR_OF:
55                  run(OPR_NODE(op, 0));
56                  if ( DEPTH() == (depth+1) )
57                  {
58                      ip0 = TOP();
59                      attrib *attr = ip0->attr;
60                      switch ( ip0->type )
61                      {
62                          case ID_ITEM:     ip0->type = IMMD_ITEM;  return;
63                          case INDIR_ITEM:ip0->type = TEMP_ITEM;  return;
64                          case PID_ITEM:    ip0->type = ID_ITEM;    return;
65                          case DIR_ITEM:    ip0->type = CON_ITEM;   return;
66
67                          case IMMD_ITEM:
68                              if ( attr && attr->dimVect )
69                              {
70 //                              warningPrint("improper address specifying!", NULL);
71                                  free(attr->dimVect),
72                                  attr->dimVect = NULL;
73                                  return;
74                              }
75                              break;
76
77                          case TEMP_ITEM:
78                          case ACC_ITEM:
79                              if ( ptrWeight(attr) > 0 )
80                              {
81                                  if ( attr->dimVect )
82                                      decDim(attr);
83                                  else
84                                      reducePtr(attr);
85                                  return;
86                              }
87                          default:
88                              break;
89                      }
90                      DEL();
91                      errPrint("can't locate address!");
92                  }
93              break;
```
/p16ecc/cc_source_6.5/pcoder2.cpp   Code-6.19

- #55: generate the item for the data variable and push it onto the stack.
- #58~59: get the top item in the stack.
- #62~65: for some item types, alter the type for the item as indicated below:
  ```
  ID_ITEM         → IMMD_ITEM
  INDIR_ITEM      → TEMP_ITEM
  PID_ITEM        → ID_ITEM
  DIR_ITEM        → CON_ITEM
  ```
- #67~74: if it's an `IMMD_ITEM` item and it has an array attribute, then delete the array attributes. This situation can be illustrated in Example 6.9 above.
- #77~86: it it's a `TEMP_ITEM` or `ACC_ITEM`, validate if the item has the 'pointer' characteristics.

6.5.3 P-code for POS_OF

Like the operator '`&`', the operator '`*`' fulfils the indirect access operation.

C language grammar:	$*X$
Parsing rules:	`'*' cast_expr   { $$ = oprNode(POS_OF, 1, $2); }`

X should possess a 'pointer' characteristic to fulfil the operation. Code-6.20a~Code-6.20b shows the P-code generation.

```
 95              case POS_OF:     // indirect access using pointer - *p
 96                 run(OPR_NODE(op, 0));
 97                 if ( DEPTH() == (depth+1) )
 98                 {
 99                     ip0 = TOP();
100                     attrib *attr = ip0->attr;
101                     switch ( ip0->type )
102                     {
103                         case CON_ITEM:
104                         case TEMP_ITEM:
105                         case ID_ITEM:
106                         case ACC_ITEM:
107                             if ( ptrWeight(attr) > 0 )
108                             {
109                                 switch ( ip0->type )
110                                 {
111                                     case CON_ITEM:  ip0->type = DIR_ITEM;    break;
112                                     case TEMP_ITEM: ip0->type = INDIR_ITEM; break;
113                                     case ID_ITEM:   ip0->type = PID_ITEM;    break;
114                                     default:        pnp = new Pnode('=');
115                                                     addPnode(&mainPcode, pnp);
116                                                     pnp->items[1] = POP();
117                                                     pnp->items[0] = makeTemp(attr);
118                                                     PUSH(pnp->items[0]->clone());
119                                                     TOP()->type = INDIR_ITEM;
120                                 }
121                                 return;
122                             }
123                             break;
124                         case IMMD_ITEM:
125                             if ( dimDepth(attr) == 0 )
126                             {
127                                 ip0->type = ID_ITEM;
128                                 return;
129                             }
130                             break;
```
/p16ecc/cc_source_6.5/pcoder2.cpp   Code-6.20a

- #99~100: generate the item for the data variable and push it onto the stack.
- #103~123: generate the P-code if the data variable X is CON_ITEM, TEMP_ITEM, ID_ITEM, or ACC_ITEM (they should have the 'pointer' characteristic).
- #124~129: generate the P-code if the data variable X is IMMD_ITEM. Example 6.10 below explains the situation.

Example 6.10:

<table>
<tr><td>Original C code</td><td>Equivalent code</td></tr>
</table>

```
int seed;                          int seed;
. . .                              . . .
int getSeed() {                    int getSeed() {
    return *&list;                     return seed;
}                                  }
```

```
131              case INDIR_ITEM:
132              case PID_ITEM:
133              case DIR_ITEM:
134                  if ( ptrWeight(attr) > 1 )
135                  {
136                      pnp = new Pnode('=');
137                      addPnode(&mainPcode, pnp);
138                      pnp->items[1] = POP();
139                      ip0 = makeTemp(attr);
140                      reducePtr(ip0->attr);
141                      pnp->items[0] = ip0;
142
143                      PUSH(ip0->clone());
144                      TOP()->type = INDIR_ITEM;
145                      return;
146                  }
```
/p16ecc/cc_source_6.5/pcoder2.cpp   Code-6.20b

- #131~146: if the data variable X is INDIR_ITEM, PID_ITEM, or DIR_ITEM, then it involves a P-code that fetches the target operand indirectly, and pushes it onto the stack. This is the most common operation for the operator '*'.

<u>6.5.4 P-code for '['</u>

The operator `'['` is to fulfil the operation of offset access.

C language grammar:	$X[Y]$
Parsing rules:	```postfix_expr '[' expr ']' {``` ``` $$ = oprNode('[', 2, $1, $3); }```

X is an array or a pointer, and Y is the offset index. Note: 'offset index' is the bias count of X basic units. The actual offset value is 'offset index' times the unit size. The following gives examples and explanations.

C language code	Calculation for actual offset

```char a[10], x;```
```void foo(int n)```
```{```
```   x = a[n];```
```}```
	The actual offset value is **n**.
	The actual access location is (a + n).

```long a[10], x;```
```void foo(int n)```
```{```
```   x = a[n];```
```}```
	The actual offset value is **n*4**.
	The actual access location is (a + n*4).

```long a[10][20], x;```
```void foo(int n, int m)```
```{```
```   x = a[n][m];```
```}```
	The actual offset value is **n*20*4 + m*4**.
	The actual access location is (a + n*20*4 + m*4), or (a + n*80 + m*4).

Table-6.1

The P-code generation for the operator, shown in Code-6.21a~Code-6.21d, appears complicated because it must handle different forms or types of X.

```
155          case '[':
156              run(OPR_NODE(op, 0));
157              run(OPR_NODE(op, 1));
158              if ( DEPTH() == (depth+2) )
159              {
160                  ip1 = POP();
161                  ip0 = POP();
162                  attrib *attr = ip0->attr;
163                  if ( !ip1->isMonoVal() || !attr )
164                  {
165                      delete ip0; delete ip1;
166                      errPrint("illegal index value!");
167                      return;
168                  }
```
`/p16ecc/cc_source_6.5/pcoder2.cpp`

Code-6.21a

- #156: generate the item for X, and push it into the stack.
- #157: generate the item for Y, and push it into the stack.
- #162~168: validation for Y.

When X is a regular instance with pointer characteristics, it needs to generate the P-code that adds up the pointer's value with the actual offset.

```
170                         switch ( ip0->type )
171                         {
172                             case ID_ITEM:
173                             case TEMP_ITEM:
174                             case CON_ITEM:
175                             case ACC_ITEM:
176                                 if ( ptrWeight(attr) > 0 )
177                                 {
178                                     attrib *ap = cloneAttr(attr);
179                                     int mem_type = -1;
180                                     if ( dimDepth(ap) > 0 )
181                                     {
182                                         if ( ip0->type == TEMP_ITEM && ptrWeight(ap) )
183                                             mem_type = reducePtr(ap);
184
185                                         ip1 = makeOffset(ip1, sizer.size(ap, SUBDIM_SIZE));
186                                         decDim(ap);
187                                     }
188                                     else // input_dim <= 0
189                                     {
190                                         mem_type = reducePtr(ap);
191                                         ip1 = makeOffset(ip1, sizer.size(ap, ATTR_SIZE));
192                                     }
193
194                                     if ( mem_type != -1 )
195                                         insertPtr(ap, mem_type);
196
197                                     pnp = new Pnode('+');
198                                     addPnode(&mainPcode, pnp);
199                                     pnp->items[2] = ip1;
200                                     pnp->items[1] = ip0;
201                                     pnp->items[0] = makeTemp();
202                                     pnp->items[0]->attr = ap;
203                                     PUSH(pnp->items[0]->clone());
204
205                                     if ( !ap->dimVect ) TOP()->type = INDIR_ITEM;
206                                     return;
207                                 }
208                                 break;
/p16ecc/cc_source_6.5/pcoder2.cpp                                    Code-6.21b
```

- #180~187: *X* has the array characteristic; the offset calculation should take its sub-array dimension(s) into consideration.
- #189~192: *X* has no array characteristic; the offset calculation is based on *X*'s type.
- #197~203: generate the P-code that adds up *X* and the actual offset.

When *X* is a variable's address, things become simple for calculating the actual offset.

```
209                             case IMMD_ITEM:
210                                 if ( dimDepth(attr) > 0 )
211                                 {
212                                     attrib *ap = cloneAttr(attr);
213                                     ip1 = makeOffset(ip1, sizer(ap, SUBDIM_SIZE));
214                                     decDim(ap);
215                                     insertPtr(ap, attr->dataBank);
216
217                                     pnp = new Pnode('+');
218                                     addPnode(&mainPcode, pnp);
219                                     pnp->items[2] = ip1;
220                                     pnp->items[1] = ip0;
221                                     pnp->items[0] = makeTemp();
222                                     pnp->items[0]->attr = ap;
223                                     PUSH(pnp->items[0]->clone());
224
225                                     if ( !ap->dimVect ) TOP()->type = INDIR_ITEM;
226                                     return;
227                                 }
228                                 break;
/p16ecc/cc_source_6.5/pcoder2.cpp                                    Code-6.21c
```

Finally, for the rest of the types of *X*, the actual offset is calculated based on *X*'s array or pointer characteristic, similar to above.

```
229                         case INDIR_ITEM:
230                         case PID_ITEM:
231                         case DIR_ITEM:
232                             if ( dimDepth(attr) > 0 )
233                             {
234                                 attrib *ap = cloneAttr(attr);
235                                 int mem_type = reducePtr(ap);
236                                 ip1 = makeOffset(ip1, sizer.size(ap, SUBDIM_SIZE));
237                                 decDim(ap);
238                                 insertPtr(ap, mem_type);
239
240                                 pnp = new Pnode('+');
241                                 addPnode(&mainPcode, pnp);
242                                 if ( ip0->type == INDIR_ITEM ) ip0->type = TEMP_ITEM;
243                                 if ( ip0->type == PID_ITEM )   ip0->type = ID_ITEM;
244                                 if ( ip0->type == DIR_ITEM )   ip0->type = CON_ITEM;
245                                 pnp->items[2] = ip1;
246                                 pnp->items[1] = ip0;
247                                 pnp->items[0] = makeTemp();
248                                 pnp->items[0]->attr = ap;
249                                 PUSH(pnp->items[0]->clone());
250
251                                 TOP()->type = INDIR_ITEM;
252                                 return;
253                             }
254                             if ( ptrWeight(ip0->attr) > 1 )
255                             {
256                                 attrib *ap = cloneAttr(attr);
257                                 reducePtr(ap);
258                                 ip1 = makeOffset(ip1, sizer.size(ap, INDIR_SIZE));
259
260                                 pnp = new Pnode('+');
261                                 addPnode(&mainPcode, pnp);
262                                 pnp->items[2] = ip1;
263                                 pnp->items[1] = ip0;
264                                 pnp->items[0] = makeTemp();
265                                 pnp->items[0]->attr = ap;
266                                 PUSH(pnp->items[0]->clone());
267
268                                 TOP()->type = INDIR_ITEM;
269                                 return;
270                             }
```

/p16ecc/cc_source_6.5/pcoder2.cpp	Code-6.21d

The following (Table-6.2) shows the test results for this section.

C language code	P-code generated

`char a[10], x;`
`void foo(int n)`
`{`
`   x = a[n];`
`}`
	`F:\p16ecc\tests>cc16e ch6_t1.c`
	`P-CODE: P_FUNC_BEG: foo : {void}`
	`P-CODE: SRC ... ch6_t1.c #4: x = a[n];`
	`P-CODE: '+'   %1{char *}, #a{char[10]}, foo_$_n{int}`
	`P-CODE: '='   x{char}, [%1]{char *}`
	`P-CODE: ';'`
	`P-CODE: LABEL   _$L1`
	`P-CODE: P_FUNC_END: foo`

`long a[10], x;`
`void foo(int n)`
`{`
`   x = a[n];`
`}`
	`F:\p16ecc\tests>cc16e ch6_t2.c`
	`P-CODE: P_FUNC_BEG: foo : {void}`
	`P-CODE: SRC ... ch6_t2.c #4: x = a[n];`
	`P-CODE: '*'   %1{unsigned int}, foo_$_n{int}, 4{int}`
	`P-CODE: '+'   %2{long *}, #a{long[10]}, %1{unsigned int}`
	`P-CODE: '='   x{long}, [%2]{long *}`
	`P-CODE: ';'`
	`P-CODE: LABEL   _$L1`
	`P-CODE: P_FUNC_END: foo`

`long a[10][20], x;`
`void foo(int n, int m)`
`{`
`   x = a[n][m];`
`}`
	`F:\p16ecc\tests>cc16e ch6_t3.c`
	`P-CODE: P_FUNC_BEG: foo : {void}`
	`P-CODE: SRC ... ch6_t3.c #4: x = a[n][m];`
	`P-CODE: '*'   %1{unsigned int}, foo_$_n{int}, 80{int}`
	`P-CODE: '+'   %2{long *[20]}, #a{long[10][20]}, %1{unsigned int}`
	`P-CODE: '*'   %3{unsigned int}, foo_$_m{int}, 4{int}`
	`P-CODE: '+'   %4{long *}, %2{long *[20]}, %3{unsigned int}`
	`P-CODE: '='   x{long}, [%4]{long *}`
	`P-CODE: ';'`
	`P-CODE: LABEL   _$L1`
	`P-CODE: P_FUNC_END: foo`

Table-6.2

6.6 P-code Generation For Oper Nodes (3)

<u>Project directory</u>: /p16ecc/cc_source_6.6
<u>New source files</u>: pcoder3.cpp
<u>Modified source files</u>: pcoder.h

6.6.1 P-code for '.'

The operator '.' directly accesses a data member in a struct/union instance. The operator's P-code generation is to add data member Y's offset to the starting address of X, resulting in biased indirect access.

C language grammar:	$X.Y$
Parsing rules:	`postfix_expr '.' identifier {` `                  $$ = oprNode('.', 2, $1, $3); }`

Where X can be a variety of forms, as shown in the following examples, which present an instance of struct/union, and Y is simply a data member name.

```
typedef struct {
   int x, y;
   char a[10];
} Data;

Data d, *dp;

void foo()
{
   d.y = 0;
   dp[0].x = 0;
   (*dp).y = 0;
}
```

P-code generation for operator '.' is biased access to a struct/union entity, so there are a lot of similarities to that of the operator '[' in generating P-code. Code-6.22a~Code-6.22b shows the procedure.

```cpp
29         case '.':    // struct/union direct addressing
30             run(OPR_NODE(op, 0));
31             if ( DEPTH() == (depth+1) )
32             {
33                 node *np = OPR_NODE(op, 1);
34                 if ( np->type != NODE_ID )
35                 {
36                     DEL();
37                     errPrint("illegal member name!");
38                     return;
39                 }
40                 ip0 = POP();
41                 attrib *ap, *attr = cloneAttr(ip0->attr);
42                 int data_bank = 0, offset, i0_type;
43                 const char *error_msg = NULL;
44                 switch ( i0_type = ip0->type )
45                 {
46                     case ID_ITEM:   case DIR_ITEM:  case INDIR_ITEM:    case PID_ITEM:
47                         if ( !(attr->type == STRUCT || attr->type == UNION) )
48                             error_msg = "not a struct/union data1!";
49                         else if ( attr->dimVect )
50                             error_msg = "no array allowed struct/union!";
51                         else if ( i0_type != ID_ITEM  )
52                         {
53                             if ( ptrWeight(attr) != 1 )
54                                 error_msg = "improper multi-pointer!";
55                         }
56                         else if ( ptrWeight(attr) != 0 )
57                             error_msg = "improper pointer!";
58
59                         if ( error_msg )
60                         {
61                             errPrint(error_msg);
62                             delAttr(attr);  delete ip0;
63                             return;
64                         }
```
    Code-6.22a

- #30: generate the item for X, and push it into the stack.
- #40: pop out X from the stack.
- #44~64: validate the properties of X.

```
66      data_bank = (i0_type == ID_ITEM)? attr->dataBank:
67                                         reducePtr(attr);
68
69      ap = sizer.memberAttrClone(attr, np->id.name);   // member's attr.
70      if ( ap == NULL )
71      {
72          errPrint("struct/union member not found!", np->id.name);
73          delAttr(attr);   delete ip0;
74          return;
75      }
76
77      CLONE_ST_UN_NEWDATA(ap, attr);
78      offset = sizer.memberOffset(attr, np->id.name);
79      insertPtr(ap, data_bank);
80      delAttr(attr);
81
82      pnp = new Pnode('+');
83      addPnode(&mainPcode, pnp);
84      if ( i0_type == DIR_ITEM   ) ip0->type = CON_ITEM;
85      if ( i0_type == ID_ITEM    ) ip0->type = IMMD_ITEM;
86      if ( i0_type == INDIR_ITEM ) ip0->type = TEMP_ITEM;
87      if ( i0_type == PID_ITEM   ) ip0->type = ID_ITEM;
88      pnp->items[2] = intItem(offset);
89      pnp->items[1] = ip0;
90      pnp->items[0] = makeTemp();
91      pnp->items[0]->attr = ap;
92
93      PUSH(pnp->items[0]->clone());
94      if ( !ap->dimVect ) TOP()->type = INDIR_ITEM;
95      break;
```

`/p16ecc/cc_source_6.6/pcoder3.cpp`   Code-6.22b

- #66~67: get the data member storage property.
- #69: get the data member attributes.
- #77: retrieve the attributes (data structure) if it's a self-referenced member (see in Example 5.7).
- #78: calculate the offset for the data member.
- #82~91: generate the P-code for the operator.

6.6.2 P-code for PTR_OP

The operator `PTR_OP` indirectly accesses a data member in a struct/union instance.

C language grammar:	X->Y
Parsing rules:	`postfix_expr PTR_OP identifier {` `                $$ = oprNode(PTR_OP, 2, $1, $3); }`

Where *X* can be in different forms, as shown in the following examples, which shall be a pointer or equivalent pointing to a struct/union, and *Y* is simply a data member name. The operator's P-code generation is to add data member *Y*'s offset to the address pointed by *X*, resulting in biased indirect access.

```
typedef struct {
  int x, y;
  char a[10];
} Data;

void foo(Data *dp)
{
  dp->x = 0;
  ((Data *)0x100)->y = 0;
}
```

P-code generation for operator PTR_OP is similar to that of '.', as shown in Code-6.23.

```
105     case PTR_OP:      // struct/union indirect addressing
106         run(OPR_NODE(op, 0));
107         if ( DEPTH() == (depth+1) )
108         {
109             node *np = OPR_NODE(op, 1);
110             if ( np->type != NODE_ID )
111             {
112                 DEL();
113                 errPrint("illegal member name!");
114                 return;
115             }
116             ip0 = POP();
117             attrib *ap, *attr = cloneAttr(ip0->attr);
118             int data_bank = 0, offset, i0_type;
119             const char *error_msg = NULL;
120             switch ( i0_type = ip0->type )
121             {
122                 case IMMD_ITEM:  case ID_ITEM:  case TEMP_ITEM: case ACC_ITEM:  case CON_ITEM:
123                 case INDIR_ITEM: case PID_ITEM: case DIR_ITEM:
124                     if ( !(attr->type == STRUCT || attr->type == UNION) )
125                     {
126                         error_msg = "not a struct/union data2!";
127                     }
128                     else if ( i0_type == INDIR_ITEM || i0_type == PID_ITEM || i0_type == DIR_ITEM )
129                     {
130                         if ( ptrWeight(attr) != 2 ) error_msg = "improper multi-pointer!";
131                         reducePtr(attr);
132                     }
133                     else if ( i0_type == IMMD_ITEM )
134                     {
135                         if ( ptrWeight(attr) != 0 ) error_msg = "improper address!";
136                     }
137                     else
138                     {
139                         if ( ptrWeight(attr) != 1 ) error_msg = "improper pointer!";
140                     }
141
142                     if ( !error_msg && attr->dimVect )
143                         error_msg = "no array allowed struct/union!";
144
145                     if ( error_msg )
146                     {
147                         errPrint(error_msg);
148                         delAttr(attr);  delete ip0;
149                         return;
150                     }
151
152                     data_bank = (i0_type == IMMD_ITEM)? attr->dataBank:
153                                                         reducePtr(attr);
154
155                     ap = sizer.memberAttrClone(attr, np->id.name);  // member's attr.
156                     if ( ap == NULL )
157                     {
158                         errPrint("struct/union member not found!", np->id.name);
159                         delAttr(attr);  delete ip0;
160                         return;
161                     }
162
163                     CLONE_ST_UN_NEWDATA(ap, attr);
164                     offset = sizer.memberOffset(attr, np->id.name);
165                     insertPtr(ap, data_bank);
166                     delAttr(attr);
167
168                     pnp = new Pnode('+');
169                     addPnode(&mainPcode, pnp);
170                     pnp->items[2] = intItem(offset);
171                     pnp->items[1] = ip0;
172                     pnp->items[0] = makeTemp();
173                     pnp->items[0]->attr = ap;
174
175                     PUSH(pnp->items[0]->clone());
176                     if ( !ap->dimVect ) TOP()->type = INDIR_ITEM;
177                     break;
```


- #106: generate the item for X, and push it into the stack.
- #116: pop out X from the stack.
- #124~150: validate the properties of X.
- #155: get the data member attributes.

122

- #157: get the offset for the data member.
- #163: retrieve the attributes (data structure) if it's a self-referenced member (see in Example 5.7).
- #168~173: generate the P-code for the operator.

6.6.3 P-code for PRE_INC and PRE_DEC

Pre-increment and pre-decrement are frequently used in C programming. They increase or decrease a variable by 1 unit, and the new values are then used in the expression.

C language grammar:	$++X$ $--X$
Parsing rules:	`INC_OP unary_expr    { $$ = oprNode(PRE_INC, 1, $2); }` `DEC_OP unary_expr    { $$ = oprNode(PRE_DEC, 1, $2); }`

It is essential to understand that the actual increment or decrement amount is based on the X's type. For examples

```
int n, *p;

void foo()
{
   ++n;   // n  increases by 1
   --p;   // p  decreases by 2 since every int variable takes 2 bytes
}
```

Code-6.24 shows the P-code generation for the operators.

```
180          case PRE_INC:
181          case PRE_DEC:
182              run(OPR_NODE(op, 0));
183              if ( DEPTH() == (depth+1) )
184              {
185                  ip0 = POP();
186                  attrib *attr = ip0->attr;
187                  int code = (op->oper == PRE_INC)? INC_OP: DEC_OP;
188                  int size = ip0->stepSize();
189                  if ( attr->dimVect || !ip0->isWritable() || size == 0 ||
190                       ip0->type == TEMP_ITEM || ip0->type == ACC_ITEM )
191                  {
192                      delete ip0;
193                      errPrint("inc/dec data type error!");
194                      return;
195                  }
196                  pnp = new Pnode(code);
197                  addPnode(&mainPcode, pnp);
198                  pnp->items[0] = ip0;
199                  pnp->items[1] = intItem(size);
200
201                  PUSH(pnp->items[0]->clone());
202                  return;
203              }
204              break;
```
`/p16ecc/cc_source_6.6/pcoder3.cpp`   Code-6.24

- #180: generate the item for X, and push it into the stack.
- #185: pop out X from the stack.
- #188: determine the actual increment/decrement amount.
- #196~199: generate the P-code for the operator.
- #201: push updated X into the stack

The following (Table-6.3) shows the test results for this section.

C language code	P-code generated

`int n, *p, x;`
 	
`void foo()`
`{`
`    x = ++n;`
`    x = *--p;`
`}`
	`F:\p16ecc\tests>cc16e ch6_t6.c`
	`P-CODE: P_FUNC_BEG: foo : {void}`
	`P-CODE: SRC ... ch6_t6.c #5: x = ++n;`
	`P-CODE: '++'  n{int}, 1{int}`
	`P-CODE: '='   x{int}, n{int}`
	`P-CODE: ';'`
	`P-CODE: SRC ... ch6_t6.c #6: x = *--p;`
	`P-CODE: '--'  p{int *}, 2{int}`
	`P-CODE: '='   x{int}, [p:0]{int *}`
	`P-CODE: ';'`
	`P-CODE: LABEL   _$L1`
	`P-CODE: P_FUNC_END: foo`

Table-6.3

6.6.4 P-code for POST_INC and POST_DEC

Post-increment and post-decrement are also frequently used in C programming. They increase or decrease a variable by 1 unit, but the old values are used in the expression instead. For this reason, X needs to be saved in a temporary variable before the increment/decrement operation.

C language grammar:	X++ X--
Parsing rules:	`postfix_expr INC_OP   { $$ = oprNode(POST_INC, 1, $1); }` `postfix_expr DEC_OP   { $$ = oprNode(POST_DEC, 1, $1); }`

Based on the X's type, the actual increment or decrement amount is the same as that for pre-increment/pre-decrement. Code-6.25 shows the P-code generation for the operators.

```
206         case POST_INC:
207         case POST_DEC:
208             run(OPR_NODE(op, 0));
209             if ( DEPTH() == (depth+1) )
210             {
211                 ip0 = POP();
212                 attrib *attr = ip0->attr;
213                 int code = (op->oper == POST_INC)? INC_OP: DEC_OP;
214                 int size = ip0->stepSize();
215
216                 if ( attr->dimVect || !ip0->isWritable() || size == 0 ||
217                      ip0->type == TEMP_ITEM || ip0->type == ACC_ITEM )
218                 {
219                     delete ip0;
220                     errPrint("inc/dec data type error!");
221                     return;
222                 }
223
224                 pnp = new Pnode('=');
225                 addPnode(&mainPcode, pnp);
226                 pnp->items[1] = ip0;
227
228                 if ( ip0->type == DIR_ITEM || ip0->type == PID_ITEM || ip0->type == INDIR_ITEM )
229                 {
230                     attrib *ap = cloneAttr(attr);
231                     reducePtr(ap);
232                     pnp->items[0] = makeTemp(ap);
233                     delAttr(ap);
234                 }
235                 else
236                     pnp->items[0] = makeTemp(attr);
237
238                 PUSH(pnp->items[0]->clone());
239
240                 pnp = new Pnode(code);
241                 addPnode(&mainPcode, pnp);
242                 pnp->items[0] = ip0->clone();
243                 pnp->items[1] = intItem(size);
244             }
245             break;
```
`/p16ecc/cc_source_6.6/pcoder3.cpp`

Code-6.25

- #224~236: generate the P-code to save the old value of X into a temporary variable.

- #238: push the temporary variable into the stack.

The following (Table-6.4) shows the test results for this section.

C language code	P-code generated

<pre>int n, *p, x;
 	
void foo()
{
    x = n++;
    x = *p--;
}</pre>
	<pre>F:\p16ecc\tests>cc16e ch6_t7.c
	P-CODE: P_FUNC_BEG: foo : {void}
	P-CODE: SRC ... ch6_t7.c #5: x = n++;
	P-CODE: '=' %1{int}, n{int}
	P-CODE: '++' n{int}, 1{int}
	P-CODE: '=' x{int}, %1{int}
	P-CODE: ';'
	P-CODE: SRC ... ch6_t7.c #6: x = *p--;
	P-CODE: '=' %1{int *}, p{int *}
	P-CODE: '--' p{int *}, 2{int}
	P-CODE: '=' x{int}, [%1]{int *}
	P-CODE: ';'
	P-CODE: LABEL _$L1
	P-CODE: P_FUNC_END: foo</pre>

Table-6.4

6.7 P-code Generation For Oper Nodes (4)

<u>Project directory</u>: /p16ecc/cc_source_6.7
<u>New source files</u>: pcoder4.cpp
<u>Modified source files</u>: pcoder.h

6.7.1 P-code for SIZEOF

SIZEOF operator is for compile-time function sizeof(X), where X can be either a data type or a data variable. sizeof(X) returns the size of X in bytes, which occurs during the pre-scan procedure if X is a data type. Otherwise, the conversion happens during P-code generation if X is a data variable name. A CON_ITEM type item will be generated and pushed into the stack, as shown in Code-6.26.

The data item X could be a virtual array, like

```
    extern int table[]; // virtual array.
```

In this case, sizeof(X) cannot give the actual value at the moment. Instead, a special symbol or ID name, **table$sizeof$**, will be created and left over for being solved until the linking process.

```
28      case SIZEOF:
29          run(OPR_NODE(op, 0));
30          if ( DEPTH() == (depth+1) )
31          {
32              ip0 = POP();
33              int n;
34
35              if ( ip0->type == IMMD_ITEM && !ip0->attr->dimVect )
36                  n = 2;
37              else
38                  n = sizer.size(ip0->attr, TOTAL_SIZE);
39
40              if ( n <= 0 )    // external virtual array
41              {
42                  Dnode *dp = (Dnode*)ip0->home;
43                  if ( ip0->type == IMMD_ITEM && dp && dp->func == NULL )
44                  {
45                      std::string s = ip0->val.s;
46                      s += "$sizeof$";
47                      ip0->updateName((char*)s.c_str());
48                      PUSH(ip0);
49                      return;
50                  }
51              }
52
53              delete ip0;
54              if ( n > 0 )
55                  PUSH(intItem(n));
56              else
57                  errPrint("can't size the data");
58          }
```
/p16ecc/cc_source_6.7/pcoder4.cpp

Code-6.26

- #29: generate the item for X, and push it into the stack.
- #32: pop out X from the stack.
- #35~36: if X is the address of a variable, size is fixed to 2.
- #38: otherwise calculate the size for the variable.
- #40~51: a virtual external array detected, rename the item with a special symbol.
- #55: generate a CON_ITEM item and push it into the stack.

The following (Table-6.5) shows the test results for this section.

C language code	P-code generated

```int array[10];```
```int n, x;```
 	
```void foo()```
```{```
```   x = sizeof(n);```
```   x = sizeof(array);```
```}```
	```F:\p16ecc\tests>cc16e ch6_t8.c```
	```P-CODE: P_FUNC_BEG: foo : {void}```
	```P-CODE: SRC ... ch6_t8.c #6: x = sizeof(n);```
	```P-CODE: '=' x{int}, 2{int}```
	```P-CODE: ';'```
	```P-CODE: SRC ... ch6_t8.c #7: x = sizeof(array);```
	```P-CODE: '=' x{int}, 20{int}```
	```P-CODE: ';'```
	```P-CODE: LABEL _$L1```
	```P-CODE: P_FUNC_END: foo```

Table-6.5

6.7.2 P-code for LABEL and GOTO

These two operators have the same structure in the parsing tree.

C language grammar:	*label* : **goto** *label*;
Parsing rules:	```identifier ':' { $$ = oprNode(LABEL, 1, $1); }``` ```GOTO identifier ';' { $$ = oprNode(GOTO, 1, $2); }```

P-code generation for these two operators is shown in Code-6.27.

```
61        case LABEL:
62        case GOTO:
63            np = OPR_NODE(op, 0);
64            pnp = new Pnode(code, lblItem(np->id.name));
65            addPnode(&mainPcode, pnp);
66            break;
```
/p16ecc/cc_source_6.7/pcoder4.cpp
Code-6.27

- #63~65: generate P-code {LABEL, [*label*]} or {GOTO, [*label*]}

6.7.3 P-code for unary operators

There are three unary operators covered here: NEG_OF, '!', and '~'.

C language grammar:	$-X$ $!X$ $\sim X$

Parsing rules:	```'-' cast_expr { $$ = oprNode(NEG_OF, 1, $2); }```
	```'!' cast_expr { $$ = oprNode('!', 1, $2); }```
	```'~' cast_expr { $$ = oprNode('~', 1, $2); }```

Usually, the P-code generated will be

$$\{op\text{-}code, [t, X]\}$$

where *t* is a temporary variable. The following (Code-6.28) shows the P-code generation.

127

```
 77                 switch ( ip0->type )
 78                 {
 79                     case ID_ITEM:
 80                     case TEMP_ITEM:
 81                     case ACC_ITEM:
 82                         pnp = new Pnode(code);
 83                         addPnode(&mainPcode, pnp);
 84                         pnp->items[1] = ip0;
 85                         pnp->items[0] = makeTemp(attr);
 86                         PUSH(pnp->items[0]->clone());
 87                         return;

 89                     case IMMD_ITEM:
 90                     case LBL_ITEM:
 91                         if ( code == '!' )
 92                         {
 93                             delete ip0;
 94                             ip0 = intItem(0);
 95                         }
 96                         else if ( code == '~' )
 97                         {
 98                             std::string s = "(~";
 99                             s += ip0->val.s;    s += ")";
100                             ip0->updateName((char*)s.c_str());
101                         }
102                         else // NEG_OF
103                         {
104                             std::string s = "(0-";
105                             s += ip0->val.s; s += ")";
106                             ip0->updateName((char*)s.c_str());
107                         }
108                         PUSH(ip0);
109                         return;

111                     case DIR_ITEM:
112                     case PID_ITEM:
113                     case INDIR_ITEM:
114                         pnp = new Pnode(code);
115                         addPnode(&mainPcode, pnp);
116                         pnp->items[1] = ip0;
117                         pnp->items[0] = makeTemp();
118                         pnp->items[0]->attr = cloneAttr(attr);
119                         reducePtr(pnp->items[0]->attr);

121                         PUSH(pnp->items[0]->clone());
122                         return;
123                     default:
124                         break;
125                 }
126             delete ip0;
127         }
```

/p16ecc/cc_source_6.7/pcoder4.cpp

Code-6.28

- #79~87: generate the P-code when X is a directly accessible variable.
- #111~122: generate the P-code when X is an indirectly accessible variable.

Note: it has been converted during the pre-scan if X is a constant.

6.7.4 P-code for CAST

There are three unary operators covered here: NEG_OF, '!', and '~'.

C language grammar:	$(data\text{-}type)\,X$

Parsing rules:	`'(' type_specifier2 ')' cast_expr`
	`        { $$ = oprNode(CAST, 1, $4);`
	`          $$->opr.attr = $2;     }`

Also, the P-code generated will be
{ '=', [t, X]}

128

where *t* is a temporary variable of the required data type, the following code (Code-6.29) shows the P-code generation.

```
131         case CAST:
132             run(OPR_NODE(op, 0));
133             if ( DEPTH() == (depth+1) )
134             {
135                 ip0 = POP();
136                 attrib *ap = op->attr;        // casting attribute
137
138                 if ( !ip0->isOperable() )
139                 {
140                     delete ip0;
141                     errPrint("can't be casted!");
142                     return;
143                 }
144                 if ( ip0->type == CON_ITEM )
145                 {
146                     ip0->updateAttr(cloneAttr(ap));
147                     PUSH(ip0);
148                     return;
149                 }
150                 pnp = new Pnode('=');
151                 addPnode(&mainPcode, pnp);
152                 pnp->items[1] = ip0;
153                 pnp->items[0] = makeTemp(ap);
154                 PUSH(pnp->items[0]->clone());
155                 return;
156             }
157             break;
```
/p16ecc/cc_source_6.7/pcoder4.cpp   Code-6.29

- 144~149: it simply modifies the attributes if *X* is a constant, no P-code generated.
- 150~154: generate the P-code for the operator.

6.8 P-code Generation For Oper Nodes (5)

<u>Project directory</u>: /p16ecc/cc_source_6.8
<u>New source files</u>: pcoder5.cpp
<u>Modified source files</u>: pcoder.h

6.8.1 P-code for ADD_ASSIGN and SUB_ASSIGN

`ADD_ASSIGN` and `SUB_ASSIGN` are for C operators **+=** and **−=**, which are compound assignment operators. They provide a shorthand way to add or subtract a value to a variable and assign the result back to the same variable.

C language grammar:	$X += Y$ $X -= Y$

Parsing rules:	`unary_expr ADD_ASSIGN assignment_expr`
	`            { $$ = oprNode(ADD_ASSIGN, 2, $1, $3); }`
	`unary_expr SUB_ASSIGN assignment_expr`
	`            { $$ = oprNode(SUB_ASSIGN, 2, $1, $3); }`

In P-code generation, they are similar to a regular assignment ('=') operation. They usually have the P-code generated as:

$$\{op\text{-}code, [X, Y]\}$$

There might be an extra consideration/operation needed for the P-code generation if X is a pointer or equivalent, where Y's value might need to be rescaled, depending on X's unit size, for the addition or subtraction, such as

$$\{ \text{'*'}, [t, Y, scaling_value]\}$$
$$\{op\text{-}code, [X, t]\}$$

For example,

```
int n, *p;

void foo(int m)
{
    n += m;   // n increases by m
    p -= m;   // p decreases by 2*m since every int variable takes 2 bytes
}
```

The following, Code-6.30, shows the P-code generation for the section.

```
25    case ADD_ASSIGN:
26    case SUB_ASSIGN:
27        run(OPR_NODE(op, 0));
28        run(OPR_NODE(op, 1));
29        if ( DEPTH() == (depth+2) )
30        {
31            ip1 = POP();
32            ip0 = POP();
33            int size = ip0->stepSize();// incremental size
34
35            if ( size > 0 && ip0->isWritable() && ip0->attr->type != SBIT &&
36                 ip0->isOperable() && ip1->isOperable() && ip1->attr->type != SBIT )
37            {
38                if ( ip0->isAccePtr() && ip1->isMonoVal() ) // modify a pointer?
39                {
40                    if ( ip1->type == CON_ITEM )
41                        ip1->val.i *= size;
42                    else if ( size > 1 )
43                    {
44                        pnp = new Pnode('*');
45                        addPnode(&mainPcode, pnp);
46                        pnp->items[2] = intItem(size);
47                        pnp->items[1] = ip1;
48                        pnp->items[0] = makeTemp();
49                        pnp->items[0]->attr = newAttr(INT);
50                        ip1 = pnp->items[0]->clone();
51                    }
52                }
53
54                pnp = new Pnode(code);
55                addPnode(&mainPcode, pnp);
56                pnp->items[1] = ip1;
57                pnp->items[0] = ip0;
58                PUSH(pnp->items[0]->clone());
59                break;
60            }
61            delete ip0;
62            delete ip1;
63            errPrint("invalid +=/-= operation!");
64        }
```
`/p16ecc/cc_source_6.8/pcoder5.cpp`   Code-6.30

- #27~28: generate the items for X and Y, and push them into the stack.
- #31~32: pop out X and Y from the stack.
- #38~52: an extra P-code might be needed if X is a pointer.
- #54~58: P-coder generation for ADD_ASSIGN or SUB_ASSIGN.

The following (Table-6.6) shows the test results for this section.

C language code	P-code generated
`int n, *p;`	`F:\p16ecc\tests>cc16e ch6_t9.c` `P-CODE: P_FUNC_BEG: foo : {void}`
`void foo(int m)`	`P-CODE: SRC ... ch6_t9.c #5: n += m;`

{
 	
  n += m;
 	
  p -= m;
 	
}
	P-CODE: '+=' n{int}, foo_$_m{int}
	P-CODE: ';'
	P-CODE: SRC ... ch6_t9.c #6: p -= m;
	P-CODE: '*' %1{int}, foo_$_m{int}, 2{int}
	P-CODE: '-=' p{int *}, %1{int}
	P-CODE: ';'
	P-CODE: LABEL _$L1
	P-CODE: P_FUNC_END: foo

Table-6.6

6.8.2 P-code for '+' and '-'

'+' and '-' are among the most common arithmetic operators.

C language grammar:	$X + Y$ $X - Y$

Parsing rules:	```additive_expr '+' multiplicative_expr {```
	``` $$ = oprNode('+', 2, $1, $3); }```
	```additive_expr '-' multiplicative_expr {```
	``` $$ = oprNode('-', 2, $1, $3); }```

Usually, they will end up with the P-code

$$\{\,\text{'+'},\,[t, X, Y]\}\ \text{or}\ \{\,\text{'-'},\,[t, X, Y]\}$$

Where t is a temporary variable, a TEMP_ITEM item that holds the operation result.

Similar to ADD_ASSIGN or SUB_ASSIGN, scaling may happen if either X or Y is a pointer or equivalent. Moreover, in some cases, the size of the temporary variable t should be extended to avoid result overflow.

The following, Code-6.31, shows the P-code generation for the operators.

```
67        case '+':
68        case '-':
69            run(OPR_NODE(op, 0));
70            run(OPR_NODE(op, 1));
71            if ( DEPTH() == (depth+2) )
72            {
73                ip1 = POP();
74                ip0 = POP();
75
76                if ( !(ip0->isOperable() && ip1->isOperable()) )
77                {
78                    delete ip0;
79                    delete ip1;
80                    errPrint("invalid operand(s)!");
81                    return;
82                }
83                if ( (ip0->isAccePtr() || ip0->type == IMMD_ITEM) && ip1->isMonoVal() )
84                {
85                    ptrBiasing(code, ip0, ip1);
86                    return;
87                }
88                if ( (ip1->isAccePtr() || ip1->type == IMMD_ITEM) && ip0->isMonoVal() && code == '+' )
89                {
90                    ptrBiasing(code, ip1, ip0);
91                    return;
92                }
```

```
 93
 94                    pnp = new Pnode(code);
 95                    addPnode(&mainPcode, pnp);
 96                    pnp->items[1] = ip0;
 97                    pnp->items[2] = ip1;
 98                    Item *tmp = makeTemp();
 99                    tmp->attr = (ip0->isMonoVal() && ip1->isMonoVal())? maxMonoAttr(ip0, ip1, code):
100                                                              maxSizeAttr(ip0, ip1);
101                    pnp->items[0] = tmp;
102                    PUSH(pnp->items[0]->clone());
103                    return;
104                }
105            break;
```
/p16ecc/cc_source_6.8/pcoder5.cpp   Code-6.31

- #83~92: if either X or Y is a pointer or equivalent, and the other is a mono value, calling the function `ptrBiasing()` to generate the P-code, in which the scaling may happen.
- #94~101: generate P-code, { '+' , [t, X, Y]} or { '-' , [t, X, Y]}.
- #98~100: generate t and determine its size.
- #102: duplicate t and push it into the stack.

The following (Table-6.7) shows the test results for this section.

C language code	P-code generated

`int n, *p;`
 	
`void foo(int m)`
`{`
`   int x;`
 	
`   x = n + m;`
`   x = (int)(p - m);`
`}`
	`F:\p16ecc\tests>cc16e ch6_t10.c`
	`P-CODE: P_FUNC_BEG: foo : {void}`
	`P-CODE: ';'`
	`P-CODE: SRC ... ch6_t10.c #7: x = n + m;`
	`P-CODE: '+'    %1{short}, n{int}, foo_$_m{int}`
	`P-CODE: '='    foo_$1_x{int}, %1{short}`
	`P-CODE: ';'`
	`P-CODE: SRC ... ch6_t10.c #8: x = (int)(p - m);`
	`P-CODE: '*'    %1{int}, foo_$_m{int}, 2{int}`
	`P-CODE: '-'    %2{int *}, p{int *}, %1{int}`
	`P-CODE: '='    %3{int}, %2{int *}`
	`P-CODE: '='    foo_$1_x{int}, %3{int}`
	`P-CODE: ';'`
	`P-CODE: LABEL    _$L1`
	`P-CODE: P_FUNC_END: foo`

Table-6.7

6.8.3 P-code for bitwise operators

Three bitwise operators ('&' , '|' , '^') are covered here.

C language grammar:	$X \mid Y$ $X \wedge Y$ X & Y

Parsing rules:	`inclusive_or_expr '|' exclusive_or_expr    {`
	`          $$ = oprNode('|', 2, $1, $3); }`
	`exclusive_or_expr '^' and_expr            {`
	`          $$ = oprNode('^', 2, $1, $3); }`
	`and_expr '&' equality_expr                {`
	`          $$ = oprNode('&', 2, $1, $3); }`

Accordingly, they will end up with the P-code

$$\{ '|' , [t, X, Y]\} \text{ or }$$
$$\{ '^' , [t, X, Y]\} \text{ or }$$
$$\{ '\&' , [t, X, Y]\}$$

where t is a temporary variable, a `TEMP_ITEM` item that holds the operation result. Unlike that for arithmetic operators '+' and '-', the P-code generation for logic operators doesn't involve the scaling consideration. Therefore, the procedure becomes straightforward, as illustrated in Code-6.32.

```
107     case '|': case '&': case '^':
108         run(OPR_NODE(op, 0));
109         run(OPR_NODE(op, 1));
110         if ( DEPTH() == (depth+2) )
111         {
112             ip1 = POP();
113             ip0 = POP();
114             if ( ip0->isOperable() && ip1->isOperable() )
115             {
116                 pnp = new Pnode(code);
117                 addPnode(&mainPcode, pnp);
118                 pnp->items[1] = ip0;
119                 pnp->items[2] = ip1;
120
121                 Item *tmp = makeTemp();
122                 tmp->attr = maxSizeAttr(ip0, ip1);
123                 tmp->attr->isUnsigned = 1;
124                 pnp->items[0] = tmp;
125                 PUSH(pnp->items[0]->clone());
126                 return;
127             }
128             delete ip0;
129             delete ip1;
130             errPrint("invalid operation type!");
131             return;
132         }
133         break;
```
/p16ecc/cc_source_6.8/pcoder5.cpp   Code-6.32

- #121~122: generate t and determine its size which is max(X, Y).
- #125: duplicate t and push it into the stack.

6.8.4 P-code for compound assignment operators

This section addresses the following compound assignment operators.

C language grammar:	$X \mathrel{\&=} Y$ $X \mathrel{\|=} Y$ $X \mathrel{\char94=} Y$ $X \mathrel{*=} Y$ $X \mathrel{/=} Y$ $X \mathrel{\%=} Y$ $X \mathrel{<<=} Y$ $X \mathrel{>>=} Y$

Parsing rules:	`unary_expr AND_ASSIGN assignment_expr  {`
	`        $$ = oprNode(AND_ASSIGN, 2, $1, $3); }`
	`unary_expr OR_ASSIGN assignment_expr   {`
	`        $$ = oprNode(OR_ASSIGN, 2, $1, $3); }`
	`unary_expr XOR_ASSIGN assignment_expr  {`
	`        $$ = oprNode(XOR_ASSIGN, 2, $1, $3); }`
	`unary_expr MUL_ASSIGN assignment_expr  {`
	`        $$ = oprNode(MUL_ASSIGN, 2, $1, $3); }`
	`unary_expr DIV_ASSIGN assignment_expr  {`
	`        $$ = oprNode(DIV_ASSIGN, 2, $1, $3); }`
	`unary_expr MOD_ASSIGN assignment_expr  {`
	`        $$ = oprNode(MOD_ASSIGN, 2, $1, $3); }`
	`unary_expr LEFT_ASSIGN assignment_expr {`
	`        $$ = oprNode(LEFT_ASSIGN, 2, $1, $3); }`
	`unary_expr RIGHT_ASSIGN assignment_expr {`
	`        $$ = oprNode(RIGHT_ASSIGN, 2, $1, $3); }`

The P-code generation for these operators has the same structure as
> {*op-code*, [*X*, *Y*]}

The P-code generation is illustrated in Code-6.33.

```
135       case AND_ASSIGN:      case OR_ASSIGN:      case XOR_ASSIGN:
136       case MUL_ASSIGN:      case DIV_ASSIGN:      case MOD_ASSIGN:
137       case LEFT_ASSIGN:     case RIGHT_ASSIGN:
138           run(OPR_NODE(op, 0));
139           run(OPR_NODE(op, 1));
140           if ( DEPTH() == (depth+2) )
141           {
142               ip1 = POP();
143               ip0 = POP();
144               if ( ip0->isOperable() && ip1->isOperable() && ip0->isWritable() )
145               {
146                   if ( code == AND_ASSIGN || code == OR_ASSIGN || code == XOR_ASSIGN )
147                   {
148                       pnp = new Pnode(code);
149                       addPnode(&mainPcode, pnp);
150                       pnp->items[0] = ip0;
151                       pnp->items[1] = ip1;
152                       PUSH(pnp->items[0]->clone());
153                       return;
154                   }
155                   if ( ip0->isMonoVal() && ip1->isMonoVal() )      // *=, /=, %=, >>=, <<=
156                   {
157                       pnp = new Pnode(code);
158                       addPnode(&mainPcode, pnp);
159                       pnp->items[0] = ip0;
160                       pnp->items[1] = ip1;
161                       PUSH(pnp->items[0]->clone());
162                       return;
163                   }
164               }
165               delete ip0;
166               delete ip1;
167               errPrint("invalid operation type!");
168               return;
```

/p16ecc/cc_source_6.8/pcoder5.cpp   Code-6.33

- #146~154: P-code generation for AND_ASSIGN, OR_ASSIGN and XOR_ASSIGN.
- #155~163: P-code generation for the other operators.

6.9 P-code Generation For Oper Nodes (6)

Project directory: /p16ecc/cc_source_6.9
New source files: pcoder6.cpp
Modified source files: pcoder.h

6.9.1 P-code for '*', '/', '%' and shift operators

This section addresses the unsolved binary math/shift operators as listed below.

C language grammar:	$X * Y$ X / Y $X \% Y$ $X << Y$ $X >> Y$

Parsing rules:	`multiplicative_expr '*' cast_expr    {`
	`                $$ = oprNode('*', 2, $1, $3); }`
	`multiplicative_expr '/' cast_expr    {`
	`                $$ = oprNode('/', 2, $1, $3); }`
	`multiplicative_expr '%' cast_expr    {`
	`                $$ = oprNode('%', 2, $1, $3); }`

| | ```
shift_expr LEFT_OP additive_expr     {
                $$ = oprNode(LEFT_OP, 2, $1, $3); }
shift_expr RIGHT_OP additive_expr    {
                $$ = oprNode(RIGHT_OP, 2, $1, $3); }
``` |

The P-code generated for them is in the same structure as
$$\{op\text{-}code, [t, X, Y]\}$$

Where t is a temporary variable, a `TEMP_ITEM` item that holds the operation result. It is important here to set the proper size for t, noted as size(t), as shown in the following table, to avoid result overflow:

Operator (*op-code*)	Result t size
`'*'`	size(t) = size(X) + size(Y), max to 4 bytes (`long` type)
`'/'`	size(t) = size(X)
`'%'`	size(t) = size(X)
`LEFT_OP`	size(t) = 4 bytes (`long` type)
`RIGHT_OP`	size(t) = size(X)

<div align="right">Table-6.8</div>

The following, Code-6.34, illustrates the P-code generation for the operators.

```
26      case '*':   case '/':   case '%':
27      case LEFT_OP:   case RIGHT_OP:
28          run(OPR_NODE(op, 0));
29          run(OPR_NODE(op, 1));
30          if ( DEPTH() == (depth+2) )
31          {
32              ip1 = POP();
33              ip0 = POP();
34              if ( ip0->isMonoVal() && ip1->isMonoVal() )
35              {
36                  pnp = new Pnode(code);
37                  addPnode(&mainPcode, pnp);
38                  pnp->items[1] = ip0;
39                  pnp->items[2] = ip1;
40
41                  Item *tmp = makeTemp();
42                  if ( code == LEFT_OP )
43                  {
44                      tmp->attr = newAttr(LONG);
45                      tmp->attr->isUnsigned = ip0->acceSign()? 0: 1;
46                  }
47                  else
48                      tmp->attr = maxMonoAttr(ip0, ip1, code);
49
50                  pnp->items[0] = tmp;
51                  PUSH(pnp->items[0]->clone());
52                  return;
53              }
54              delete ip0;
55              delete ip1;
56              errPrint("invalid operand(s)!");
57          }
58          break;
```

<div style="display:flex; justify-content:space-between;">/p16ecc/cc_source_6.9/pcoder6.cpp<span>Code-6.34</span></div>

- #42~46: set size for t (4 bytes long) unconditionally.
- #51: duplicate t and push it into the stack.
```

The following (Table-6.8) shows the test results for this section.

C language code	P-code generated

```void foo(int n, int m)```
```{```
  ```int x;```
 	
  ```x = n * m;```
  ```x = n / m;```
  ```x = n << m;```
 	
```}```
	```F:\p16ecc\tests>cc16e ch6_t11.c```
	```P-CODE: P_FUNC_BEG: foo : {void}```
	```P-CODE: ';'```
	```P-CODE: SRC ... ch6_t11.c #5: x = n * m;```
	```P-CODE: '*' %1{long}, foo_$_n{int}, foo_$_m{int}```
	```P-CODE: '=' foo_$1_x{int}, %1{long}```
	```P-CODE: ';'```
	```P-CODE: SRC ... ch6_t11.c #6: x = n / m;```
	```P-CODE: '/' %1{int}, foo_$_n{int}, foo_$_m{int}```
	```P-CODE: '=' foo_$1_x{int}, %1{int}```
	```P-CODE: ';'```
	```P-CODE: SRC ... ch6_t11.c #7: x = n << m;```
	```P-CODE: '<<' %1{long}, foo_$_n{int}, foo_$_m{int}```
	```P-CODE: '=' foo_$1_x{int}, %1{long}```
	```P-CODE: ';'```
	```P-CODE: LABEL _$L1```
	```P-CODE: P_FUNC_END: foo```

Table-6.7

6.9.2 P-code for relational operators

This section addresses all six relational operators as listed below.

C language grammar:	$X == Y$ $X != Y$ $X > Y$ $X >= Y$ $X < Y$ $X <= Y$
Parsing rules:	```relational_expr RELATIONAL_OP shift_expr {``` ``` $$ = oprNode($2, 2, $1, $3); }```

Relational operators typically appear in `if`, `for`, and `while` statements. They can occur in some other type of statement, such as

```
x = (a > b);
```

A relational operator makes a comparison or test between operands X and Y, to check if it meets the comparison type ($==$, $!=$, $<$, $>$, …). The result will be put in a temporary variable t, which will be set to true '1' if the result meets the comparison type, otherwise it will be set to false '0'.

Moreover, all those comparison types have their complementary types, shown in the table below:

Original comparison (*comp-op*)	Complementary (*comp-op'*)
==	!=
!=	==
>	<=
>=	<
<	>=
<=	>

Table-6.8

That means it is suitable to use the complementary type for comparison if it improves efficiency.

The P-code for the operators will be as follows. (It uses the complementary option for comparison)

{ '=', [*t*, 0]}	set *t* to '0' as the default
{*comp-op'*, [*X, Y, label2*]}	compare *X* and *Y*, and skip to *label2* if the test is true
{LABEL, [*label1*]}	put a label (*label1*)
{INC_OP, [*t*]}	set *t* to '1', (Note: increment op. is more efficient for PIC16)
{LABEL, [*label2*]}	put a label (*label2*)

The following (Code-6.35) illustrates the P-code generation for the operators.

```
60      case EQ_OP: case NE_OP:
61      case LE_OP: case GE_OP:
62      case '<':   case '>':
63          run(OPR_NODE(op, 0));
64          run(OPR_NODE(op, 1));
65          if ( DEPTH() == (depth+2) )
66          {
67              ip1 = POP();
68              ip0 = POP();
69              if ( !comparable(code, ip0, ip1) )
70              {
71                  delete ip0; delete ip1;
72                  errPrint("can't compare values!");
73                  return;
74              }
75
76              tmp = makeTemp();
77              tmp->attr = newAttr(CHAR); tmp->attr->isUnsigned = 1;
78
79              pnp = new Pnode('=');
80              addPnode(&mainPcode, pnp);
81              pnp->items[0] = tmp;
82              pnp->items[1] = intItem(0);
83
84              true_lbl  = getLbl();
85              false_lbl = getLbl();
86              compareBranch(code, ip0, ip1, true_lbl, false_lbl, true_lbl);
87
88              PUT_LBL(true_lbl);
89              pnp = new Pnode(INC_OP);
90              addPnode(&mainPcode, pnp);
91              pnp->items[0] = tmp->clone();
92              pnp->items[1] = intItem(1);
93
94              PUT_LBL(false_lbl);
95              PUSH(tmp->clone());
96          }
97      break;
```

/p16ecc/cc_source_6.9/pcoder6.cpp
Code-6.35

- #76~77: generate the temporary variable *t*.
- #79~82: generate P-code that sets *t* to 0.
- #84~85: create two label items for skip.
- #86: call function `compareBranch()` that generates P-code '{*comp-op'*, [*X, Y, label2*]}'.
- #88~94: generate the rest of the P-codes.
- #95: push the item *t* into the stack.

The following code, Code-6.36, illustrates the function `compareBranch()`, which generates the complementary of the operator and will be utilized in other places.

```
160   void Pcoder :: compareBranch(int op, Item *ip0, Item *ip1, int true_lbl, int false_lbl, int next_lbl)
161   {
162       int lbl;
163       if ( true_lbl == next_lbl )
164       {
165           switch ( op )
166           {
167               case EQ_OP: op = P_JNE; break;  // '=='
168               case NE_OP: op = P_JEQ; break;  // '!='
169               case '<':   op = P_JGE; break;  // '<'
170               case '>':   op = P_JLE; break;  // '>'
171               case GE_OP: op = P_JLT; break;  // '>='
172               case LE_OP: op = P_JGT; break;  // '<='
173           }
174           lbl = false_lbl;
175       }
176       else
177       {
178           switch ( op )
179           {
180               case EQ_OP: op = P_JEQ; break;  // '=='
181               case NE_OP: op = P_JNE; break;  // '!='
182               case '<':   op = P_JLT; break;  // '<'
183               case '>':   op = P_JGT; break;  // '>'
184               case GE_OP: op = P_JGE; break;  // '>='
185               case LE_OP: op = P_JLE; break;  // '<='
186           }
187           lbl = true_lbl;
188       }
189
190       Pnode *pnp = new Pnode(op);
191       addPnode (&mainPcode, pnp);
192       pnp->items[0] = ip0;
193       pnp->items[1] = ip1;
194       pnp->items[2] = lblItem(lbl);
195   }
```
`/p16ecc/cc_source_6.9/pcoder6.cpp`   Code-6.36

- #163~175: generate the complementary operator.

The following (Table-6.9) shows the test results for this section.

C language code	P-code generated

`void foo(int n, int m)`
`{`
 	
`  int x;`
 	
`  x = (n > m);`
`}`
	`F:\p16ecc\tests>cc16e ch6_t12.c`
	`P-CODE: P_FUNC_BEG: foo : {void}`
	`P-CODE: ';'`
	`P-CODE: SRC ... ch6_t12.c #5: x = (n > m);`
	`P-CODE: '=' %1{unsigned char}, 0{int}`
	`P-CODE: JP <= foo_$_n{int}, foo_$_m{int}, _$L3`
	`P-CODE: LABEL _$L2`
	`P-CODE: '++' %1{unsigned char}, 1{int}`
	`P-CODE: LABEL _$L3`
	`P-CODE: '=' foo_$1_x{int}, %1{unsigned char}`
	`P-CODE: ';'`
	`P-CODE: LABEL _$L1`
	`P-CODE: P_FUNC_END: foo`

Table-6.9

6.9.3 P-code for logic AND

Logic AND (`'&&'`) typically appears in `if`, `for`, and `while` statements, as well.

C language grammar:	X `&&` Y
Parsing rules:	`logical_and_expr AND_OP inclusive_or_expr          {` `           $$ = oprNode(AND_OP, 2, $1, $3); }`

It returns true '1' only if both X and Y are true (non-zero). Otherwise, it returns false '0'. Importantly, Y is not evaluated if X is false, a behavior known as short-circuiting evaluation. Additionally, both X and Y can contain complex expressions, which should be treated appropriately. For example,

```c
void foo(int n, int m)
{
    int x;

    x = (n < 100 && m > 10);
}
```

The P-code generation for the operator is described below and Code-6.37.

{ '=', [t, 0]}	set t to '0' as the default
P-codes for X	evaluate X, skip to *label3* if the result is false
{LABEL, [*label1*]}	put a label (*label1*)
P-codes for Y	evaluate Y, skip to *label3* if the result is false
{LABEL, [*label2*]}	put a label (*label2*)
{INC_OP, [t]}	set t to '1'
{LABEL, [*label3*]}	put a label (*label3*)

```cpp
 99             case AND_OP:     // '&&'
100                 false_lbl = getLbl();
101
102                 pnp = new Pnode('=');
103                 addPnode(&mainPcode, pnp);
104                 tmp = makeTemp();
105                 tmp->attr = newAttr(CHAR);
106                 tmp->attr->isUnsigned = 1;
107                 pnp->items[0] = tmp;
108                 pnp->items[1] = intItem(0);
109
110                 true_lbl = getLbl();
111                 logicBranch(OPR_NODE(op, 0), true_lbl, false_lbl, true_lbl);
112                 PUT_LBL(true_lbl);
113
114                 true_lbl = getLbl();
115                 logicBranch(OPR_NODE(op, 1), true_lbl, false_lbl, true_lbl);
116                 PUT_LBL(true_lbl);
117
118                 pnp = new Pnode(INC_OP);
119                 addPnode(&mainPcode, pnp);
120                 pnp->items[0] = tmp->clone();
121                 pnp->items[1] = intItem(1);
122
123                 PUT_LBL(false_lbl);
124                 PUSH(tmp->clone());
125                 break;
```
/p16ecc/cc_source_6.9/pcoder6.cpp — Code-6.37

- #102~108: generate P-code { '=', [t, 0]}.
- #111: evaluate X, to generate the P-codes that make the jump based on the result.
- #112: put a label (*label1*).
- #115: evaluate Y, to generate the P-codes that make the jump based on the result.
- #112: put a label (*label2*).
- #118~121: generate P-code {INC_OP, [t]}
- #123: put a label (*label3*).
- #124: push the item t into the stack.

The following (Table-6.10) shows the test results for this section.

C language code	P-code generated
| ```
void foo(int n, int m)
{
   int x;


    x = (n < 100 &&
       m > 10);

}
``` | ```
F:\p16ecc\cc_source_6.9>cc16e ch6_t13.c
P-CODE: P_FUNC_BEG: foo : {void}
P-CODE: ';'
P-CODE: SRC ... ch6_t13.c #6: x = (n < 100 && m > 10);
P-CODE: '='    %1{unsigned char}, 0{int}
P-CODE: JP >=  foo_$_n{int}, 100{int}, _$L2
P-CODE: LABEL   _$L3
P-CODE: JP <=  foo_$_m{int}, 10{int}, _$L2
P-CODE: LABEL   _$L4
P-CODE: '++'   %1{unsigned char}, 1{int}
P-CODE: LABEL   _$L2
P-CODE: '='    foo_$1_x{int}, %1{unsigned char}
P-CODE: ';'
P-CODE: LABEL   _$L1
P-CODE: P_FUNC_END: foo
``` |

<div align="right">Table-6.10</div>

6.9.4 P-code for logic OR

Logic OR (`'||'`) also typically appears in `if`, `for`, and `while` statements, like logic AND. In C grammar, the OR operator has lower precedence than the AND operator.

C language grammar:	$X \mid\mid Y$
| Parsing rules: | ```
logical_or_expr OR_OP logical_and_expr    {
              $$ = oprNode(OR_OP, 2, $1, $3); }
``` |

There are several similarities between them in P-code generation, as illustrated in Code-6.38. It returns true '1' if either X or Y is true '1'. Otherwise, it returns false '0'. That means Y is not evaluated if X is true.

Table-6.11 shows the test results for this section.

```
127        case OR_OP:      // '||'
128            true_lbl  = getLbl();
129
130            pnp = new Pnode('=');
131            addPnode (&mainPcode, pnp);
132            tmp = makeTemp();
133            tmp->attr = newAttr(CHAR); tmp->attr->isUnsigned = 1;
134            pnp->items[0] = tmp;
135            pnp->items[1] = intItem(1);
136
137            false_lbl = getLbl();
138            logicBranch(OPR_NODE(op, 0), true_lbl, false_lbl, false_lbl);
139            PUT_LBL(false_lbl);
140
141            false_lbl = getLbl();
142            logicBranch(OPR_NODE(op, 1), true_lbl, false_lbl, false_lbl);
143            PUT_LBL(false_lbl);
144
145            pnp = new Pnode(DEC_OP);
146            addPnode (&mainPcode, pnp);
147            pnp->items[0] = tmp->clone();
148            pnp->items[1] = intItem(1);
149
150            PUT_LBL(true_lbl);
151            PUSH(tmp->clone());
152            break;
/p16ecc/cc_source_6.9/pcoder6.cpp
```

<div align="right">Code-6.38</div>
```

C language code	P-code generated
| ```
void foo(int n, int m)
{
  int x;

    x = (n < 100 ||
      m > 10);
}
``` | ```
F:\p16ecc\tests>cc16e ch6_t14.c
P-CODE: P_FUNC_BEG: foo : {void}
P-CODE: ';'
P-CODE: SRC ... ch6_t14.c #6: x = (n < 100 || m > 10);
P-CODE: '='    %1{unsigned char}, 1{int}
P-CODE: JP <  foo_$_n{int}, 100{int}, _$L2
P-CODE: LABEL   _$L3
P-CODE: JP >  foo_$_m{int}, 10{int}, _$L2
P-CODE: LABEL   _$L4
P-CODE: '--'   %1{unsigned char}, 1{int}
P-CODE: LABEL   _$L2
P-CODE: '='    foo_$1_x{int}, %1{unsigned char}
P-CODE: ';'
P-CODE: LABEL   _$L1
P-CODE: P_FUNC_END: foo
``` |

Table-6.11

Note that the function `logicBranch()` called in both Code-6.37 and Code-6.38 is to evaluate an item and make a jump based on the test result, in which recursive calling may happen. The function will be called elsewhere (such as during P-code generation for `if`, `for`, and `while` statements). It dispatches the process according to the data types to improve the efficiency, and also greatly ease the design. The P-code generation is illustrated in Code-6.39a~Code-6.39b.

```cpp
197    void Pcoder :: logicBranch(node *np, int true_lbl, int false_lbl, int next_lbl)
198    {
199        int depth = DEPTH();
200        int lbl;
201        if ( np && np->type == NODE_OPR ) switch ( np->opr.oper )
202        {
203            case '!':
204                takeSrc(np);
205                logicBranch(np->opr.op[0], false_lbl, true_lbl, next_lbl);
206                return;
207            case EQ_OP: case NE_OP:
208            case GE_OP: case LE_OP:
209            case '>':   case '<':
210                takeSrc(np);
211                run(np->opr.op[0]);
212                run(np->opr.op[1]);
213                if ( DEPTH() == (depth+2) )
214                {
215                    Item *ip1 = POP();
216                    Item *ip0 = POP();
217                    int  code = np->opr.oper;
218                    if ( comparable(code, ip0, ip1) )
219                        compareBranch(code, ip0, ip1, true_lbl, false_lbl, next_lbl);
220                    else
221                    {
222                        delete ip0;
223                        delete ip1;
224                        errPrint("can't compare the values!");
225                    }
226                }
227                return;
228            case AND_OP:    // x && y
229            case OR_OP:     // x || y
230                takeSrc(np);
231                lbl = getLbl();
232                if ( np->opr.oper == AND_OP )
233                    logicBranch(np->opr.op[0], lbl, false_lbl, lbl);
234                else
235                    logicBranch(np->opr.op[0], true_lbl, lbl, lbl);
236                PUT_LBL(lbl);
237                logicBranch(np->opr.op[1], true_lbl, false_lbl, next_lbl);
238                return;
239        }
```

/p16ecc/cc_source_6.9/pcoder6.cpp   Code-6.39a

- #228~238: direct recursive calling happens here.

```cpp
241          run(np);
242          if ( DEPTH() != (depth+1) )
243          {
244              errPrint("invalid branch operation!");
245              return;
246          }
247
248          Item *ip = POP();
249          Pnode *pnp;
250
251          if ( !ip->isOperable() )
252          {
253              delete ip;
254              errPrint("invalid value type!");
255              return;
256          }
257
258          if ( ip->type == CON_ITEM )
259          {
260              int n = ip->val.i;
261              delete ip;
262
263              if ( (true_lbl  == next_lbl &&  n) ||
264                   (false_lbl == next_lbl && !n) )
265                  return;
266
267              pnp = new Pnode(GOTO);
268              addPnode(&mainPcode, pnp);
269              pnp->items[0] = lblItem(n? true_lbl: false_lbl);
270          }
271          else
272          {
273              int lbl = (true_lbl == next_lbl)? false_lbl: true_lbl;
274              pnp = new Pnode((true_lbl == next_lbl)? P_JZ: P_JNZ);
275              addPnode(&mainPcode, pnp);
276              pnp->items[0] = ip;
277              pnp->items[1] = lblItem(lbl);
278          }
```

`/p16ecc/cc_source_6.9/pcoder6.cpp`

Code-6.39b

- #258~270: the procedure is simplified to evaluate a constant.

6.10 P-code Generation For Oper Nodes (7)

<u>Project directory</u>: /p16ecc/cc_source_6.10
<u>New source files</u>: pcoder7.cpp
<u>Modified source files</u>: pcoder.h

6.10.1 Expand Pcoder class

In the C language, `break` and `continue` statements can appear in loop statements (such as `for`, `while`, `do...while`). These two statements make jumps and alter the execution sequence. Furthermore, it will complicate things when nested loop statements are involved. For this reason, a class named `JumpStack` is introduced (see Code-6.40), which manages the jump points for nested break and continue statements in a FILO (First in, Last out) manner, and is instantiated (`breakStack` and `continueStack`) in the `Pcoder` class.

```
19  class JumpStack {
20      private:
21          int FILO[1024];
22          int index;
23
24      public:
25          JumpStack()      { index = 0; }
26          void reset()     { index = 0; }
27          void push(int lbl) { if (index < 128) FILO[index++] = lbl; }
28          int  pop()       { return (index > 0)? FILO[--index]: 0; }
29          int  top()       { return (index > 0)? FILO[index-1]: 0; }
30          int  depth()     { return index; }
31  };
32
33  class Pcoder {
34      public:
35          Pcoder();
36          ~Pcoder();
37          void run(node *np);
38
39          void  takeSrc(node *np);
40          Pnode *mainPcode;
41          Pnode *initPcode;
42          Pnode *constPcode;
43          src_t *src;
44          int   errorCount;
45          int   warningCount;
46
47          Dlink *curDlink;      // current data link
48          Fnode *curFnode;      // current func node
49
50          int   labelSeed;
51          int   getLbl(void) { return ++labelSeed; }
52          Item  *iStack;
53          Const *constGroup;
54
55          JumpStack continueStack;
56          JumpStack breakStack;
```
`/p16ecc/cc_source_6.10/pcoder7.cpp`   Code-6.40

6.10.2 P-code for IF statement

The whole `IF` statement consists of several parts.

C language grammar:	`if ( ` *expr* ` )`
	statement1
	`else`
	statement2

Parsing rules:	`IF '(' expr ')' statement else_statement    {` `                $$ = oprNode(IF, 3, $3, $5, $6); }`

The flowchart in Figure 6.2 illustrates the structure of the generated P-codes. Three labels (true_lbl, false_lbl, end_lbl) are generated and placed in the P-codes sequences.

Code-6.41 is the procedure of P-code generation for `IF` statement, and Table-6.12 shows the test result.

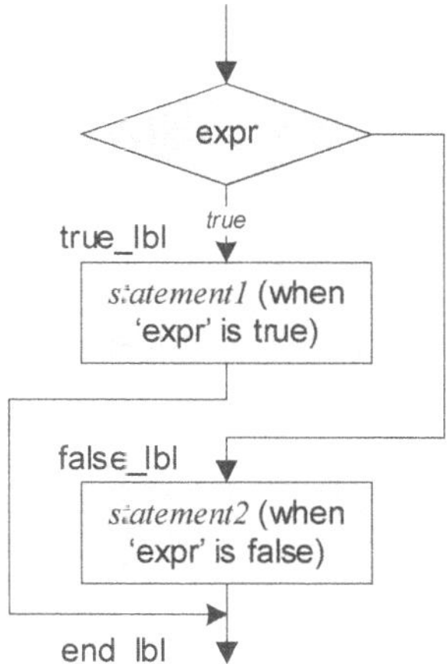

Figure-6.2

```
26      case IF:
27          true_lbl  = getLbl();
28          false_lbl = getLbl();
29          end_lbl   = getLbl();
30
31          logicBranch(op->op[0], true_lbl, false_lbl, true_lbl);
32          addPnode(&mainPcode, new Pnode(';'));
33
34          PUT_LBL(true_lbl);
35          run(op->op[1]); // true statement
36          addPnode(&mainPcode, new Pnode(GOTO, lblItem(end_lbl)));
37
38          PUT_LBL(false_lbl);
39          run(op->op[2]); // false statement(can be NULL)
40
41          PUT_LBL(end_lbl);
42          break;
```

/p16ecc/cc_source_6.10/pcoder7.cpp

Code-6.41

- #27~29: three labels (true_lbl, false_lbl, end_lbl) generated.
- #31: call function `logicBranch()` to evaluate *expr* and make the jump according the test result.
- #34: P-code for placing the label true_lbl before *statement1*.
- #35: generate the P-codes for *statement1*.
- #36: generate P-code {GOTO, [end_lbl]} that unconditionally jumps to the end of the statement.
- #38: P-code for placing the label false_lbl before *statement2*.
- #39: generate the P-codes for *statement2*.
- #41: P-code for placing the label end_lbl.

C language code	P-code generated

`void foo(int n)`
`{`
`  int x;`
 	
`  if ( n > 10 )`
`    x = 1;`
`  else`
`    x = 0;`
`}`
	`F:\p16ecc\tests>cc16e ch6_t15.c`
	`P-CODE: P_FUNC_BEG: foo : {void}`
	`P-CODE: ';'`
	`P-CODE: SRC ... ch6_t15.c #5: if ( n > 10 )`
	`P-CODE: JP <=  foo_$_n{int}, 10{int}, _$L3`
	`P-CODE: ';'`
	`P-CODE: LABEL   _$L2`
	`P-CODE: SRC ... ch6_t15.c #6: x = 1;`
	`P-CODE: '='   foo_$1_x{int}, 1{int}`
	`P-CODE: GOTO    _$L4`
	`P-CODE: LABEL   _$L3`
	`P-CODE: SRC ... ch6_t15.c #8: else    x = 0;`
	`P-CODE: '='   foo_$1_x{int}, 0{int}`
	`P-CODE: LABEL   _$L4`
	`P-CODE: ';'`
	`P-CODE: LABEL   _$L1`
	`P-CODE: P_FUNC_END: foo`

Table-6.12

145

<u>6.10.3 P-code for WHILE statement</u>

WHILE statement has a loop mechanism that allows a block of code to be repeatedly executed as long as a specified condition remains true. As mentioned earlier, break and continue statements can alter the looping sequence, which means it's necessary to create and set labels for them.

C language grammar:	**while (*expr*)** *statement*
Parsing rules:	```WHILE_'(' expr ')' statement {``` ``` $$ = oprNode(WHILE, 2, $3, $5); }```

Figure-6.3 illustrates how the P-codes generated for WHILE are structured.

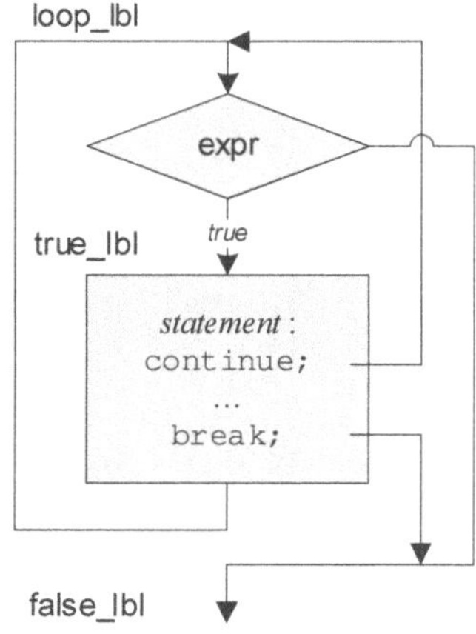

Figure-6.3

Code-6.42 illustrates the P-code generation for WHILE statement (pay attention to continueStack and breakStack operations).

```
44      case WHILE: // while (...) ...
45          loop_lbl  = getLbl();
46          true_lbl  = getLbl();
47          false_lbl = getLbl();
48
49          PUT_LBL(loop_lbl);
50          logicBranch(op->op[0], true_lbl, false_lbl, true_lbl);
51          addPnode(&mainPcode, new Pnode(';'));
52
53          PUT_LBL(true_lbl);
54
55          continueStack.push(loop_lbl);
56          breakStack.push(false_lbl);
57          run(op->op[1]);
58          continueStack.pop();
59          breakStack.pop();
60
61          addPnode(&mainPcode, new Pnode(GOTO, lblItem(loop_lbl)));
62          PUT_LBL(false_lbl);
63          break;
```
/p16ecc/cc_source_6.10/pcoder7.cpp   Code-6.42

- #45~47: three labels (loop_lbl, true_lbl, false_lbl) generated.
- #49: P-code for placing the label loop_lbl.
- #50: call function logicBranch() to evaluate *expr* and make the jump according the test result.
- #53: P-code for placing the label true_lbl.
- #55: push loop_lbl into continueStack.

- #56: push false_lbl into `breakStack`.
- #57: generate the P-codes for *statement*.
- #58: pop loop_lbl out of `continueStack`.
- #59: pop false_lbl out of `breakStack`.
- #61: generate P-code {GOTO, [loop_lbl]} that unconditionally jumps to the start of the statement.
- #62: P-code for placing the label false_lbl.

Table-6.13 shows the test result for this section.

C language code	P-code generated

<pre>void delay(int n)
{
  while (n)
  {
    if (n > 1000)
      break;
    n--;
  }
}</pre>
	<pre>F:\p16ecc\tests>cc16e ch6_t16.c
	P-CODE: P_FUNC_BEG: delay : {void}
	P-CODE: LABEL _$L2
	P-CODE: SRC ... ch6_t16.c #3: while (n)
	P-CODE: JP_Z delay_$_n{int}, _$L4
	P-CODE: ';'
	P-CODE: LABEL _$L3
	P-CODE: SRC ... ch6_t16.c #5: if (n > 1000)
	P-CODE: JP <= delay_$_n{int}, 1000{int}, _$L6
	P-CODE: ';'
	P-CODE: LABEL _$L5
	P-CODE: SRC ... ch6_t16.c #6: break;
	P-CODE: GOTO _$L4
	P-CODE: GOTO _$L7
	P-CODE: LABEL _$L6
	P-CODE: LABEL _$L7
	P-CODE: ';'
	P-CODE: SRC ... ch6_t16.c #7: --;
	P-CODE: '=' %1{int}, delay_$_n{int}
	P-CODE: '--' delay_$_n{int}, 1{int}
	P-CODE: ';'
	P-CODE: GOTO _$L2
	P-CODE: LABEL _$L4
	P-CODE: ';'
	P-CODE: LABEL _$L1
	P-CODE: P_FUNC_END: delay</pre>

Table-6.13

6.10.4 P-code for DO...WHILE statement

DO ... WHILE loop is a variant of the while loop. This loop will execute the code block once before checking if the condition is proper. Then, it will repeat the loop as long as the condition remains true.

C language grammar:	**do** *statement* **while (** *expr* **);**

Parsing rules:	<pre>DO statement WHILE_'(' expr ')' ';' { {
	             $$ = oprNode(DO, 2, $2, $5); }</pre>

Figure-6.4 illustrates how the P-codes are generated for DO...WHILE statement is structured. And Code-6.43 is the program that produces the P-code for the statement.

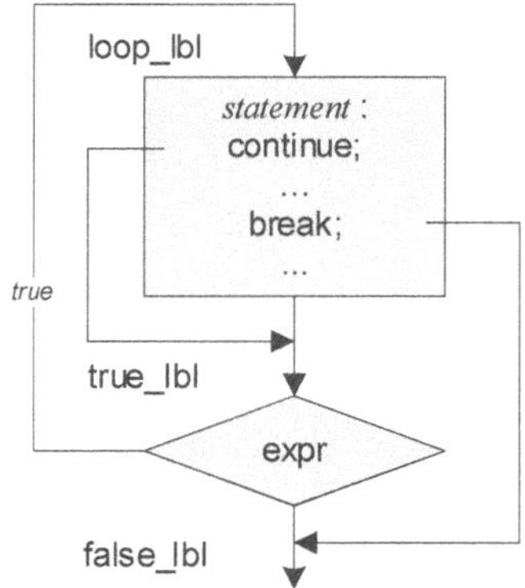

Figure-6.4

```
65              case DO:     // do ... while (...);
66                  loop_lbl  = getLbl();
67                  true_lbl  = getLbl();
68                  false_lbl = getLbl();
69
70                  PUT_LBL(loop_lbl);
71
72                  continueStack.push(true_lbl);
73                  breakStack.push(false_lbl);
74                  run(op->op[0]);
75                  continueStack.pop();
76                  breakStack.pop();
77
78                  PUT_LBL(true_lbl);
79                  logicBranch(op->op[1], loop_lbl, false_lbl, false_lbl);
80                  PUT_LBL(false_lbl);
81                  break;
```

/p16ecc/cc_source_6.10/pcoder7.cpp   Code-6.43

6.10.5 P-code for FOR statement

A FOR loop is a control flow statement that allows you to repeatedly execute a block of code a specific number of times. It is often used when you know in advance how many times you need to iterate.

C language grammar:	**for** (*expr1*; *expr2*; *expr3*) *statement*

Parsing rules:	FOR '(' opt_expr ';'
	    opt_expr ';'
	        opt_expr ')' statement {
	            $$ = oprNode(FOR, 4, $3, $5, $7, $9); }

Note, *expr1, expr2* and *expr3* are optional. And when *expr2* doesn't exist (NULL), it will become an endless loop; only break and goto can end the execution of *statement*.

Figure-6.5 illustrates how the P-codes are generated for the FOR statement in both cases. And Code-6.44 is the program that produces the P-code for the statement.

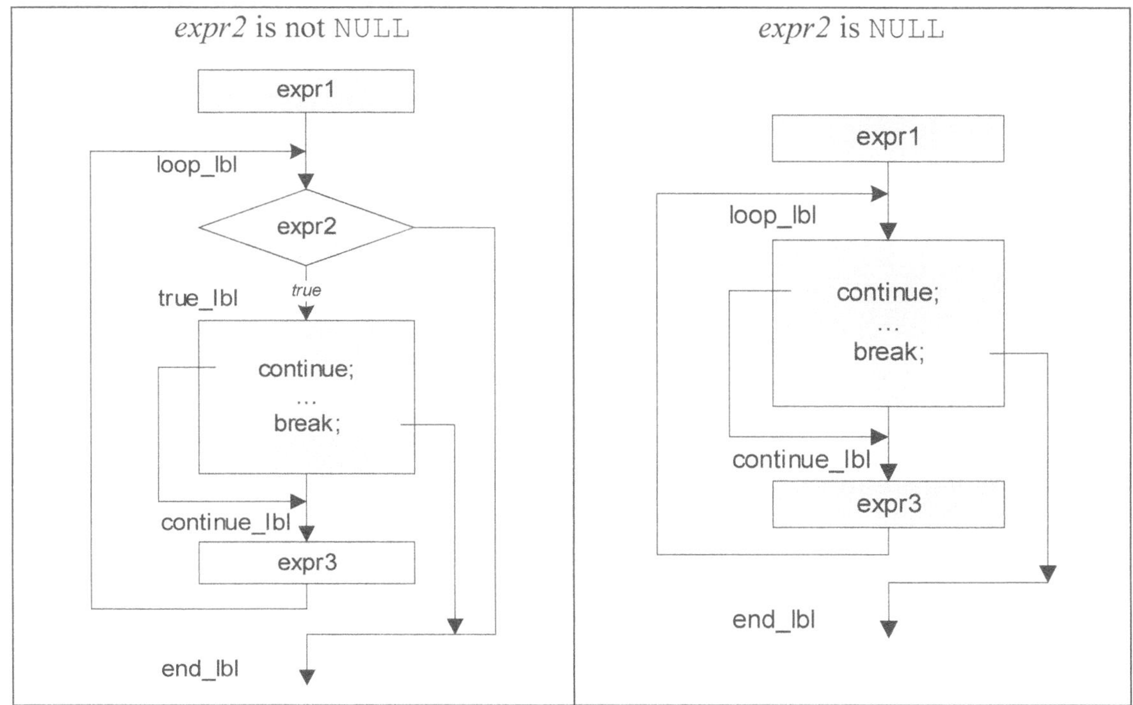

Figure-6.5

```
83              case FOR:
84                  loop_lbl      = getLbl();
85                  true_lbl      = getLbl();
86                  end_lbl       = getLbl();
87                  continue_lbl = getLbl();
88
89                  run(op->op[0]); addPnode(&mainPcode, new Pnode(';'));
90
91                  PUT_LBL(loop_lbl);
92
93                  if ( op->op[1] )
94                  {
95                      logicBranch(op->op[1], true_lbl, end_lbl, true_lbl);
96                      PUT_LBL(true_lbl);
97                  }
98
99                  continueStack.push(continue_lbl);
100                 breakStack.push(end_lbl);
101                 run(op->op[3]); addPnode(&mainPcode, new Pnode(';'));
102                 continueStack.pop();
103                 breakStack.pop();
104
105                 PUT_LBL(continue_lbl);
106                 run(op->op[2]); addPnode(&mainPcode, new Pnode(';'));
107                 addPnode(&mainPcode, new Pnode(GOTO, lblItem(loop_lbl)));
108
109                 PUT_LBL(end_lbl);
110                 break;
```

/p16ecc/cc_source_6.10/pcoder7.cpp   Code-6.44

- #93~97: check if *expr2* exists.

6.10.6 P-code for BREAK and CONTINUE statements

BREAK and CONTINUE statements are control flow mechanisms used to terminate the execution of a loop (such as for, while, or do-while) or a switch statement prematurely.

C language grammar:	`break ;` `continue ;`	
Parsing rules:	`BREAK ';'`	`{ $$ = oprNode(BREAK, 0);   }`
	`CONTINUE ';'`	`{ $$ = oprNode(CONTINUE, 0); }`

In the P-code generation, illustrated in Code-6.45, both BREAK and CONTINUE statements retrieve the label from the top of either breakStack or continueStack, and then place a GOTO P-code.

```
112        case BREAK:
113            if ( breakStack.depth() > 0 )
114                addPnode(&mainPcode, new Pnode(GOTO, lblItem(breakStack.top())));
115            else
116                errPrint("illegal 'break' statement!");
117            break;
118
119        case CONTINUE:
120            if ( continueStack.depth() > 0 )
121                addPnode (&mainPcode, new Pnode(GOTO, lblItem(continueStack.top())));
122            else
123                errPrint("illegal 'continue' statement!");
124            break;
```
/p16ecc/cc_source_6.10/pcoder7.cpp   Code-6.45

6.10.7 P-code for SWITCH statement

The SWITCH statement in C programming provides a structured way to execute different code blocks based on the value of a single expression. It serves as an alternative to long chains of if-else statements when comparing a variable against multiple possible values.

A typical SWITCH statement is shown as follows.

```
        void foo(int x)
        {
            .  .  .

            switch ( x )
            {
              case N1:
                statement block1
              case N2:
                statement block2
              .  .  .
              case Nn:
                statement blockn
              default:
                statement block for default
            }


            .  .  .

        }
```

The corresponding parsing tree is shown in Figure-6.6.

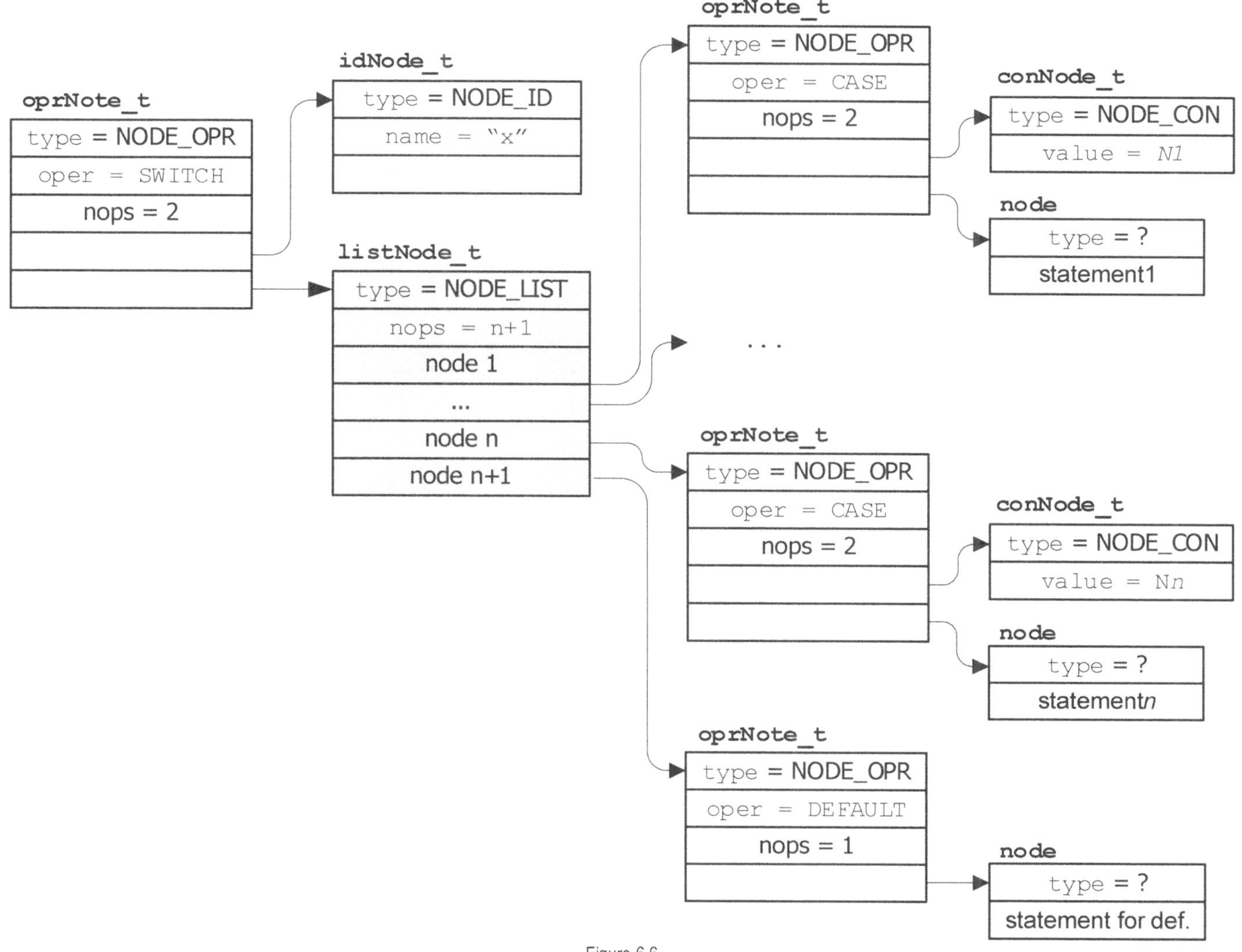

Figure-6.6

There are several issues that should be clarified for the P-code generation:
- The whole statement will be divided into two sequences of P-codes for SWITCH statement: (1) compare-jump sequence; (2) execution code blocks sequence. They are connected with jumping labels.
- DEFAULT statement is optional. And it can only appear once in the sequence.
- All the P-code execution code blocks are arranged in back-to-back sequential order, exactly the same as those that appear in the source code.
- The DEFAULT statement can appear in any place in the sequence. It is only executed when all the CASE tests fail, regardless of its position in the sequence.
- Figure-6.7 shows the flowchart for the implementation of P-code. Note that it shows some discrepancies compared to those that appear in most textbooks.

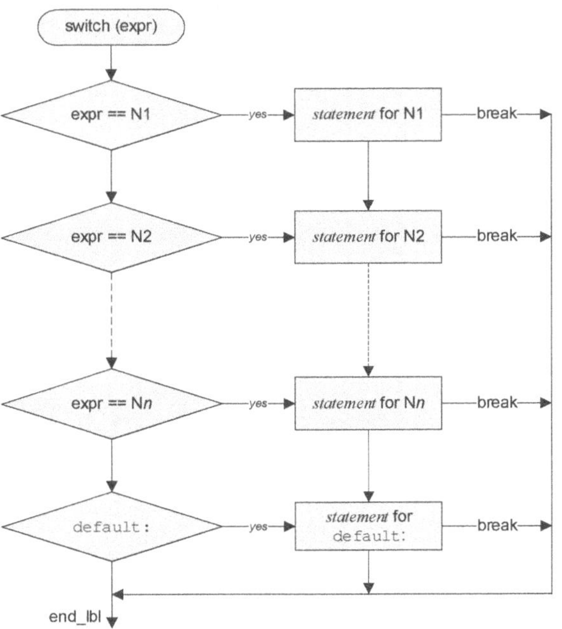

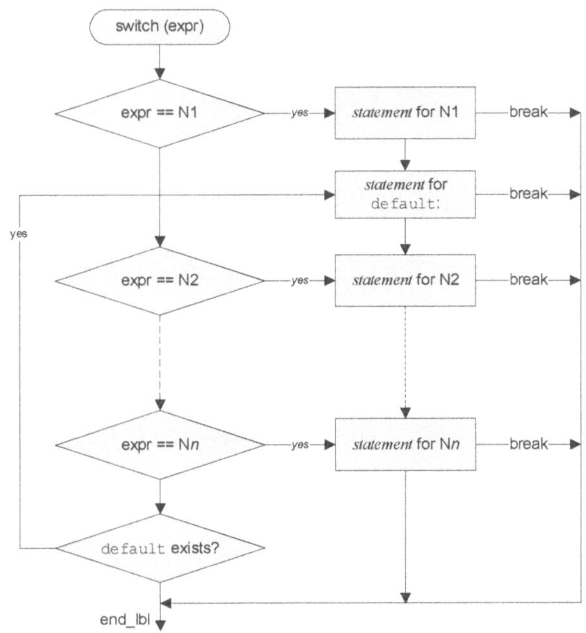

Figure-6.7

The P-code generation for SWITCH statement is illustrated in Code-6.46 and Code-6.47a~Code-6.47d.

```
126         case SWITCH:
127             run(op->op[0]);
128
129             if ( DEPTH() == (depth+1) )
130             {
131                 Item *ip = POP();
132                 if ( ip->isMonoVal() )
133                     switchListProc(ip, op->op[1]);
134                 else
135                 {
136                     errPrint("can't evaluate for SWITCHing!");
137                     delete ip;
138                 }
139             }
140             break;
```
/p16ecc/cc_source_6.10/pcoder7.cpp   Code-6.46

- #127: generate the item that holds 'x'.
- #133: call the function switchListProc() that generates the P-code.

```
148     void Pcoder :: switchListProc(Item *ip, node *list)
149     {
150         Pnode *pnp;
151         int length = LIST_LENGTH(list);
152         int *lbl_list = new int[length];       // case labels
153         int *val_list = new int[length];       // case values
154         int  end_lbl  = getLbl();              // end label
155         int  default_idx = -1;                 // default case label index
156
157         if ( !(ip->type == ID_ITEM || ip->type == TEMP_ITEM || ip->type == ACC_ITEM) )
158         {
159             pnp = new Pnode('=');    addPnode (&mainPcode, pnp);
160             pnp->items[1] = ip;
161             pnp->items[0] = makeTemp();
162             pnp->items[0]->attr = newAttr((ip->acceSize() == 1)? CHAR :
163                                           (ip->acceSize() == 2)? INT   :
164                                           (ip->acceSize() == 3)? SHORT: LONG);
165             ip = pnp->items[0]->clone();
166         }
```
/p16ecc/cc_source_6.10/pcoder7.cpp   Code-6.47a

- #152: create a list, lbl_list[], that holds all the labels that are for entries of the execution code blocks.

- #153: create a list, `val_list[]`, that holds the comparing values (N_1, N_2, …).
- #154: create `end_lbl`.
- #155: sequence index for `default`.
- #157~166: P-code that moves 'x' to temporary variable if needed.

```
168        for (int i = 0; i < length; i++)
169        {
170            node *np = LIST_NODE(list, i);
171            lbl_list[i] = getLbl();
172
173            if ( np->opr.nops == 1 )      // is DEFAULT case?
174            {
175                if ( default_idx >= 0 )
176                {
177                    errPrint("multiple 'default'!");
178                    return;
179                }
180                default_idx = i;
181            }
182            else
183            {
184                int depth = DEPTH();
185                run(np->opr.op[0]);        // get 'case' value...
186                if ( DEPTH() == (depth+1) )
187                {
188                    Item *cip = POP();   // case N node
189                    if ( cip->type != CON_ITEM )
190                        errPrint("contant expected for 'case'!");
191                    else if ( caseValCheck(val_list, i, cip->val.i, default_idx) )
192                        errPrint("duplicated switch case value!");
193                    else
194                    {
195                        val_list[i] = cip->val.i;
196                        pnp = new Pnode(CASE);
197                        addPnode (&mainPcode, pnp);
198                        pnp->items[0] = ip->clone();
199                        pnp->items[1] = intItem(cip->val.i);
200                        pnp->items[2] = lblItem(lbl_list[i]);
201                    }
202                    delete cip;              // delete case N node
203                }
204            }
205        }
```
/p16ecc/cc_source_6.10/pcoder7.cpp   Code-6.47b

- #171: generate the label, *lbl*, for the sequence.
- #173~181: detect `default` statement, and mark the sequence index.
- #183~204: P-code, {CASE, [x, N*i*, *lbl*]}, that does compare-jump, for the whole CASE sequence.

```
207        pnp = new Pnode(GOTO);   addPnode (&mainPcode, pnp);
208        if ( default_idx < 0 )   // if no 'default' applied?
209            pnp->items[0] = lblItem(end_lbl);
210        else
211            pnp->items[0] = lblItem(lbl_list[default_idx]);
212
213        addPnode(&mainPcode, new Pnode(';'));
```
/p16ecc/cc_source_6.10/pcoder7.cpp   Code-6.47c

- #207~211: generate P-code {GOTO, [*label*]} that jumps to the end of the statement or DEFAULT execution code block.

```
215         breakStack.push(end_lbl);
216         for (int i = 0; i < length; i++)
217         {
218             node *np = LIST_NODE(list, i);
219
220             PUT_LBL(lbl_list[i]);
221
222             if ( np->opr.nops == 1 )      // default:
223                 run(np->opr.op[0]);
224             else                          // case N:
225                 run(np->opr.op[1]);
226
227             addPnode(&mainPcode, new Pnode(';'));
228         }
229         breakStack.pop();
230         PUT_LBL(end_lbl);
```

/p16ecc/cc_source_6.10/pcoder7.cpp   Code-6.47d

- #215: push `end_lbl` into `breakStack`, which will be used for `break` statements in all the execution code blocks.
- #218: place the entry label in front of each execution code block.
- #216~228: generate P-codes for all the execution code blocks, one by one.
- #229: pop `end_lbl` out of `breakStack`.

Table-6.14 shows the test result for this section.

C language code	P-code generated

``void foo(int x)
{
 int n;
 switch (x)
 {
  case 1: n = 0;
        break;
  case 2: n = 10;
        break;
  default: n = 100;
        break;
 }
}```
	``F:\p16ecc\tests>cc16e ch6_t18.c
	P-CODE: P_FUNC_BEG: foo : {void}
	P-CODE: ';'
	P-CODE: SRC ... ch6_t18.c #5: switch (x)
	P-CODE: SRC ... ch6_t18.c #6: case 1:
	P-CODE: CASE foo_$_x{int}, 1{int}, _$L3
	P-CODE: SRC ... ch6_t18.c #8: case 2:
	P-CODE: CASE foo_$_x{int}, 2{int}, _$L4
	P-CODE: GOTO _$L5
	P-CODE: ';'
	P-CODE: LABEL _$L3
	P-CODE: SRC ... ch6_t18.c #6: n = 0;
	P-CODE: '=' foo_$1_n{int}, 0{int}
	P-CODE: SRC ... ch6_t18.c #7: break;
	P-CODE: GOTO _$L2
	P-CODE: ';'
	P-CODE: LABEL _$L4
	P-CODE: SRC ... ch6_t18.c #8: n = 10;
	P-CODE: '=' foo_$1_n{int}, 10{int}
	P-CODE: SRC ... ch6_t18.c #9: break;
	P-CODE: GOTO _$L2
	P-CODE: ';'
	P-CODE: LABEL _$L5
	P-CODE: SRC ... ch6_t18.c #10: n = 100;
	P-CODE: '=' foo_$1_n{int}, 100{int}
	P-CODE: SRC ... ch6_t18.c #11: break;
	P-CODE: GOTO _$L2
	P-CODE: ';'
	P-CODE: LABEL _$L2
	P-CODE: ';'
	P-CODE: LABEL _$L1
	P-CODE: P_FUNC_END: foo```

Table-6.14

6.11 P-code Generation For Oper Nodes (8)

<u>Project directory</u>: /p16ecc/cc_source_6.11
<u>New source files</u>: pcoder8.cpp, p16e/pic16e_inst.h
<u>Modified source files</u>: pcoder.h

<u>6.11.1 P-code for CALL statement</u>

CALL statement is to make a function call and get its returning value. It has the following formats.

C language grammar:	*function_name* **(** *argument_list* **)** *function_name* **(** **)**

Parsing rules:	`postfix_expr '(' argument_expr_list ')' {`
	`                oprNode(CALL, 2, $1, $3); }`
	`postfix_expr '(' ')'  { oprNode(CALL, 2, $1, NULL); }`

The P-code generation for CALL is illustrated in Code-6.48a~Code-6.48d.

There is a compile-time function, asm(), which inserts a line of assembly code and requires special treatment during P-code generation. Additionally, in **p16ecc**, more assembly code is supplied for the PIC16F MCUs.

```
30      case CALL:
31          np = op->op[0];
32          if ( np->type == NODE_ID && strcmp(np->id.name, "asm") == 0 )
33          {   // it's an in-line assembly code...
34              if ( op->nops > 1 )
35              {
36                  np = op->op[1]; // fetch parameter list
37                  if ( np->type == NODE_LIST && LIST_LENGTH(np) == 1 )
38                  {
39                      np = LIST_NODE(np, 0);
40                      if ( np->type == NODE_STR )
41                      {
42                          pp = new Pnode(AASM);
43                          addPmode (&mainPcode, pp);
44                          pp->items[0] = strItem(np->str.str);
45                          return;
46                      }
47                  }
48              }
49              errPrint("asm(...) format error!");
50              return;
51          }
52
53          if ( np->type == NODE_ID && buildInAsm((node*)op) )
54              return;
```
/p16ecc/cc_source_6.11/pcoder8.cpp   Code-6.48a

- #32~51: P-code generation for asm().
- #53~54: P-code generation for other types of assembly code insertion.

There are two types of function calls: direct call and indirect call. The latter implies using a function pointer. *function_name* shall be validated (in **pcoder.cpp**) to confirm the calling type, and obtain its attributes (argument list length/type and returning type).

```
56                      run(np);
57                      if ( DEPTH() == (depth+1) )
58                      {
59                          Item *ip = POP();
60                          node *par_list= (op->nops > 1 && op->op[1])? op->op[1]: NULL;
61                          int   par_cnt = (op->nops > 1 && op->op[1])? LIST_LENGTH(par_list): 0;
62                          int   f_par_count;
63                          attrib *r_attr = NULL;
64                          node *parnp = NULL;
65                          bool par_elipsis;
66
67                          switch ( ip->type )
68                          {
69                              case ID_ITEM:    // it's function pointer
70                              case TEMP_ITEM:
71                              case ACC_ITEM:
72                              case DIR_ITEM:
73                              case PID_ITEM:
74                              case INDIR_ITEM:
75                                  if ( !(ip->attr && ip->attr->isFptr) )
76                                  {
77                                      delete ip;
78                                      errPrint("invalid function pointer!");
79                                      return;
80                                  }
81                                  parnp = (node*)ip->attr->parList;
82                                  f_par_count = LIST_LENGTH(parnp);
83                                  par_elipsis = (parnp && parnp->list.elipsis);
84
85                                  r_attr = cloneAttr(ip->attr);// function return attr.
86                                  r_attr->isFptr = 0;
87                                  if ( ip->type == INDIR_ITEM || ip->type == DIR_ITEM || ip->type == PID_ITEM )
88                                      reducePtr(r_attr);
89                                  break;
90
91                              case LBL_ITEM:   // it's a regular function call
92                                  fp = (Fnode*)ip->home;
93                                  if ( fp == NULL )
94                                  {
95                                      delete ip;
96                                      errPrint("invalid function name!");
97                                      return;
98                                  }
99                                  f_par_count = fp->parCount();
100                                  par_elipsis = fp->elipsis? true: false;
101
102                                  r_attr = cloneAttr(fp->attr);
103                                  break;
104
105                              default:
106                                  delete ip;
107                                  errPrint("invalid function name!");
108                                  return;
109                          }
110
```

`/p16ecc/cc_source_6.11/pcoder8.cpp`   Code-6.48b

- #69~89: it's an indirect function call (based on a function pointer)
- #91~103: it's a direct (regular) function call.

In P-code generation, the function parameters or arguments must be passed (ideally, push in a hardware stack) before the call is made.

```
119                      // push parameter(s) into stack...
120                      for (int i = 0; i < par_cnt; i++)
121                      {
122                          attrib *par_attr = fp? fp->parAttr(i): LIST_NODE(parnp, i)->id.attr;
123                          passParameter(par_attr, LIST_NODE(par_list, i), i, f_par_count);
124                      }
125
126                      pp = fp? new Pnode(CALL): new Pnode(P_CALL);
127                      pp->items[0] = ip;
128                      addPnode(&mainPcode, pp);
```

`/p16ecc/cc_source_6.11/pcoder8.cpp`   Code-6.48c

- #120~124: call function `passParameter()`, to generate P-code that passes function parameters, in the sequence from start to end.
- #126~128: P-code that makes a call (direct or indirect).

Note, `passParameter()` will also take care of the situation where a variable number of arguments in a function happens (denoted by the ellipsis operator '…' that indicates extra arguments might be appended at the end of the parameter list).

Finally, two more things need to be done, as shown in Code-6.48d.
1. Clear up the parameters (if they are in the stack) that correspond to the ellipsis operator '…'.
2. Generate P-code that moves the return value (in the accumulator) to a temporary variable.

```
130     // clear up extra parameters' space ...
131     if ( par_cnt > f_par_count )
132     {
133         pp = new Pnode(P_ARG_CLEAR);
134         addPnode (&mainPcode, pp);
135         pp->items[0] = intItem((par_cnt - f_par_count)*4);
136     }
137     // return value stored in ACC
138     if ( r_attr && (r_attr->type != VOID || r_attr->ptrVect) )
139     {
140         pp = new Pnode('=');
141         addPnode (&mainPcode, pp);
142         pp->items[0] = makeTemp(r_attr);
143         pp->items[1] = accItem(r_attr);
144         PUSH(pp->items[0]->clone());
145     }
```
`/p16ecc/cc_source_6.11/pcoder8.cpp`   Code-6.48d

Table-6.15 shows the test result for this section.

C language code	P-code generated

`void func(int);`
 	
`void foo(int x,`
`    int (*f)(int))`
`{`
`  func(x);`
`  f(x);`
`}`
	`F:\p16ecc\tests>cc16e ch6_t19.c`
	`P-CODE:  P_FUNC_BEG: foo : {void}`
	`P-CODE:  SRC ... ch6_t19.c #6: func(x);`
	`P-CODE:  PASS_ARG    0{int}, foo_$_x{int}, 1{int}`
	`P-CODE:  CALL  func`
	`P-CODE:  ';'`
	`P-CODE:  SRC ... ch6_t19.c #7: f(x);`
	`P-CODE:  PASS_ARG    0{int}, foo_$_x{int}, 1{int}`
	`P-CODE:  I_CALL foo_$_f{int}`
	`P-CODE:  '='    %1{int}, ACC{int}`
	`P-CODE:  ';'`
	`P-CODE:  LABEL    _$L1`
	`P-CODE:  P_FUNC_END: foo`

Table-6.15

6.11.2 P-code for '?' operator

The '?' operator is known as the conditional or ternary operator. It provides a concise way to express a simple if-else condition.

C language grammar:	*expr1* **?** *expr2* **:** *expr3*
Parsing rules:	`logical_or_expr '?' expr ':' conditional_expr    {` `                $$ = oprNode('?', 3, $1, $3, $5); }`

Actually, in the implementation of the operator, it is more complicated than it looks in the grammar. The operation of '?' returns *expr2*'s value if the test of *expr1* is true (not zero); otherwise, *expr3*'s value is returned. That means, in P-code generation, both *expr2* and *expr3* should be assigned to the same data holder as the result of the operator, which usually is a temporary variable. Additionally, determining the size of the temporary variable t is necessary, as *expr2* and *expr3* may differ in size. That means,

$$\text{size}(t) = \max\ (\text{size}\ (\textit{expr2}),\ \text{size}\ (\textit{expr3}))$$

Figure-6.8, which follows, shows the flowchart of the P-code structure, and Code-6.49 illustrates the P-code generation for the operator.

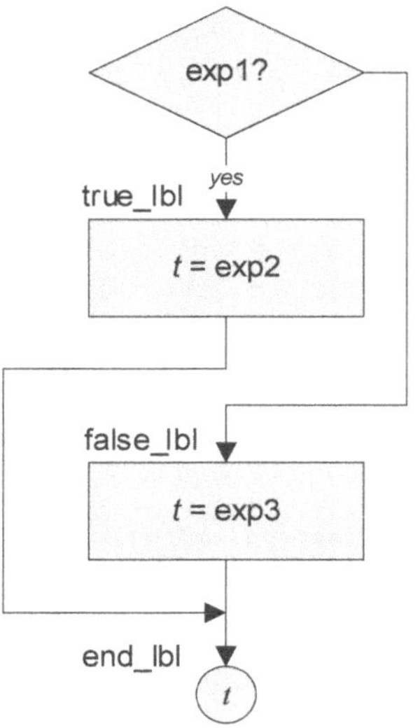

Figure-6.8

```
151        case '?':
152            true_lbl  = getLbl();
153            false_lbl = getLbl();
154            end_lbl   = getLbl();
155            logicBranch(op->op[0], true_lbl, false_lbl, true_lbl);
156            ip0 = ip1 = NULL;
157
158            PUT_LBL(true_lbl);
159            run(op->op[1]);
160            if ( DEPTH() == (depth+1) )
161            {
162                pp0 = new Pnode(P_MOV);
163                addPnode(&mainPcode, pp0);
164                pp0->items[1] = ip0 = POP();
165
166                pp = new Pnode(GOTO);
167                addPnode(&mainPcode, pp);
168                pp->items[0] = lblItem(end_lbl);
169            }
170
171            PUT_LBL(false_lbl);
172            run(op->op[2]);
173            if ( DEPTH() == (depth+1) )
174            {
175                pp1 = new Pnode(P_MOV);
176                addPnode(&mainPcode, pp1);
177                pp1->items[1] = ip1 = POP();
178            }
179            PUT_LBL(end_lbl);
180
181            if ( ip0 && ip1 && ip0->isOperable() && ip1->isOperable() )
182            {
183                int size0 = ip0->acceSize();
184                int size1 = ip1->acceSize();
185                if ( size0 <= 4 && size1 <= 4 )
186                {
187                    int max_size = (size0 >= size1)? size0: size1;
188                    if ( max_size > 0 ) max_size--;
189                    Item *ip = makeTemp();
190                    ip->attr = newAttr(max_size+CHAR);
191                    ip->attr->isUnsigned = (ip0->acceSign() && ip1->acceSign())? 0: 1;
192                    pp0->items[0] = ip->clone();
193                    pp1->items[0] = ip->clone();
194                    PUSH(ip);
195                    break;
196                }
197            }
```
`/p16ecc/cc_source_6.11/pcoder8.cpp`

Code-6.49

- #155: P-code that evaluates *expr1* and makes the jumps.
- #159~164: P-code that assigns the result of *expr2* to the temporary variable *t*.
- #166~168: P-code that jumps to the end of operation.
- #172~177: P-code that assigns the result of *expr3* to the temporary variable *t*.
- #179: P-code that places `end_lbl`.
- #187~188: determine the size for item *t*.
- #189~191: generate item *t*.
- #192~193: fill *t* back in the assignments.
- #194: push *t* into the stack as the result for further operations.

Table-6.16 shows the test result for this section.

C language code	P-code generated
```void foo() { char n, m; int x; x = n? m: m + 1000; }```	```F:\p16ecc\tests>cc16e ch6_t20.c P-CODE: P_FUNC_BEG: foo : {void} P-CODE: ';' P-CODE: ';' P-CODE: SRC ... quest.c #6: x = n? m: m + 1000; P-CODE: JP_Z foo_$1_n{char}, _$L3 P-CODE: LABEL _$L2 P-CODE: P_MOV %2{short}, foo_$2_m{char} P-CODE: GOTO _$L4 P-CODE: LABEL _$L3 P-CODE: '+' %1{short}, foo_$2_m{char}, 1000{int} P-CODE: P_MOV %2{short}, %1{short} P-CODE: LABEL _$L4 P-CODE: '=' foo_$3_x{int}, %2{short} P-CODE: ';' P-CODE: LABEL _$L1 P-CODE: P_FUNC_END: foo```

Table-6.16

6.11.3 P-code for ',' operator

The comma operator ',' is a binary operator that evaluates its operands from left to right and returns the value of the rightmost operand. It implies that both operands are concatenated into an execution block.

C language grammar:	*expr1* , *expr2*
Parsing rules:	```expr ',' assignment_expr { $$ = oprNode(',', 2, $1, $3); }```

The P-code generation for the operator is straightforward, as shown in Code-6.50.

```
206        case ',':
207            run(op->op[0]); if ( DEPTH() > depth ) DEL(DEPTH() - depth);
208            run(op->op[1]);
209            break;
```
/p16ecc/cc_source_6.11/pcoder8.cpp

Code-6.50

The following, Table-6.17, is the test results that illustrate the effect when the operator ´ used (against '; ').

C language code	P-code generated
| ```
void foo(int x)
{
   int a, b;

   if ( x ) a++; b++;
}
``` | ```
F:\p16ecc\tests>cc16e ch6_t21.c
P-CODE: P_FUNC_BEG: foo : {void}
P-CODE: ';'
P-CODE: SRC ... ch6_t21.c #5: if ( x )
P-CODE: JP_Z  foo_$_x{int}, _$L3
P-CODE: ';'
P-CODE: LABEL   _$L2
P-CODE: SRC ... ch6_t21.c #5: a++;
P-CODE: '='   %1{int}, foo_$1_a{int}
P-CODE: '++'  foo_$1_a{int}, 1{int}
P-CODE: GOTO   _$L4
P-CODE: LABEL   _$L3
P-CODE: LABEL   _$L4
P-CODE: ';'
P-CODE: SRC ... ch6_t21.c #5: ++;
P-CODE: '='   %1{int}, foo_$2_b{int}
P-CODE: '++'  foo_$2_b{int}, 1{int}
P-CODE: ';'
P-CODE: LABEL   _$L1
P-CODE: P_FUNC_END: foo
``` |
| ```
void foo(int x)
{
   int a, b;

   if ( x ) a++, b++;
}
``` | ```
F:\p16ecc\tests>cc16e ch6_t22.c
P-CODE: P_FUNC_BEG: foo : {void}
P-CODE: ';'
P-CODE: SRC ... ch6_t22.c #5: if ( x )
P-CODE: JP_Z  foo_$_x{int}, _$L3
P-CODE: ';'
P-CODE: LABEL   _$L2
P-CODE: SRC ... ch6_t22.c #5: a++, b++;
P-CODE: '='   %1{int}, foo_$1_a{int}
P-CODE: '++'  foo_$1_a{int}, 1{int}
P-CODE: '='   %2{int}, foo_$2_b{int}
P-CODE: '++'  foo_$2_b{int}, 1{int}
P-CODE: GOTO   _$L4
P-CODE: LABEL   _$L3
P-CODE: LABEL   _$L4
P-CODE: ';'
P-CODE: LABEL   _$L1
P-CODE: P_FUNC_END: foo
``` |

Table-6.17

6.11.4 P-code for RETURN statement

RETURN statement happens in functions, which terminates the execution of the function, with or without a returning value.

C language grammar:	`return ;` `return` *expr* `;`
| Parsing rules: | ```
RETURN ';'          { $$ = oprNode(RETURN, 0);   }
RETURN expr ';'     { $$ = oprNode(RETURN, 1, $2); }
``` |

In the P-code generated for RETURN statement, a {GOTO, [*lbl*]} P-code is generated that jumps to the end of the function. If *expr* exists, the returning value will be stored in a virtual accumulator, a global data holder, before the GOTO.
```

The following, Code-6.51, illustrates the P-code generation for RETURN statement.

```
211         case RETURN:
212             fp = curFnode;
213             if ( op->nops == 0 ) // return without value
214             {
215                 if ( !ISR_FUNC(fp) && fp->attr && (fp->attr->type != VOID || fp->attr->ptrVect) )
216                     errPrint("missing return value!");
217                 else
218                 {   // jump to the end of function
219                     pp = new Pnode(GOTO);
220                     addPnode(&mainPcode, pp);
221                     pp->items[0] = lblItem(fp->endLbl);
222                 }
223             }
224             else                    // return with a value
225             {
226                 if ( !ISR_FUNC(fp) && fp->attr && (fp->attr->type != VOID || fp->attr->ptrVect) )
227                 {
228                     run(op->op[0]);
229                     if ( DEPTH() == (depth+1) )
230                     {
231                         ip0 = POP();
232                         if ( ip0->isOperable() )
233                         {   // put the returning value into ACC
234                             pp = new Pnode('=');
235                             addPnode (&mainPcode, pp);
236                             pp->items[0] = accItem(cloneAttr(fp->attr));
237                             pp->items[1] = ip0;
238
239                             // jump to the end of function
240                             pp = new Pnode(GOTO);
241                             addPnode (&mainPcode, pp);
242                             pp->items[0] = lblItem(fp->endLbl);
243                         }
244                         else
245                         {
246                             delete ip0;
247                             errPrint("invalid return value!");
248                         }
249                     }
250                 }
251                 else
252                 {
253                     errPrint("invalid 'return'!");
254                 }
255             }
```

/p16ecc/cc_source_6.11/pcoder8.cpp   Code-6.51

- #213~223: P-code for RETURN without returning a value.
    - #219~221: P-code for GOTO.
- #226~250: P-code for RETURN with returning value.
    - #228: generate returning item.
    - #234~237: P-code that moves the returning value to the (virtual) accumulator (ACC).
    - #240~242: P-code for GOTO.

6.11.5 P-code for AASM statement

AASM statements are originally from source code #asm ... #endasm, generated during the pre-scan procedure. Each AASM node carries a string which is an assembly code. It is unconditionally converted to a P-code, shown as in Code-6.52.

{AASM, ["*asm_code*"]}

```
258         case AASM:
259             pp = new Pnode(AASM);
260             addPnode(&mainPcode, pp);
261             pp->items[0] = strItem(op->op[0]->str.str);
262             break;
```

/p16ecc/cc_source_6.11/pcoder8.cpp   Code-6.52

<u>6.11.6 P-code for PRAGMA statement</u>

PRAGMA statements are used to control the compiling activities. It has the format, in **p16ecc**, shown below

<table>
<tr><td rowspan="2">C language grammar:</td><td>**#pragma** *id*</td></tr>
<tr><td>**#pragma** *id* *expr*</td></tr>
<tr><td rowspan="2">Parsing rules:</td><td>PRAGMA identifier EOL   {
            $$ = oprNode(PRAGMA, 1, $2); }</td></tr>
<tr><td>PRAGMA identifier conditional_expr EOL   {
            $$ = oprNode(PRAGMA, 2, $2, $3); }</td></tr>
</table>

The P-code generation for PRAGMA simply generates the items for *id* and *expr*, and them form the P-code

$$\{\text{PRAGMA}, [id, expr]\}$$

```
264        case PRAGMA:
265            pp = new Pnode(PRAGMA);
266            addPnode(&mainPcode, pp);
267            pp->items[0] = idItem(op->op[0]->id.name);
268            if ( op->nops > 1 )
269            {
270                run(op->op[1]);
271                if ( DEPTH() == (depth+1) )
272                    pp->items[1] = POP();
273                else
274                    errPrint("unknown item in pragma!");
275            }
276            break;
```
/p16ecc/cc_source_6.11/pcoder8.cpp   Code-6.53

Chapter-7

P-code Optimization

The test results in Chapter 6 show that the P-codes generated are verbal and 'bulky', far from efficient. There are many places that have inefficient P-codes, unused labels, and so on. The following table, Table-7.1, shows the differences before and after optimization.

```
void delay(int n)
{
  while ( n )
  {
    if ( n > 1000 )
      break;
    n--;
  }
}
```

Before optimization	After optimization
P-CODE: P_FUNC_BEG: delay : {void} P-CODE: LABEL _$L2 P-CODE: SRC … ch6_t16.c #3: while (n) P-CODE: JP_Z delay_$_n{int}, _$L4 P-CODE: ';' P-CODE: LABEL _$L3 P-CODE: SRC … ch6_t16.c #5: if (n > 1000) P-CODE: JP <= delay_$_n{int}, 1000{int}, _$L6 P-CODE: ';' P-CODE: LABEL _$L5 P-CODE: SRC … ch6_t16.c #6: break; P-CODE: GOTO _$L4 P-CODE: GOTO _$L7 P-CODE: LABEL _$L6 P-CODE: LABEL _$L7 P-CODE: ';' P-CODE: SRC … ch6_t16.c #7: --; P-CODE: '=' %1{int}, delay_$_n{int} P-CODE: '−' delay_$_n{int}, 1{int} P-CODE: ';' P-CODE: GOTO _$L2 P-CODE: LABEL _$L4 P-CODE: ';' P-CODE: LABEL _$L1 P-CODE: P_FUNC_END: delay	P-CODE: P_FUNC_BEG: delay : {void} P-CODE: LABEL _$L2 P-CODE: SRC … ch6_t16.c #3: while (n) P-CODE: JP_Z delay_$_n{int}, _$L4 P-CODE: ';' P-CODE: SRC … ch6_t16.c #5: if (n > 1000) P-CODE: JP > delay_$_n{int}, 1000{int}, _$L4 P-CODE: ';' P-CODE: SRC … ch6_t16.c #6: break; P-CODE: ';' P-CODE: SRC … ch6_t16.c #7: --; P-CODE: '−' delay_$_n{int}, 1{int} P-CODE: ';' P-CODE: GOTO _$L2 P-CODE: LABEL _$L4 P-CODE: ';' P-CODE: P_FUNC_END: delay

Table-7.1

Code optimization is a crucial phase in compiler design that aims to enhance the performance and efficiency of the executable code. By improving the quality of the generated machine code, optimizations can reduce execution time, minimize code size, and enhance overall performance. This process involves the various techniques and strategies applied during compilation to produce more efficient code without altering the program's functionality.

Code optimization can occur at various stages during the compilation process. This chapter only covers the optimization that happens to P-code (intermediate code), which is independent of the target (CPU).

As mentioned above, numerous books, papers, and studies have addressed the topic. Most of the strategies and algorithms presented in them are profound and complicated to implement. Moreover, code optimization may introduce risks to the results, leading to unexpected errors.

This chapter presents a simple, straightforward strategy for P-code optimization that focuses on the P-code group corresponding to a single sentence in the source code. Matching the code patterns in the P-code series, it will replace them with the optimized codes. Numerous pattern recognition methods, organized into groups, are presented in the chapter and applied to scan the P-codes. Once a pattern is recognized and optimization happens, the whole optimization process will start again from the beginning.

The P-code optimization operation is implemented using a class, `Optimizer`, as shown in Code-7.1.

```
19   class Optimizer {
20       public:
21           Optimizer(Pcoder *pcoder);
22           void run(void);
23
24       protected:
25           int optiCount;
26           Pcoder *pcoder;
27           Pnode  *head, *funcPtr;
28           Item   *ip0, *ip1, *ip2;
29
30           Pnode *next(Pnode *p, int offset);
31           bool updateLbl(char *old_lbl, char *new_lbl, Pnode *p);
32           bool updateGoto(char *old_lbl, char *new_lbl, Pnode *p);
33           bool unusedLbl(Pnode *pnp, char *lbl, Pnode *p);
34           bool unusedTmp(Pnode *pnp, int tmp_idx);
35           bool endOfScope(Pnode *pnp);
36           bool assignmentCode(int code, int types);
37           Pnode *indirReferenced(Pnode *p, Item *ip, int *index);
38           void replaceTmp(Pnode *pnp, int tmp_old, int tmp_new);
39           bool reduceTmpSize(Pnode *pnp, Item *tmp, int size);
40           bool accReferenced(Pnode *pnp);
41           bool replaceIndir(Pnode *pnp, Item *ip, int temp_index);
42           bool bitSelect(Item *ip, int *n);
43           bool bitDeselect(Item *ip, int *n, int size);
44
45           bool group1(Pnode *pnp);    // simplify
46           bool group2(Pnode *pnp);    // merge
47           bool group3(Pnode *pnp);    // jump
48           bool group4(Pnode *pnp);    // special constance
49           bool group5(Pnode *pnp);    // temporary variable
50           bool group6(Pnode *pnp);    // others
51   };
```
/p16ecc/cc_source_7.1/popt.h   Code-7.1

- #21: the constructor of the class. It gets the P-code pointer here.
- #22: function to start the optimization.
- #30~43: functions that serve for optimization.
- #45~50: a group of optimizing functions that perform pattern matching and P-code replacement.

The optimization is started in `main()`, after P-code generation, as shown in following, Code-7.2.

```
14   int main(int argc, char *argv[])
15   {
     . . .
```

```
30 ⊟                    {
31                          Nlist nlist;
32                          PreScan preScan(&nlist);
33                          progUnit = preScan.scam(progUnit);
34
35                          Pcoder pcoder;            // P-code generation
36                          pcoder.run(progUnit);     // run it now.
37                          display(pcoder.mainPcode);
38
39                          if ( pcoder.errorCount == 0 )
40 ⊟                        {
41                              Optimizer opt(&pcoder);
42                              opt.run();
43                              display(pcoder.mainPcode);
44                          }
45                    }
```

`/p16ecc/cc_source_7.1/main.cpp`   Code-7.2

- #41: instantiate the class `Optimizer`.
- #42: start optimization.

When `Optimizer` is running, it will go through all pattern recognitions, as illustrated in Code-7.3.

```
14 Optimizer :: Optimizer(Pcoder *_pcoder)
15 ⊟{
16     memset(this, 0, sizeof(Optimizer));
17     pcoder = _pcoder;
18 }
19
20 void Optimizer :: run(void)
21 ⊟{
22     head = pcoder->mainPcode;
23     bool done = false;
24     while ( !done )
25 ⊟    {
26         done = true;
27         for(Pnode *pnp = head; pnp; pnp = pnp->next)
28 ⊟        {
29             ip0 = pnp->items[0];
30             ip1 = pnp->items[1];
31             ip2 = pnp->items[2];
32             if ( pnp->type == P_FUNC_BEG ) funcPtr = pnp;
33             if ( pnp->type == P_FUNC_END ) funcPtr = NULL;
34
35             if ( group1(pnp) ) { done = false; break; }
36 //          if ( group2(pnp) ) { done = false; break; }
37 //          if ( group3(pnp) ) { done = false; break; }
38 //          if ( group4(pnp) ) { done = false; break; }
39 //          if ( group5(pnp) ) { done = false; break; }
40 //          if ( group6(pnp) ) { done = false; break; }
41         }
42     }
43 }
```

`/p16ecc/cc_source_7.1/popt.cpp`   Code-7.3

- #29~31: fetch the items from the current P-code.
- #35~40: go through the pattern recognition and P-code replacement, in all groups.

7.1 P-code Optimization (1)

```
Project directory: /p16ecc/cc_source_7.1
New source files: popt.cpp, popt.h, popt1.cpp
Modified source files: main.cpp
```

7.1.1 Remove LABEL redundancy

When two lines of LABEL P-code are in sequence, it shall remove one of them after the replacement.

$$\begin{array}{l}\{LABEL,\ [Lx]\}\\ \{LABEL,\ [Ly]\}\end{array} \qquad \rightarrow \qquad \{LABEL,\ [Lx]\}$$

In addition, P-code {LABEL, [*Lx*]} will be removed if label *Lx* is not in use.

This part of the optimization does not seem to reduce the final code size. But it will trigger other optimizations.

```
16    bool Optimizer :: group1(Pnode *pnp)
17    {
18        int ptype = pnp->type;
19        Pnode *p1 = next(pnp, 1);
20
21        if ( ptype == LABEL && p1 && p1->type == LABEL )      // {LABEL, [Lx]}
22        {                                                      // {LABEL, [Ly]}
23            updateLbl(p1->items[0]->val.s, ip0->val.s, p1);
24            delete p1;
25            return true;
26        }
27
28        if ( ptype == LABEL && unusedLbl(head, ip0->val.s, pnp) )   // {LABEL, [Lx]}
29        {
30            delete pnp;
31            return true;
32        }
```
/p16ecc/cc_source_7.1/popt1.cpp   Code-7.4

- #23: replace *Ly* with *Lx*,
- #24: and then remove {LABEL, [*Ly*]}.
- #28~32: remove unused label line.

Table-7.2 shows the result.

```
void foo(char a, char *p)
{
  if ( a > 0 )  (*p)++;
}
```

Before optimization	After optimization
P-CODE: P_FUNC_BEG: foo : {void}	P-CODE: P_FUNC_BEG: foo : {void}
P-CODE: SRC ... ch7_t1.c #3: if (a > 0)	P-CODE: SRC ... ch7_t1.c #3: if (a > 0)
P-CODE: JP <= foo_$_a{char}, 0{int}, _$L3	P-CODE: JP <= foo_$_a{char}, 0{int}, _$L3
P-CODE: ';'	P-CODE: ';'
P-CODE: LABEL _$L2	P-CODE: SRC ... ch7_t1.c #3: (*p)++;
P-CODE: SRC ... ch7_t1.c #3: (*p)++;	P-CODE: '=' %1{char}, [foo_$_p:0]{char *}
P-CODE: '=' %1{char}, [foo_$_p:0]{char *}	P-CODE: '++' [foo_$_p:0]{char *}, 1{int}
P-CODE: '++' [foo_$_p:0]{char *}, 1{int}	P-CODE: GOTO _$L3
P-CODE: GOTO _$L4	P-CODE: LABEL _$L3
P-CODE: LABEL _$L3	P-CODE: ';'
P-CODE: LABEL _$L4	P-CODE: P_FUNC_END: foo
P-CODE: ';'	
P-CODE: LABEL _$L1	
P-CODE: P_FUNC_END: foo	

Table-7.2

7.1.2 Remove GOTO redundancy

Several P-code segments with GOTO instructions can be optimized.

1) When a GOTO code jumps to the label line that follows. It shall be removed.

$$\begin{matrix} \{GOTO, [Lx]\} \\ \{LABEL, [Lx]\} \end{matrix} \quad \rightarrow \quad \{LABEL, [Lx]\}$$

```
34   if ( ptype == GOTO && p1 && p1->type == LABEL &&        // {GOTO, [Lx]}
35        !strcmp(ip0->val.s, p1->items[0]->val.s) )         // {LABEL, [Lx]}
36   {
37       delete pnp;
38       return true;
39   }
```
/p16ecc/cc_source_7.1/popt1.cpp   Code-7.5

Table-7.3 shows the result, continued from Table-7.2.

Before optimization	After optimization
P-CODE: P_FUNC_BEG: foo : {void}	P-CODE: P_FUNC_BEG: foo : {void}
P-CODE: SRC ... ch7_t1.c #3: if (a > 0)	P-CODE: SRC ... ch7_t1.c #3: if (a > 0)
P-CODE: JP <= foo_$_a{char}, 0{int}, _$L3	P-CODE: JP <= foo_$_a{char}, 0{int}, _$L3
P-CODE: ';'	P-CODE: ';'
P-CODE: SRC ... ch7_t1.c #3: (*p)++;	P-CODE: SRC ... ch7_t1.c #3: (*p)++;
P-CODE: '=' %1{char}, [foo_$_p:0]{char *}	P-CODE: '=' %1{char}, [foo_$_p:0]{char *}
P-CODE: '++' [foo_$_p:0]{char *}, 1{int}	P-CODE: '++' [foo_$_p:0]{char *}, 1{int}
P-CODE: GOTO _$L3	P-CODE: LABEL _$L3
P-CODE: LABEL _$L3	P-CODE: ';'
P-CODE: ';'	P-CODE: P_FUNC_END: foo
P-CODE: P_FUNC_END: foo	

Table-7.3

2) If a label line is followed by a GOTO code, as shown below, then a replacement of the label name will occur, which may trigger other optimizations to take place.

$$\begin{matrix} \{jump_code, [\ldots, Lx]\} \\ \ldots \\ \{LABEL, [Lx]\} \\ \{GOTO, [Ly]\} \end{matrix} \quad \rightarrow \quad \begin{matrix} \{jump_code, [\ldots, \boldsymbol{Ly}]\} \\ \ldots \\ \{LABEL, [Lx]\} \\ \{GOTO, [Ly]\} \end{matrix}$$

```
41   if ( ptype == LABEL && p1 && p1->type == GOTO     &&       // {LABEL, [Lx]}
42        strcmp(ip0->val.s, p1->items[0]->val.s) != 0  )        // {GOTO, [Ly]}
43   {   // make all 'jump to LABELx' to 'jump to LABELy'
44       if ( updateGoto(ip0->val.s, p1->items[0]->val.s, p1) )
45           return true;
46   }
```
/p16ecc/cc_source_7.1/popt1.cpp   Code-7.6

3) Remove the P-code that follows GOTO, if it's neither a label nor P_FUNC_END.

$$\begin{matrix} \{GOTO, [Lx]\} \\ \{\ldots\} \end{matrix} \quad \rightarrow \quad \{GOTO, [Lx]\}$$

```
48   if ( ptype == GOTO && p1 && !(p1->type == LABEL     ||      // {GOTO, [Lx]}
49                                  p1->type == P_FUNC_END) )     // {...}
50   {
51       delete p1;
52       return true;
53   }
```
/p16ecc/cc_source_7.1/popt1.cpp   Code-7.7

4) Remove the GOTO if it follows RETURN.

{RETURN, [?]}
{GOTO, [Lx]}   →   {RETURN, [?]}

```
56       if ( ptype == RETURN && p1 && p1->type == GOTO )        // {RETURN, [?]}
57       {   // remove useless GOTO                              // {GOTO, [Lx]}
58           delete p1;
59           return true;
60       }
```
/p16ecc/cc_source_7.1/popt1.cpp   Code-7.8

7.2 P-code Optimization (2)

Project directory: /p16ecc/cc_source_7.2
New source files: popt2.cpp
Modified source files: popt.cpp

7.2.1 Simplify P-code for IMMD_ITEM

IMMD_ITEM type items are equivalent to constants. So, some operations over the item can be simplified.

{ '+', [X, #immd_item, N]}   →   { '=', [X, #(immd_item + N)]}

```
22       if ( assignmentCode(ptype, MATH_OP | LOGIC_OP) &&        // {'+', [X, #Y, N]}
23            ip1->type == IMMD_ITEM && ip2->type == CON_ITEM )
24       {
25           if ( ip2->val.i != 0 )
26           {
27               char buf[32];
28               sprintf(buf, "%c%d)", ptype, ip2->val.i);
29               std::string str = ip1->val.s;
30               str.insert(0, "("); str += buf;
31               pnp->updateName(1, (char*)str.c_str());
32           }
33           pnp->type = '=';
34           pnp->updateItem(2, NULL);
35           return true;
36       }
```
/p16ecc/cc_source_7.2/popt2.cpp   Code-7.9

Table-7.4 shows the result.

```
int array[20];

void foo(int *p)
{
    *p = array[3];
}
```

Before optimization	After optimization
P-CODE: P_FUNC_BEG: foo : {void}	P-CODE: P_FUNC_BEG: foo : {void}
P-CODE: SRC ... ch7_t3.c #5: *p = array[3];	P-CODE: SRC ... ch7_t3.c #5: *p = array[3];
P-CODE: '+' %1{int *}, #array{int[20]}, 6{int}	P-CODE: '=' %1{int *}, #(array+6){int[20]}
P-CODE: '=' [foo_$_p:0]{int *}, [%1]{int *}	P-CODE: '=' [foo_$_p:0]{int *}, [%1]{int *}
P-CODE: ';'	P-CODE: ';'
P-CODE: LABEL _$L1	P-CODE: P_FUNC_END: foo
P-CODE: P_FUNC_END: foo	

Table-7.4

The Optimization result above shows that further optimization can be done by merging P-code lines as follows.

$$\begin{array}{ccc} \{\,`='\,,\,[t,\,\#immd]\} & & \\ \{\,`='\,,\,[X,\,(t)]\} & \rightarrow & \{\,`='\,,\,[X,\,immd_name]\} \end{array}$$

```
38    if ( ptype == '=' &&
39        ip0->type == TEMP_ITEM && ip1->type == IMMD_ITEM )          // {'=', [t, #X]}
40    {
41        int i;
42        Pnode *p1 = indirReferenced(pnp->next, ip0, &i);
43        if ( p1 )                                                    // {op, [.. (t) ..]}
44        {
45            if ( p1->items[i]->type == INDIR_ITEM )
46            {
47                attrib *attr = cloneAttr(p1->items[i]->attr);
48                reducePtr(attr);
49                attr->dataBank = ip1->attr->dataBank;
50                ip1->updateAttr(attr);
51                ip1->type = ID_ITEM;
52            }
53            p1->updateItem(i, ip1);
54            pnp->items[1] = NULL;
55            delete pnp;
56            return true;
57        }
58    }
```
`/p16ecc/cc_source_7.2/popt2.cpp`   Code-7.10

Thus, the further optimization will become as shown in Table-7.5,

Before optimization	After optimization
P-CODE: P_FUNC_BEG: foo : {void}	P-CODE: P_FUNC_BEG: foo : {void}
P-CODE: SRC ... ch7_t3.c #5: *p = array[3];	P-CODE: SRC ... ch7_t3.c #5: *p = array[3];
P-CODE: '=' %1{int *}, #(array+6){int[20]}	P-CODE: '=' [foo_$_p:0]{int *}, (array+6){int}
P-CODE: '=' [foo_$_p:0]{int *}, [%1]{int *}	P-CODE: ';'
P-CODE: ';'	P-CODE: P_FUNC_END: foo
P-CODE: P_FUNC_END: foo	

Table-7.5

7.2.3 Simplify the P-code operation

Usually, following math and logic operations can be simplified to their corresponding compound assignments if the pattern is met:

$$\begin{array}{ccc} \{\,`+'\,,\,[X,X,Y]\} & \rightarrow & \{\,`+='\,,\,[X,Y]\} \\ \{\,`-'\,,\,[X,X,Y]\} & \rightarrow & \{\,`-='\,,\,[X,Y]\} \\ \{\,`*'\,,\,[X,X,Y]\} & \rightarrow & \{\,`*='\,,\,[X,Y]\} \\ \{\,`/'\,,\,[X,X,Y]\} & \rightarrow & \{\,`/='\,,\,[X,Y]\} \\ \{\,`\%'\,,\,[X,X,Y]\} & \rightarrow & \{\,`\%='\,,\,[X,Y]\} \\ \{\,`\&'\,,\,[X,X,Y]\} & \rightarrow & \{\,`\&='\,,\,[X,Y]\} \\ \{\,`|'\,,\,[X,X,Y]\} & \rightarrow & \{\,`|='\,,\,[X,Y]\} \\ \{\,`{\wedge}'\,,\,[X,X,Y]\} & \rightarrow & \{\,`{\wedge}='\,,\,[X,Y]\} \end{array}$$

```
60    int new_type;
61    if ( shortenOper(ptype, &new_type) &&                           // {'+', [X, X, Y]}
62        same(ip0, ip1) && !same(ip1, ip2) && !same(ip0, ip2) )
63    {
64        pnp->type = new_type;
65        pnp->updateItem(1, ip2);
66        pnp->items[2] = NULL;
67        return true;
68    }
```
`/p16ecc/cc_source_7.2/popt2.cpp`   Code-7.11

7.2.4 Remove temporary item

A temporary item is generated due to shift-reduce operation in P-code generation. Most of time, it is just used for once right after. That means that it can be eliminated as

$$\begin{array}{l} \{\text{`+'}, [t, X, Y]\} \\ \{\text{`='}, [Z, t]\} \end{array} \quad \rightarrow \quad \{\text{`+'}, [Z, X, Y]\}$$

$$\begin{array}{l} \{\text{`--'}, [t, X, Y]\} \\ \{\text{`='}, [Z, t]\} \end{array} \quad \rightarrow \quad \{\text{`--'}, [Z, X, Y]\}$$

$$\dots$$

$$\begin{array}{l} \{\text{`='}, [t, X]\} \\ \{\text{`='}, [Z, t]\} \end{array} \quad \rightarrow \quad \{\text{`='}, [Z, X]\}$$

```
70    Pnode *p1 = next(pnp, 1);
71    if ( assignmentCode(ptype, ALL_OP) && ip0->type == TEMP_ITEM &&    // {'+', [t, x, y]}
72         p1 && p1->type == '=' && same(ip0, p1->items[1])        &&    // {'=', [z, t]}
73         unusedTmp(p1->next, ip0->val.i)                         )
74    {
75         pnp->updateItem(0, p1->items[0]);                             // {'+', [z, x, y]}
76         p1->items[0] = NULL;
77         delete p1;
78         return true;
79    }
```
/p16ecc/cc_source_7.2/popt2.cpp   Code-7.12

7.2.5 Alter P-code lines for ++ and --

The following code pattern occurs very often in everyday programming.

```
x = y++;
```

And its P-code is generated, as well as its optimization can be

$$\begin{array}{l} \{\text{`='}, [t, Y]\} \\ \{\text{INC_OP}, [Y, 1]\} \\ \{\text{`='}, [X, t]\} \end{array} \quad \rightarrow \quad \begin{array}{l} \{\text{`='}, [t, Y]\} \\ \{\text{`='}, [X, t]\} \\ \{\text{INC_OP}, [Y, 1]\} \end{array}$$

Further, after using the optimization described in Section 7.2.4, the final optimization result will be

$$\begin{array}{l} \{\text{`='}, [t, Y]\} \\ \{\text{`='}, [X, t]\} \\ \{\text{INC_OP}, [Y, 1]\} \end{array} \quad \rightarrow \quad \begin{array}{l} \{\text{`='}, [X, Y]\} \\ \{\text{INC_OP}, [Y, 1]\} \end{array}$$

```
81    Pnode *p2 = next(pnp, 2);
82    if ( ptype == '=' && ip0->type == TEMP_ITEM && p1 && p2 &&      // {'=', [t, x]}
83         (p1->type == INC_OP || p1->type == DEC_OP)        &&       // {INC_OP, [x, N]}
84         same(ip1, p1->items[0])                           &&       // {'=', [y, t]}
85         p2->type == '=' && related(ip0, p2->items[1])     &&
86                           !related(ip1, p2->items[0])     &&
87         unusedTmp(p2->next, ip0->val.i)                   )
88    {
89         swap(p1, p2);    // swap line p1 and p2.
90         return true;
91    }
```
/p16ecc/cc_source_7.2/popt2.cpp   Code-7.13

The test result is shown in Table-7.6.

```
int x, y;

void foo()
{
   x = y++;
}
```

Before optimization	After optimization
P-CODE: P_FUNC_BEG: foo : {void}	P-CODE: P_FUNC_BEG: foo : {void}
P-CODE: SRC ... ch7_t4.c #5: x = y++;	P-CODE: SRC ... ch7_t4.c #5: x = y++;
P-CODE: '=' %1{int}, y{int}	P-CODE: '=' x{int}, y{int}
P-CODE: '++' y{int}, 1{int}	P-CODE: '++' y{int}, 1{int}
P-CODE: '=' x{int}, %1{int}	P-CODE: ';'
P-CODE: ';'	P-CODE: P_FUNC_END: foo
P-CODE: LABEL _$L1	
P-CODE: P_FUNC_END: foo	

Table-7.6

7.2.6 Merge multi addition/subtraction of constants

For the C code, like

$$x = y + N - M;$$

Where N and M are constants. Those multiple additions/subtractions will result in the P-code and can be simplified as

$$\{`+', [t1, Y, N]\}$$
$$\{`-', [t2, t1, M]\} \quad \rightarrow \quad \{`+', [t1, Y, (N - M)]\} \quad \rightarrow \quad \{`+', [X, Y, (N - M)]\}$$
$$\{`=', [X, t2]\} \qquad\qquad \{`=', [X, t1]\}$$

```
 93     if ( (ptype == '+' || ptype == '-') &&
 94         ip0->type == TEMP_ITEM && ip2->type == CON_ITEM )        // {'+', [t, x, N]}
 95     {                                                            // {'+', [y, t, M]}
 96         if ( p1 && (p1->type == '+' || p1->type == '-') &&
 97             same(ip0, p1->items[1]) && p1->items[2]->type == CON_ITEM &&
 98             (related(p1->items[0], ip0) || unusedTmp(p1->next, ip0->val.i)) )
 99         {
100             if ( ptype == p1->type )                             // {'+', [y, x, N+M]}
101                 ip2->val.i += p1->items[2]->val.i;
102             else if ( ip2->val.i >= p1->items[2]->val.i )
103                 ip2->val.i -= p1->items[2]->val.i;
104             else {
105                 ip2->val.i = p1->items[2]->val.i - ip2->val.i;
106                 pnp->type  = p1->type;
107             }
108
109             pnp->updateItem(0, p1->items[0]);
110             p1->items[0] = NULL;
111             delete p1;
112             return true;
113         }
114     }
/p16ecc/cc_source_7.2/popt2.cpp
```

Code-7.14

The test result is shown in Table-7.7.

```
void foo(int x, int y)
{
   x = y + 10 - 5;
}
```

Before optimization	After optimization
P-CODE: P_FUNC_BEG: foo : {void} P-CODE: SRC ... ch7_t5.c #3: x = y + 10 - 5; P-CODE: '+' %1{short}, foo_$_y{int}, 10{int} P-CODE: '-' %2{long}, %1{short}, 5{int} P-CODE: '=' foo_$_x{int}, %2{long} P-CODE: ';' P-CODE: LABEL _$L1 P-CODE: P_FUNC_END: foo	P-CODE: P_FUNC_BEG: foo : {void} P-CODE: SRC ... ch7_t5.c #3: x = y + 10 - 5; P-CODE: '+' foo_$_x{int}, foo_$_y{int}, 5{int} P-CODE: ';' P-CODE: P_FUNC_END: foo

Table-7.7

Note, this optimization can be extended for other situations, as long as the operator meets the 'associative property', such as

$$\begin{array}{l} \{\,`*\text{'},\,[t1,\,Y,\,N]\} \\ \{\,`*\text{'},\,[t2,\,Y,\,M]\} \\ \{\,`=\text{'},\,[X,\,t2]\} \end{array} \quad\longrightarrow\quad \{\,`*\text{'},\,[X,\,Y,\,(N * M)]\}$$

$$\begin{array}{l} \{\,`\&\text{'},\,[t1,\,Y,\,N]\} \\ \{\,`\&\text{'},\,[t2,\,Y,\,M]\} \\ \{\,`=\text{'},\,[X,\,t2]\} \end{array} \quad\longrightarrow\quad \{\,`\&\text{'},\,[X,\,Y,\,(N \& M)]\}$$

. . .

7.2.7 Alter the P-code operation for += and -=

For C code like

```
x += N;
```

Where N is a constant, the P-coder operator generated is ADD_ASSIGN or SUB_ASSIGN. It can be changed to INC_OP or DEC_OP, which sometimes is more efficient.

```
116     if ( (ptype == ADD_ASSIGN || ptype == SUB_ASSIGN) && ip1->type == CON_ITEM )    // {'+=', [X, N]}
117     {
118         pnp->type = (ptype == ADD_ASSIGN)? INC_OP: DEC_OP;
119         return true;
120     }
```

/p16ecc/cc_source_7.2/popt2.cpp

Code-7.15

7.3 P-code Optimization (3)

<u>Project directory</u>: /p16ecc/cc_source_7.3
<u>New source files</u>: popt3.cpp
<u>Modified source files</u>: popt.cpp

This part covers the P-code optimizations that relate to conditional jumps.

7.3.1 Alter the jump types in P-code

As discussed in Section 6.9.2, every conditional jump is based on its comparison type. Changing the jump type will optimize the P-code if the pattern matches, as follows.

$$\begin{array}{l}\{P_JZ, [X, Lx]\} \\ \{GOTO, [Ly]\} \\ \{LABEL, [Lx]\}\end{array} \qquad \rightarrow \qquad \begin{array}{l}\{P_JNZ, [X, Ly]\} \\ \{LABEL, [Lx]\}\end{array}$$

$$\begin{array}{l}\{P_JEQ, [X, Y, Lx]\} \\ \{GOTO, [Ly]\} \\ \{LABEL, [Lx]\}\end{array} \qquad \rightarrow \qquad \begin{array}{l}\{P_JNE, [X, Y, Ly]\} \\ \{LABEL, [Lx]\}\end{array}$$

. . .

```
21        int ptype = pnp->type;
22        Pnode *p1 = next(pnp, 1);
23        Pnode *p2 = next(pnp, 2);
24        int rev_type, n;
25
26        if ( convertJump(ptype, &rev_type, &n) && p1 && p2 )      // {J_Z, [X, Lx]}
27        {                                                         // {GOTO, [Ly]}
28            if ( p1->type == GOTO && p2->type == LABEL &&         // {LABEL, [Lx]}
29                 same(pnp->items[n-1], p2->items[0]) )
30            {
31                pnp->type = rev_type;
32                pnp->updateName(n-1, p1->items[0]->val.s);
33                delete p1;
34                return true;
35            }
36        }
```
/p16ecc/cc_source_7.3/popt3.cpp   Code-7.16

```
void foo(int x)
{
  while ( x != 0 )
  {
    if ( x < 0 ) break;
    x--;
  }
}
```

Before optimization	After optimization
P-CODE: P_FUNC_BEG: foo : {void}	P-CODE: P_FUNC_BEG: foo : {void}
P-CODE: LABEL _$L2	P-CODE: LABEL _$L2
P-CODE: SRC ... ch7_t6.c #3: while (x != 0)	P-CODE: SRC ... ch7_t6.c #3: while (x != 0)
P-CODE: JP == foo_$_x{int}, 0{int}, _$L4	P-CODE: JP == foo_$_x{int}, 0{int}, _$L4
P-CODE: ';'	P-CODE: ';'
P-CODE: SRC ... ch7_t6.c #5: if (x < 0)	P-CODE: SRC ... ch7_t6.c #5: if (x < 0)
P-CODE: JP >= foo_$_x{int}, 0{int}, _$L6	P-CODE: JP < foo_$_x{int}, 0{int}, _$L4
P-CODE: ';'	P-CODE: ';'
P-CODE: SRC ... ch7_t6.c #5: break;	P-CODE: SRC ... ch7_t6.c #5: break;
P-CODE: GOTO _$L4	P-CODE: ';'
P-CODE: LABEL _$L6	P-CODE: SRC ... ch7_t6.c #6: --;

```
P-CODE: ';'
P-CODE: SRC ... ch7_t6.c #6: --;
P-CODE: '='    %1{int}, foo_$_x{int}
P-CODE: '--'   foo_$_x{int}, 1{int}
P-CODE: ';'
P-CODE: GOTO   _$L2
P-CODE: LABEL  _$L4
P-CODE: ';'
P-CODE: P_FUNC_END: foo
```

```
P-CODE: '='    %1{int}, foo_$_x{int}
P-CODE: '--'   foo_$_x{int}, 1{int}
P-CODE: ';'
P-CODE: GOTO   _$L2
P-CODE: LABEL  _$L4
P-CODE: ';'
P-CODE: P_FUNC_END: foo
```

Table-7.7

7.3.2 Replace '&' operation with bit test

C code operation like

$$\text{if } (\ (X \ \& \ 128)\ ) \ . \ . \ .$$

Where the constant $128 = 2^7$. The P-code of the test-jump operation can be optimized as

$$\{\text{`\&'}, [t, X, 128]\}$$
$$\{\text{P_JZ}, [t, Lx]\} \qquad \rightarrow \qquad \{\text{P_JBZ}, [X, 7, Lx]\}$$
$$\ldots \qquad\qquad\qquad\qquad\qquad \ldots$$

```
38  if ( ptype == '&' && ip0->type == TEMP_ITEM && p1 &&      // {'&', [t, X, 2^N]}
39       bitSelect(ip2, &n)                             &&      // {J_Z, [t, Lx]}
40       (p1->type == P_JZ || p1->type == P_JNZ)        &&
41       same(ip0, p1->items[0])                        &&
42       unusedTmp(p1->next, ip0->val.i)                )
43  {
44       p1->type = (p1->type == P_JZ)? P_JBZ: P_JBNZ;
45       p1->updateItem(0, ip1);
46       p1->items[2] = p1->items[1];
47       p1->items[1] = intItem(n);
48       pnp->items[1] = NULL;
49       delete pnp;
50       return true;
51  }
/p16ecc/cc_source_7.3/popt3.cpp
```

Code-7.17

7.3.3 Optimization of comparing

Some comparing operations, which compare against with '0' value, can be optimized.

$$\{\text{P_JEQ}, [X, 0, Lx]\} \qquad \rightarrow \qquad \{\text{P_JZ}, [X, Lx]\}$$
$$\ldots \qquad\qquad\qquad\qquad\qquad \ldots$$

P-code $\{\text{P_JZ}, [X, Lx]\}$ usually is more efficient.

```
53  if ( (ptype == P_JEQ || ptype == P_JNE) &&          // {P_JEQ, [X, 0, Lx]}
54       ip1->type == CON_ITEM && ip1->val.i == 0 )
55  {
56       pnp->type = (ptype == P_JEQ)? P_JZ: P_JNZ;
57       pnp->updateItem(1, ip2);
58       pnp->items[2] = NULL;
59       return true;
60  }
/p16ecc/cc_source_7.3/popt3.cpp
```

Code-7.18

174

7.3.4 Swap item positions in P-code

Swapping item positions in P-code, as well as the operation, may trigger the optimizations described above. Or, it will bring convenience for the assembly code generation.

A swapping of the item positions is described below, which may involve changing the P-code operator.

$$\{P_JEQ, [N, X, Lx]\} \quad \rightarrow \quad \{P_JEQ, [X, N, Lx]\}$$
$$\{P_JGT, [N, X, Lx]\} \quad \rightarrow \quad \{P_JLT, [X, N, Lx]\}$$
$$\ldots \quad \rightarrow \quad \ldots$$

Where N is a constant.

```
62      int swap_type;
63      if ( swapJump(pnp->type, &swap_type) &&              // {P_JEQ, [N, X, Lx]}
64          ip0->type == CON_ITEM && ip1->type != CON_ITEM )
65      {
66          pnp->type = swap_type;
67          pnp->items[0] = ip1;
68          pnp->items[1] = ip0;
69          return true;
70      }
```
`/p16ecc/cc_source_7.3/popt3.cpp`   Code-7.19

- #63: a new operator may be generated when calling the function `swapJump()`.

7.3.5 New P-code type for combining P-codes

A group of new P-code types (`P_JZ_INC`, `P_JZ_DEC`, `P_JNZ_INC`, `P_JNZ_DEC`) is created, combining P-codes into a single instruction for optimization.

$\{`=', [t, X]\}$ $\{P_INC, [X, N]\}$ $\{P_JZ, [t, Lx]\}$	$\rightarrow$	$\{P_JZ_INC, [X, N, Lx]\}$
$\{`=', [t, X]\}$ $\{P_DEC, [X, N]\}$ $\{P_JNZ, [t, Lx]\}$	$\rightarrow$	$\{P_JNZ_DEC, [X, N, Lx]\}$
$\ldots$	$\rightarrow$	$\ldots$

```
72      if ( ptype == '=' && ip0->type == TEMP_ITEM && p1 && p2 )   // {'=', [t, X]}
73      {                                                          // {P_INC, [X, N]}
74          if ( (p1->type == INC_OP || p1->type == DEC_OP) &&     // {P_JZ, [t, Lx]}
75              same(ip1, p1->items[0])                    &&
76              (p2->type == P_JZ   || p2->type == P_JNZ)  &&
77              same(ip0, p2->items[0])                    &&
78              unusedTmp(p2->next, ip0->val.i)            )
79          {
80              if ( p2->type == P_JZ )
81                  p2->type = (p1->type == INC_OP)? P_JZ_INC: P_JZ_DEC;
82              else
83                  p2->type = (p1->type == INC_OP)? P_JNZ_INC: P_JNZ_DEC;

85              p2->items[2] = p2->items[1];
86              p2->items[1] = p1->items[1];
87              p2->updateItem(0, p1->items[0]);
88              p1->items[0] = p1->items[1] = NULL;
89              delete pnp;
90              delete p1;
91              return true;
92          }
93      }
```
`/p16ecc/cc_source_7.3/popt3.cpp`   Code-7.20

Table-7.8 shows the test result.

```
void foo(int x)
{
   while ( x-- );
}
```

Before optimization	After optimization
P-CODE: P_FUNC_BEG: foo : {void} P-CODE: LABEL _$L2 P-CODE: SRC ... ch7_t8.c #3: while (x--) P-CODE: '=' %1{int}, foo_$_x{int} P-CODE: '--' foo_$_x{int}, 1{int} P-CODE: JP_Z %1{int}, _$L4 P-CODE: ';' P-CODE: LABEL _$L3 P-CODE: SRC ... ch7_t8.c #3: ; P-CODE: ';' P-CODE: GOTO _$L2 P-CODE: LABEL _$L4 P-CODE: ';' P-CODE: LABEL _$L1 P-CODE: P_FUNC_END: foo	P-CODE: P_FUNC_BEG: foo : {void} P-CODE: LABEL _$L2 P-CODE: SRC ... ch7_t8.c #3: while (x--) P-CODE: JZ -- foo_$_x{int}, 1{int}, _$L4 P-CODE: ';' P-CODE: SRC ... ch7_t8.c #3: ; P-CODE: ';' P-CODE: GOTO _$L2 P-CODE: LABEL _$L4 P-CODE: ';' P-CODE: P_FUNC_END: foo

Table-7.8

7.4 P-code Optimization (4)

<u>Project directory</u>: /p16ecc/cc_source_7.4
<u>New source files</u>: popt4.cpp
<u>Modified source files</u>: popt.cpp

This part covers the P-code optimizations that relate to 'special' constants.

<u>7.4.1 Math and logic operations with constant '0'</u>

1) Simplify P-code if ip1 is constant '0'. Such as

$$\{\,'+',[X,0,Y]\} \rightarrow$$
$$\{\,'-',[X,0,Y]\} \rightarrow$$
$$\{\,'|',[X,0,Y]\} \rightarrow \quad \{\,'=',[X,Y]\}$$
$$\{\,'\wedge',[X,0,Y]\} \rightarrow$$

$$\{\,'+=',[X,0]\} \rightarrow$$
$$\{\,'-=',[X,0]\} \rightarrow$$
$$\{\,'|=',[X,0]\} \rightarrow \quad \text{(remove the P-code)}$$
$$\{\,'\wedge=',[X,0]\} \rightarrow$$

$$\{\,'*',[X,0,Y]\} \rightarrow$$
$$\{\,'/',[X,0,Y]\} \rightarrow$$
$$\{\,'\%',[X,0,Y]\} \rightarrow$$
$$\{\,'\&',[X,0,Y]\} \rightarrow \quad \{\,'=',[X,0]\}$$
$$\{\,'<<',[X,0,Y]\} \rightarrow$$
$$\{\,'>>',[X,0,Y]\} \rightarrow$$

$$\{\,'<<=',[X,0]\} \rightarrow$$
$$\{\,'>>=',[X,0]\} \rightarrow$$
$$\{INC_OP,[X,0]\} \rightarrow \quad \text{(remove the P-code)}$$
$$\{DEC_OP,[X,0]\} \rightarrow$$

```
20        if ( IS_CONST(ip1, 0) )        // ip1 is constant '0'
21        {
22            switch (ptype)
23            {
24                case '+':               // {'+', [x, 0, y]}
25                case '-':               // {'+', [x, 0, y]}
26                case '|':               // {'|', [x, 0, y]}
27                case '^':               // {'^', [x, 0, y]}
28                    pnp->type = '=';
29                    pnp->updateItem(1, ip2);
30                    pnp->items[2] = NULL;
31                    return true;
32
33                case '*':               // {'*', [x, 0, y]}
34                case '/':               // {'/', [x, 0, y]}
35                case '%':               // {'%', [x, 0, y]}
36                case '&':               // {'&', [x, 0, y]}
37                case LEFT_OP:           // {'<<', [x, 0, y]}
38                case RIGHT_OP:          // {'>>', [x, 0, y]}
39                    pnp->type = '=';
40                    pnp->updateItem(2, NULL);
41                    return true;
42
43                case ADD_ASSIGN:        // {'+=', [x, 0]}
44                case SUB_ASSIGN:        // {'-=', [x, 0]}
45                case OR_ASSIGN:         // {'|=', [x, 0]}
46                case XOR_ASSIGN:        // {'^=', [x, 0]}
47                case LEFT_ASSIGN:       // {'<<=', [x, 0]}
48                case RIGHT_ASSIGN:      // {'>>=', [x, 0]}
49                case INC_OP:            // {INC_OP, [x, 0]}
50                case DEC_OP:            // {DEC_OP, [x, 0]}
51                    delete pnp;
52                    return true;
53            }
54        }
```

/p16ecc/cc_source_7.4/popt4.cpp   Code-7.21

2) Simplify P-code if `ip2` is constant '0'. Such as

$$\{ `+', [\,X, Y, 0\,] \} \rightarrow$$
$$\{ `-', [\,X, Y, 0\,] \} \rightarrow$$
$$\{ `|', [\,X, Y, 0\,] \} \rightarrow \quad \{ `=', [\,X, Y\,] \}$$
$$\{ `^', [\,X, Y, 0\,] \} \rightarrow$$
$$\{ `<<', [\,X, Y, 0\,] \} \rightarrow$$
$$\{ `>>', [\,X, Y, 0\,] \} \rightarrow$$

$$\{ `*', [\,X, Y, 0\,] \} \rightarrow \quad \{ `=', [\,X, 0\,] \}$$
$$\{ `\&', [\,X, Y, 0\,] \} \rightarrow$$

```
56        if ( IS_CONST(ip2, 0) )        // ip2 is constant '0'
57        {
58            switch (ptype)
59            {
60                case '+':               // {'+', [x, y, 0]}
61                case '-':               // {'-', [x, y, 0]}
62                case '|':               // {'|', [x, y, 0]}
63                case '^':               // {'^', [x, y, 0]}
64                case LEFT_OP:           // {'<<', [x, y, 0]}
65                case RIGHT_OP:          // {'>>', [x, y, 0]}
66                    pnp->type = '=';
67                    pnp->updateItem(2, NULL);
68                    return true;
69
70                case '*':               // {'*', [x, y, 0]}
71                case '&':               // {'&', [x, y, 0]}
72                    pnp->type = '=';
73                    pnp->updateItem(1, ip2);
74                    pnp->items[2] = NULL;
75                    return true;
76            }
77        }
```

/p16ecc/cc_source_7.4/popt4.cpp   Code-7.22

7.4.2 Math and logic operations with constant '1'

Similar to that in Section 7.4.1,

1) Simplify P-code if `ip1` is constant '1'. Such as

$$\{\,`*\text{'},[X,1,Y]\} \qquad \rightarrow \qquad \{\,`=\text{'},[X,Y]\}$$

$$\{\,`*=\text{'},[X,1\,]\} \qquad \rightarrow$$
$$\{\,`/=\text{'},[X,1\,]\} \qquad \rightarrow \qquad \text{(remove the P-code)}$$

$$\{\,`\%=\text{'},[X,1\,]\} \qquad \rightarrow \qquad \{\,`=\text{'},[X,0\,]\}$$

```
79      if ( IS_CONST(ip1, 1) )      // ip1 is constant '1'
80      {
81          switch (ptype)
82          {
83              case '*':            // {'*', [x, 1, y]}
84                  pnp->type = '=';
85                  pnp->updateItem(1, ip2);
86                  pnp->items[2] = NULL;
87                  return true;
88
89              case MUL_ASSIGN:     // {'*=', [x, 1]}
90              case DIV_ASSIGN:     // {'/=', [x, 1]}
91                  delete pnp;
92                  return true;
93
94              case MOD_ASSIGN:     // {'%=', [x, 1]}
95                  pnp->type = '=';
96                  ip1->val.i = 0;
97                  return true;
98          }
99      }
```
/p16ecc/cc_source_7.4/popt4.cpp   Code-7.23

2) Simplify P-code if `ip2` is constant '1'. Such as

$$\{\,`*\text{'},[X,Y,1\,]\} \qquad \rightarrow$$
$$\{\,`/\text{'},[X,Y,1\,]\} \qquad \rightarrow \qquad \{\,`=\text{'},[X,Y]\}$$

$$\{\,`\%\text{'},[X,Y,1\,]\} \qquad \rightarrow \qquad \{\,`=\text{'},[X,0\,]\}$$

```
101     if ( IS_CONST(ip2, 1) )      // ip2 is constant '1'
102     {
103         switch (ptype)
104         {
105             case '*':            // {'*', [x, y, 1]}
106             case '/':            // {'/', [x, y, 1]}
107                 pnp->type = '=';
108                 pnp->updateItem(2, NULL);
109                 return true;
110             case '%':            // {'%', [x, y, 1]}
111                 pnp->type = '=';
112                 ip2->val.i = 0;
113                 pnp->updateItem(1, ip2);
114                 pnp->items[2] = NULL;
115                 return true;
116         }
117     }
```
/p16ecc/cc_source_7.4/popt4.cpp   Code-7.24

7.4.3 Math operations with a 'magic' number

The 'magic' number here means that a constant has a value of two's power, $N = 2^n$.

1) In this case, the following optimizations can be applied if `ip1` is a magic number.

$$\{\,`*=',\,[\,X,\,N=2^n\,]\}\qquad\rightarrow\qquad\{\,`<<=',\,[\,X,\,n\,]\}$$
$$\{\,`/=',\,[\,X,\,N=2^n\,]\}\qquad\rightarrow\qquad\{\,`>>=',\,[\,X,\,n\,]\}$$
$$\{\,`\%=',\,[\,X,\,N=2^n\,]\}\qquad\rightarrow\qquad\{\,`=',\,[\,X,\,N-1\,]\}$$

```
119     int n;
120     if ( bitSelect(ip1, &n) )    // ip1 is constant (1 << N)
121     {
122         switch (ptype)
123         {
124             case MUL_ASSIGN:     // {'*=', [x, 2^N]}
125                 pnp->type = LEFT_ASSIGN;
126                 ip1->val.i = n;
127                 return true;
128
129             case DIV_ASSIGN:     // {'/=', [x, 2^N]}
130                 if ( ip0->acceSign() ) break;
131                 pnp->type = RIGHT_ASSIGN;
132                 ip1->val.i = n;
133                 return true;
134
135             case MOD_ASSIGN:     // {'%=', [x, 2^N]}
136                 if ( ip0->acceSign() ) break;
137                 pnp->type = AND_ASSIGN;
138                 ip1->val.i = (1 << n) - 1;
139                 return true;
140         }
141     }
```
`/p16ecc/cc_source_7.4/popt4.cpp`   Code-7.25

- #130, #136: the optimization only applies to unsigned values.

2) The following optimizations can be applied if `ip2` is a magic number.

$$\{\,`*',\,[\,X,\,Y,\,N=2^n\,]\}\qquad\rightarrow\qquad\{\,`<<',\,[\,X,\,Y,\,n\,]\}$$
$$\{\,`/',\,[\,X,\,Y,\,N=2^n\,]\}\qquad\rightarrow\qquad\{\,`>>',\,[\,X,\,Y,\,n\,]\}$$
$$\{\,`\%',\,[\,X,\,Y,\,N=2^n\,]\}\qquad\rightarrow\qquad\{\,`\&',\,[\,X,\,Y,\,N-1\,]\}$$

```
143     if ( bitSelect(ip2, &n) )    // ip2 is constant (1 << N)
144     {
145         switch (ptype)
146         {
147             case '*':            // {'*', [x, y, 2^N]}
148                 pnp->type = LEFT_OP;
149                 ip2->val.i = n;
150                 return true;
151
152             case '/':            // {'/', [x, y, 2^N]}
153                 if ( ip1->acceSign() ) break;
154                 pnp->type = RIGHT_OP;
155                 ip2->val.i = n;
156                 return true;
157
158             case '%':            // {'%', [x, y, 2^N]}
159                 if ( ip1->acceSign() ) break;
160                 pnp->type = '&';
161                 ip2->val.i = (1 << n) - 1;
162                 return true;
163         }
164     }
```
`/p16ecc/cc_source_7.4/popt4.cpp`   Code-7.26

- #153, #159: the optimization only applies to unsigned values.

7.5 P-code Optimization (5)

<u>Project directory</u>: /p16ecc/cc_source_7.5
<u>New source files</u>: popt5.cpp
<u>Modified source files</u>: popt.cpp

This part covers the P-code optimizations for temporary variables. Temporary variables are generated during P-code generation and used to hold intermediate results that are not visible to users.

1. Temporary variables have a transient period that only lasts within one sentence of C source code.
2. Temporary variables will be reused throughout by all other sentences within a source code function.
3. Temporary variables can be assigned to CPU registers or memory RAM space, not accessible from the source code.

It is an essential aspect for a compiler to minimize both the number of temporary variables used and their size.

7.5.1 Remove unused temporary variables

In the following C code, the shift-reduce operation during P-code generation generates an assignment P-code that assigns data to a temporary variable t, but t is not used later. So, this P-code line can be removed.

```
X++;
```

$$\{\text{`='}, [t, X]\}$$
$$\{\text{INC_OP}, [X, 1]\} \qquad \rightarrow \qquad \{\text{INC_OP}, [X, 1]\}$$

```
41      if ( TERMINATE_ASSIGN(ptype) && ip0->type == TEMP_ITEM &&   // {'=', [t, x]}
42          unusedTmp(pnp->next, ip0->val.i)                    )   // 't' not used any more
43      {
44          delete pnp;
45          return true;
46      }
```
/p16ecc/cc_source_7.5/popt5.cpp   Code-7.27

7.5.2 Reuse of temporary variables

Multiple temporary variables can be generated within a single source file. A temporary variable can be reused if it's used only once, thereby minimizing the total number of variables. For example,

```
a = b + c + d + e;
```

$$\{\text{`+'}, [t1, \text{b}, \text{c}]\} \qquad\qquad \{\text{`+'}, [t1, \text{b}, \text{c}]\}$$
$$\{\text{`+'}, [t2, t1, \text{d}]\} \qquad\qquad \{\text{`+'}, [t1, t1, \text{d}]\} \qquad\qquad \{\text{`+'}, [t1, \text{b}, \text{c}]\}$$
$$\{\text{`+'}, [t3, t2, \text{e}]\} \quad\rightarrow\quad \{\text{`+'}, [t1, t1, \text{e}]\} \quad\rightarrow\quad \{\text{`+'}, [t1, t1, \text{d}]\}$$
$$\{\text{`='}, [\text{a}, t3]\} \qquad\qquad \{\text{`='}, [\text{a}, t1]\} \qquad\qquad \{\text{`+'}, [\text{a}, t1, \text{e}]\}$$

```
48          if ( TERMINATE_ASSIGN(ptype) && ip0->type == TEMP_ITEM )      // {'=', [tn, x]} or {'*', [tn, x, y]} ...
49          {
50              for (int i = 0; i < ip0->val.i; i++)
51              {
52                  if ( unusedTmp(pnp->next, i) )
53                  {
54                      replaceTmp(pnp->next, ip0->val.i, i);
55                      ip0->val.i = i;
56                      return true;
57                  }
58              }
59          }
/p16ecc/cc_source_7.5/popt5.cpp                                                          Code-7.28
```

Table-7.9 shows the test result.

```
void foo()
{
   int a, b, c, d, e;

   a = b + c + d + e;
}
```

Before optimization	After optimization
P-CODE: P_FUNC_BEG: foo : {void}	P-CODE: P_FUNC_BEG: foo : {void}
P-CODE: ';'	P-CODE: ';'
P-CODE: SRC ... ch7_t10.c #5: a = b + c + d + e;	P-CODE: SRC ... ch7_t10.c #5: a = b + c + d + e;
P-CODE: '+' %1{short}, foo_$2_b{int}, foo_$3_c{int}	P-CODE: '+' %1{short}, foo_$2_b{int}, foo_$3_c{int}
P-CODE: '+' %2{long}, %1{short}, foo_$4_d{int}	P-CODE: '+' %1{long}, %1{short}, foo_$4_d{int}
P-CODE: '+' %3{long}, %2{long}, foo_$5_e{int}	P-CODE: '+' foo_$1_a{int}, %1{long}, foo_$5_e{int}
P-CODE: '=' foo_$1_a{int}, %3{long}	P-CODE: ';'
P-CODE: ';'	P-CODE: P_FUNC_END: foo
P-CODE: LABEL _$L1	
P-CODE: P_FUNC_END: foo	

Table-7.9

7.5.3 Minimize the size of temporary variables

The test results in Table-7.9 show that the temporary variable $t1$ has been used twice to hold the operation result, and its size changes from `short` (24-bit) to `long` (32-bit). However, only its lower 16 bits are used at the end. That means $t1$ can be an `int` (16-bit) variable in both places, without affecting the final result.

In the same case above, a further optimization will happen if $t1$'s size, at all the places, changes to `int` (16-bit):

```
a = b + c + d + e;
```

$$\begin{array}{l}\{\text{'}+\text{'}, [t1, b, c]\} \\ \{\text{'}+\text{'}, [t2, t1, d]\} \\ \{\text{'}+\text{'}, [t3, t2, e]\} \\ \{\text{'}=\text{'}, [a, t3]\}\end{array} \rightarrow \begin{array}{l}\{\text{'}+\text{'}, [t1, b, c]\} \\ \{\text{'}+\text{'}, [t1, t1, d]\} \\ \{\text{'}+\text{'}, [t1, t1, e]\} \\ \{\text{'}=\text{'}, [a, t1]\}\end{array} \rightarrow \begin{array}{l}\{\text{'}+\text{'}, [t1, b, c]\} \\ \{\text{'}+\text{'}, [t1, t1, d]\} \\ \{\text{'}+\text{'}, [a, t1, e]\}\end{array} \rightarrow \begin{array}{l}\{\text{'}+\text{'}, [t1, b, c]\} \\ \{\text{'}+=\text{'}, [t1, d]\} \\ \{\text{'}+\text{'}, [a, t1, e]\}\end{array}$$

The optimization operation is made as the flowchart shows below:

P-code line	P-code operation
P-code line$_{n-1}$	$\{\ '+' ,\ [t,\ ?,\ ?]\}$
P-code line$_n$	$\{\ '+' ,\ [X,\ t,\ Y]\}$

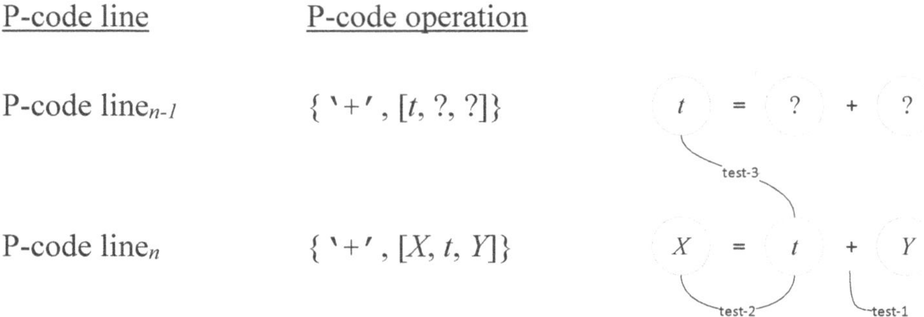

When the P-code line$_n$ is reached for optimization, it will check

1) P-code line$_n$ operation type.
2) `ip1` is a temp item, and size(`ip1`) > size(`ip0`).
3) `ip1` is same as `ip0` in P-code line$_{n-1}$.

Resizing `ip1`, size(`ip1`) = size(`ip0`), will take place if all these conditions are met, as shown in Code-7.29.

```
61     if ( TERMINATE_ASSIGN(ptype) || COMPOUND_ASSIGN(ptype) )    // {'+', [z, x, y]} or {'+=', [x, y]} ...
62     {
63         int size0 = ip0->acceSize();
64         int size1 = ip1->acceSize();
65         int size2 = ip2? ip2->acceSize(): 0;
66         if ( size0 == 0 || size0 >= 4 ) return false;
67
68         if ( ip1 && ip1->type == TEMP_ITEM && !VOLATILE_OP1(ptype) )
69         {
70             if ( size1 > size0 && reduceTmpSize(pnp->last, ip1, size0) )
71             {
72                 resizeTmpItem(ip1, size0);
73                 return true;
74             }
75         }
76
77         if ( ip2 && ip2->type == TEMP_ITEM && !VOLATILE_OP1(ptype) )
78         {
79             if ( size2 > size0 && reduceTmpSize(pnp->last, ip2, size0) )
80             {
81                 resizeTmpItem(ip2, size0);
82                 return true;
83             }
84         }
85     }
```
`/p16ecc/cc_source_7.5/popt5.cpp`   Code-7.29

Table-7.10 shows the test result.

Before optimization	After optimization
P-CODE: P_FUNC_BEG: foo : {void}	P-CODE: P_FUNC_BEG: foo : {void}
P-CODE: ';'	P-CODE: ';'
P-CODE: SRC ... ch7_t10.c #5: a = b + c + d + e;	P-CODE: SRC ... ch7_t10.c #5: a = b + c + d + e;
P-CODE: '+' %1{short}, foo_$2_b{int}, foo_$3_c{int}	P-CODE: '+' %1{int}, foo_$2_b{int}, foo_$3_c{int}
P-CODE: '+' %2{long}, %1{short}, foo_$4_d{int}	P-CODE: '+=' %1{int}, foo_$4_d{int}
P-CODE: '+' %3{long}, %2{long}, foo_$5_e{int}	P-CODE: '+' foo_$1_a{int}, %1{int}, foo_$5_e{int}
P-CODE: '=' foo_$1_a{int}, %3{long}	P-CODE: ';'
P-CODE: ';'	P-CODE: P_FUNC_END: foo
P-CODE: LABEL _$L1	
P-CODE: P_FUNC_END: foo	

Table-7.10

Note, operation type of P-code line n matters. Please see the case in Table-7.11.

```
void foo()
{
  int a, b, c, d, e;

  a = (b + c + d) / e;
}
```

Before optimization	After optimization
P-CODE: P_FUNC_BEG: foo : {void}	P-CODE: P_FUNC_BEG: foo : {void}
P-CODE: ';'	P-CODE: ';'
P-CODE: SRC ... ch7_t11.c #5: a = (b + c + d) / e;	P-CODE: SRC ... ch7_t11.c #5: a = (b + c + d) / e;
P-CODE: '+' %1{short}, foo_$2_b{int}, foo_$3_c{int}	P-CODE: '+' %1{short}, foo_$2_b{int}, foo_$3_c{int}
P-CODE: '+' %2{long}, %1{short}, foo_$4_d{int}	P-CODE: '+' %1{long}, %1{short}, foo_$4_d{int}
P-CODE: '/' %3{long}, %2{long}, foo_$5_e{int}	P-CODE: '/' foo_$1_a{int}, %1{long}, foo_$5_e{int}
P-CODE: '=' foo_$1_a{int}, %3{long}	P-CODE: ';'
P-CODE: ';'	P-CODE: P_FUNC_END: foo
P-CODE: LABEL _$L1	
P-CODE: P_FUNC_END: foo	

Table-7.11

7.6 P-code Optimization (6)

<u>Project directory</u>: /p16ecc/cc_source_7.6
<u>New source files</u>: popt6.cpp
<u>Modified source files</u>: popt.cpp

7.6.1 Swap operands positions

For the operations that meet exchangeable law, change operands positions and simplify the operation types.

$$\{\text{'+'}, [X, Y, X]\} \quad \rightarrow \quad \{\text{'+='}, [X, Y]\}$$
$$\{\text{'*'}, [X, Y, X]\} \quad \rightarrow \quad \{\text{'*='}, [X, Y]\}$$
$$\dots$$
$$\{\text{'\^{}'}, [X, Y, X]\} \quad \rightarrow \quad \{\text{'\^{}='}, [X, Y]\}$$

```
23       if ( exchaneableOpr(ptype, &new_type) && same(ip0, ip2) )    // {'+', [X, Y, X]}
24       {
25           pnp->type = new_type;
26           pnp->updateItem(2, NULL);
27           return true;
28       }
```
/p16ecc/cc_source_7.6/popt6.cpp
Code-7.30

7.6.2 Remove assignments from ACC to the temporary

The temporary variable (TEMP_ITEM) and the accumulator (ACC_ITEM) are both considered transient data holders. As described in Section 6.11.1, the returned value of a function call will be assigned to a temporary variable as a default operation. The strategies presented in Section 7.2.4 discuss how to eliminate the use of temporary variables, which covers most situations, but the following case

```
if ( func() == 0 ) ...
```

The P-code generated and optimization are

$$\{\text{'='}, [t, \text{ACC}]\} \quad \rightarrow \quad \{\text{P_JNZ}, [\text{ACC}, Lx]\}$$

183

$$\{P_JNZ, [t, Lx]\}$$
...
...

```
30    if ( ptype == '=' &&                                        // {'=', [t, ACC]}
31         ip0->type == TEMP_ITEM && ip1->type == ACC_ITEM &&
32         cmpAttr(ip0->attr, ip1->attr) == 0 && !accReferenced(p1) )
33    {
34        bool updated = false;
35        for (bool done = false; !endOfScope(p1) && !done;)
36        {
37            for (int i = 3; i-- && !done;)
38            {
39                Item *ip = p1->items[i];
40                if ( ip && same(ip0, ip) )
41                {
42                    if ( i > 0 || !assignmentCode(p1->type, ALL_OP) )
43                    {
44                        ip->type = ACC_ITEM;
45                        updated  = true;
46                    }
47                    else
48                        done = true;     // stop here!
49                }
50            }
51            p1 = p1->next;
52        }
53        if ( updated )
54        {
55            delete pnp;
56            return true;
57        }
58        p1 = next(pnp, 1);
59    }
```
/p16ecc/cc_source_7.6/popt6.cpp   Code-7.31

7.6.3 New P-code type to mimic DJNZ instruction

There is an 'efficient' instruction in the Intel 8051 CPU instruction set, **DJNZ**, which integrates multiple operations as follows.

$$\text{DJNZ } R, Lx \qquad \text{decrease register } R \text{ by 1, and jump to address } Lx \text{ if } R \text{ is not zero.}$$

This section introduces a pair of new P-code operation types that mimic DJNZ operations.

$$\begin{array}{l}\{\text{DEC_P}, [\,X, 1\,]\} \\ \{\text{P_JNZ}, [\,X, Lx\,]\}\end{array} \quad \rightarrow \quad \{\text{P_DJNZ}, [\,X, Lx\,]\} \quad \text{decrease } X, \text{ and jump if } X\,!= 0$$

$$\begin{array}{l}\{\text{INC_P}, [\,X, 1\,]\} \\ \{\text{P_JNZ}, [\,X, Lx\,]\}\end{array} \quad \rightarrow \quad \{\text{P_IJNZ}, [\,X, Lx\,]\} \quad \text{increase } X, \text{ and jump if } X\,!= 0$$

Note that this optimization applies only when X is an 8-bit value.

```
61    if ( (ptype == DEC_OP || ptype == INC_OP) && p1 && p1->type == P_JNZ )   // {DEC_OP, [X, 1]}
62    {                                                                        // {P_JNZ, [X, Lx]}
63        Item *ip = p1->items[0];
64        if ( ip0->acceSize() == 1 && ip1->val.i == 1 && same(ip0, ip) )
65        {
66            pnp->type = (ptype == DEC_OP)? P_DJNZ: P_IJNZ;
67            pnp->updateItem(1, p1->items[1]);
68            p1->items[1] = NULL;
69            delete p1;
70            return true;
71        }
72    }
```
/p16ecc/cc_source_7.6/popt6.cpp   Code-7.32

7.6.4 Optimization for offset addressing

In struct/union data structure, accessing a data member usually involves offset addressing. For example,

```
typedef struct {
   int x, y, z;
} Data;

int foo(Data *dp)
{
   dp->y = 100;
}
```

The data member y has the offset value = 2, since `int` type has size = 2. The P-codes generated are

$$\{ \,'+',[\,t,\mathrm{dp},2\,]\,\} \qquad\qquad \text{P-CODE: '+'}\quad \%1\{int *\},\ foo_\$_dp\{STRUCT *\},\ 2\{int\}$$
$$\{ \,'=',[\,(t),100\,]\,\} \qquad\qquad \text{P-CODE: '='}\quad [\%1]\{int *\},\ 100\{int\}$$

The optimization involves (1) eliminating the temporary variable t; (2) moving the offset value into dp's modifier (bias); (3) merging the P-codes.

$$\{ \,'+',[\,t,\mathrm{dp},2\,]\,\}$$
$$\{ \,'=',[\,(t),100\,]\,\} \qquad\rightarrow\qquad \{ \,'=',[\,(\mathrm{dp}:2),100\,]\,\}$$

```
86         if ( (ptype == '=' || (ptype == '+' && ip2->type == CON_ITEM)) && ip0->type == TEMP_ITEM && p1 )
87         {
88             bool ret_code;
89
90             if ( ip1->type == ID_ITEM && ip1->attr && ip1->attr->ptrVect )        // {'+', [t, *X, N]}
91             {                                                                      // {op, [.. (t) ..]}
92                 Item *ip = ip1->clone();
93                 if ( ptype == '+' ) ip->bias += ip2->val.i;
94
95                 ip->type = PID_ITEM;
96                 ret_code = replaceIndir(p1, ip, ip0->val.i);
97                 delete ip;
98
99                 if ( ret_code )
100                {
101                    delete pnp;
102                    return true;
103                }
104            }
/p16ecc/cc_source_7.6/popt6.cpp                                                                  Code-7.33
```

Table-7.12 shows the test for this section.

```
struct _Data {
    int a, b, c, d, e;
};

void foo(struct _Data *p)
{
   p->a = (p->b * p->c * p->d) + p->e;
}
```

Before optimization
P-CODE: SRC ... ch7_t18.c #3: struct _Data   int a, b, c, d, e;;
P-CODE: P_FUNC_BEG: foo : {void}
P-CODE: SRC ... ch7_t18.c #7: p->a = (p->b * p->c * p->d) + p->e;
P-CODE: '+'   %1{int *}, foo_$_p{(_Data) STRUCT *}, 0{int}
P-CODE: '+'   %2{int *}, foo_$_p{(_Data) STRUCT *}, 2{int}
P-CODE: '+'   %3{int *}, foo_$_p{(_Data) STRUCT *}, 4{int}
P-CODE: '*'   %4{long}, [%2]{int *}, [%3]{int *}

185

```
P-CODE: '+'    %5{int *}, foo_$_p{(_Data) STRUCT *}, 6{int}
P-CODE: '*'    %6{long}, %4{long}, [%5]{int *}
P-CODE: '+'    %7{int *}, foo_$_p{(_Data) STRUCT *}, 8{int}
P-CODE: '+'    %8{long}, %6{long}, [%7]{int *}
P-CODE: '='    [%1]{int *}, %8{long}
P-CODE: ';'
P-CODE: LABEL    _$L1
P-CODE: P_FUNC_END: foo
```

After optimization

```
P-CODE: SRC ... ch7_t18.c #3: struct _Data    int a, b, c, d, e;;
P-CODE: P_FUNC_BEG: foo : {void}
P-CODE: SRC ... ch7_t18.c #7: p->a = (p->b * p->c * p->d) + p->e;
P-CODE: '='    %1{int *}, foo_$_p{(_Data) STRUCT *}
P-CODE: '*'    %2{int}, [foo_$_p:2]{int *}, [foo_$_p:4]{int *}
P-CODE: '*='   %2{int}, [foo_$_p:6]{int *}
P-CODE: '+'    [%1]{int *}, %2{int}, [foo_$_p:8]{int *}
P-CODE: ';'
P-CODE: P_FUNC_END: foo
```

Table-7.12

※ ※ ※ ※ ※

Conclusions of the charter:

1. Optimization is an interesting and endless topic. The strategies or algorithms introduced in this chapter are simple. But the results achieved seem satisfying. Readers can develop or create their own methods to build a DIY compiler.
2. The P-code structure introduced in this book (a 3-tuple data structure) limits the creativity in developing new optimization strategies.
3. It is essential to be aware that any optimization may introduce a risk to the result.
4. In the practice of designing a compiler, an optimization option setting is typically used to control the level of optimizations.

PIC16Fxxxx Assembly Code Generation

Assembly code generation is the last process for **p16ecc** compiler. Additionally, assembly code generation is a process that converts P-codes into assembly output. The process is highly target-dependent. That means it is necessary to fully understand the architecture of the PIC16Fxxxx processor and its instruction set. Additionally, achieving higher efficiency is a crucial concern in assembly code generation.

8.1 Introduction of PIC16Fxxxx processor

PIC16Fxxxx processor is the enhanced version of PIC16F CPU family. It was released to the market in early 2000. PIC16Fxxxx keeps the basic architecture of PIC16F (14-bit instruction length, RISC structure, little-endian organization, …), as well as the mnemonics of the instructions. The enhancements include
-   Support up to 2Kbytes RAM space.
-   Support up to 32Kwords ROM space.
-   Instruction set expanded to 49 instructions.
-   Dual indirect addressing register pairs (**FSR0** and **FSR1**) that can access both RAM and ROM.

Note that a few newly released PIC16Fxxxx products offer an additional enhancement and can support up to 4Kbytes of RAM (referred to as 'extended' PIC16Fxxxx processors). They are not compatible with PIC16Fxxxx at the instruction level (they still keep the same instruction mnemonics). This issue will be covered later in this book.

8.1.1 PIC16Fxxxx memory space organization

Like the original PIC16F processors, the PIC16Fxxxx family uses a Harvard architecture. The RAM is banked (128 bytes/bank), and ROM/Flash memory is paged (2048 words/page).

For PIC16Fxxxx processor family, each RAM bank is divided into 4 blocks or segments, organized as in Table-8.1:

Address allocation in a bank	Purpose	
0x00 ~ 0x0B (12 bytes)	Core registers (see Table-3.3)	These locations are shared by all 32 banks (mapped to the same physical registers). No **BSR** register setting is needed to access any of core registers.
0x0C ~ 0x1F (20 bytes)	Special-purpose registers	BSR register setting is needed to access any of the registers in this area.
0x20 ~ 0x6F (80 bytes)	General data memory	BSR register setting is needed to access this area.
0x70 ~ 0x7F (16 bytes)	Common data memory	These locations are shared by all 32 banks (mapped to the same physical memory). No **BSR** register setting is needed to access this area.

Table-8.1

In theory, the maximum data RAM size is $80\times32+16 = 2576$ (bytes).

ROM memory is divided into 16 pages. Bits 6 through 3 in register PCLATH indicate the current page being accessed.

8.1.2 Memory space mapping

RAM and ROM spaces in PIC16Fxxxx are separate. They are mapped to a common 64Kbyte space, as shown in Table-8.2, which can be accessed in indirect addressing mode using the FSR0 or FSR1 register pair.

64 Kbytes common space		Memory mapping
0x0000 ~ 0x0FFF (4096 bytes)	Regular RAM space	All 32 RAM banks
0x2000 ~ 0x29FF (2560 bytes)	Linear data RAM space	All 32 general data RAM blocks
0x8000 ~ 0xFFFF (32768 bytes)	Common ROM space	Lower bytes of ROM words

Table-8.2

Note that all the general data RAM blocks (80 bytes/block) are arranged in a back-to-back sequence. This way, big data arrays (> 80 bytes) are supported.

8.1.3 PIC16Fxxxx core registers and instruction set

There are 12 core registers for PIC16Fxxxx processors. They are mapped to the address 0x00~0x0B in every bank, as shown in Table-8.3:

Register	Address in bank	Note
INDF0	0x00	Indirect addressing, using FSR0 pair
INDF1	0x01	Indirect addressing, using FSR1 pair
PCL	0x02	Lower byte of PC
STATUS	0x03	Status register that holds status flags
FSR0L	0x04	Register pair FSR0 for indirect addressing
FSR0H	0x05	
FSR1L	0x06	Register pair FSR1 for indirect addressing
FSR1H	0x07	
BSR	0x08	bit4~bit0 select RAM bank
WREG	0x09	
PCLATH	0x0A	bit6~bit3 select ROM page
INTCON	0x0B	

Table-8.3

Table-8.4 lists the instruction set of PIC16Fxxxx (mnemonics).

ADDWF	INCF	SUBWF	BTFSS	ANDLW	SUBWFB	BRW
ANDWF	INCFSZ	NOP	CALL	IORLW	LSLF	CALLW
CLRF	IORWF	SWAPF	GOTO	MOVLW	LSRF	ADDFSR
CLRW	MOVF	XORWF	RETFIE	SUBLW	ASRF	MOVIW
COMF	MOVWF	BCF	RETLW	XORLW	MOVLP	MOVWI
DECF	RLF	BSF	RETURN	SLEEP	MOVLB	RESET
DECFSZ	RRF	BTFSC	ADDLW	ADDWFC	BRA	CLRWDT

Table-8.4

Besides the instruction set, there is a set of pseudo-instructions or directives (shown in Table-8.5) used in the assembly code.

Directive	Purpose	Format	
.bsel	Set BSR	`.bsel addr`	Set BSR for *addr*.
		`.bsel addr1, addr2`	Set BSR for *addr2* if *addr1* and *addr2* are in different banks.
.psel	Set PCLATH	`.psel addr`	Set PCLATH for *addr*.
		`.psel addr1, addr2`	Set PCLATH for *addr2* if *addr1* and *addr2* are in different pages.
.segment	Start a segment	`.segment CONSTn, ...`	Start of a constant segment in ROM.
		`.segment CODEn, ...`	Start of a code segment in ROM.
		`.segment BANKn, …`	Start of a data segment in RAM bank.
		`.segment FUSE, …`	Start of a fuse segment.
.end	End of the code	`.end`	
.rs	Reserve RAM space	`.rs n`	Reserve *n* bytes of space in RAM
.equ	Set label value	`label .equ v`	Set value *v* to *label*.
.fcall	Function call note	`.fcall f1, f2`	Notation: function *f1* calls *f2*.
.dw	Define words in ROM	`.dw w1, w2, ...`	Define words *w1*, *w2*, …, in ROM.
.invoke	Involve library file	`.invoke "file_name"`	
.device	Specify CPU device	`.device "cpu_module"`	
.dblank	Disable RAM space	`.dblank addr, length`	Disable the usage of RAM *length* bytes starting at *addr*.
.cblank	Disable ROM space	`.cblank addr, length`	Disable the usage of ROM *length* words starting at *addr*.

Table-8.5

8.1.4 CPU HW resources used in p16ecc

This section discusses how the **p16ecc** compiler is built and runs on the PIC16Fxxxx processor hardware environment. (Frankly speaking, PIC16Fxxxx processor is not ideal for supporting the C language)

1. FSR1 register pair will be used for simulating the stack pointer (for passing function arguments, interrupt environment protection). It can be used for moving data in some situations.

2. The common data RAM of each bank (0x70~0x7F) is used for compiler internal usage as shown in Table-8.6.
3. The total size of internal variables in each function cannot exceed 80 bytes.
4. The maximum size of each function code cannot exceed 2048 words.

Bank address	Purpose
0x70 ~ 0x73	Accumulator (ACC0 ~ ACC3), to store function returning data.
0x7A ~ 0x7B	FSR1 protection holder (for interrupt).
0x7C ~ 0x7D	FSR1 protection holder.
0x7E	A flag bit (bit-0) indicates if FSR1 has been put in 0x7C ~ 0x7D.
0x7F	'0' value constant.

Table-8.6

8.2 Start Assembly Code Generation

During P-code generation procedure, there are three P-code streams generated:

```
mainPcode      Program running code
initPcode      External and static variable initialization code
constPcode     ROMed constants (initialization data images)
```

8.2.1 Classes and source code folder

Project directory: /p16ecc/cc_source_8.1
New source files: pic16e.cpp, pic16e.h, pic16e_asm.cpp, pic16e_asm.h, pic16e_reg.cpp
Modified source files: main.cpp

All new files used for assembly code generation are located in the "**/p16e**" folder of the project.

Class PIC16E (coded in **pic16e.cpp**, ..., **pic16e.h**) is the kernel for the process of assembly code generation. And it's started from main(), after P-code optimization, and illustrated as Code-8.1 below.

```
41      if ( pcoder.errorCount == 0 )
42      {
43          Optimizer opt(&pcoder);
44          opt.run();
45   //     display(pcoder.mainPcode);
46
47          str.replace(str.length()-3, 3, ".asm");
48          PIC16E asm_gen((char*)str.c_str(), &nlist, &pcoder);
49          asm_gen.run();
50      }
```
/p16ecc/cc_source_8.1/main.cpp

Code-8.1

- #43~44: P-code optimization.
- #47: make the output file name, with extension as ".asm".
- #48: instantiate the class PIC16E.
- #49: start assembly code generation.

As shown in Code-8.1, for every input C source file, a corresponding .asm file will be generated by class PIC16E.

The constructor of `PIC16E` is shown in Code-8.2.

```
25   PIC16E :: PIC16E(char *out_file, Nlist *_nlist, Pcoder *_pcoder)
26   {
27       memset(this, 0, sizeof(PIC16E));     // clean up the class data
28
29       if( out_file ) fout = fopen(out_file, "w");
30       if ( fout == NULL ) fout = stdout;
31       asm16e = new P16E_ASM(fout);
32       nlist  = _nlist;
33       pcoder = _pcoder;
34
35       regWREG    = new P16E_REG8(WREG,    asm16e);
36       regBSR     = new P16E_REG8(BSR,     asm16e);
37       regPCLATH  = new P16E_REG8(PCLATH, asm16e);
38       regFSR0    = new P16E_FSR0(asm16e);
39       accSave    = 4;
40       isrStackSet = true;
41   }
```
/p16ecc/cc_source_8.1/p16e/pic16e.cpp   Code-8.2

- #29~30: open output file (".asm").
- #31: instantiate the class `P16E_ASM`, which performs as a file writer, in the assembly code format.
- #35~38: instantiate the classes for core registers (`WREG`, `BSR`, `PCLATH` and `FSR0`) that handle register tracking.

Class `P16E_ASM`, described in Code-8.3, supplies a group of service functions that write assembly code into an output file.

```
7    class P16E_ASM {
8        private:
9            FILE *fout;
10
11       public:
12           P16E_ASM(FILE *fileout);
13
14           void output(char *s);
15           void output(const char *s)  { output((char*)s); }
16           void label(char *lbl, bool c=false, char *opr=NULL);
17
18           void code(char *inst);
19           void code(char *inst, char *opr1, char *opr2=NULL);
20
21           void code(char *inst, const char *opr1, char *opr2=NULL) { code(inst, (char*)opr1, opr2); }
22           void code(const char *inst, char *opr1, char *opr2=NULL) { code((char*)inst, opr1, opr2); }
23
24           void code(char *inst, int opr1, char *opr2=NULL);
25           void code(const char *inst, int opr) { code((char*)inst, opr); }
26
27           void code(char *inst, char *opr1, int opr2);
28           void code(char *inst, int opr1, int opr2);
29   };
```
/p16ecc/cc_source_8.1/p16e/pic16e_asm.cpp   Code-8.3

8.2.2 Running class PIC16E

The initial part of the class `PIC16E` runs as follows (Code-8.4). Essentially, it will process every P-code stream mentioned above, generate and output assembly code to the output file.

```cpp
55  void PIC16E :: run(void)
56  {
57      time_t t = time (&t);    // current time
58      char *buf = STRBUF();    // string buffer
59      Nnode *nnp = NULL;
60      int ram_size = 0;
61      int stack_addr = 0;
62      PreScan prescan(nlist);
63
64      sprintf(buf, ";******************************************************\n"
65                   ";  Microchip Enhanced PIC16F1xxx C Compiler (CC16E), %s\n"
66                   ";  %s"
67                   ";******************************************************\n",
68                   VERSION, ctime (&t));
69      ASM_OUTP(buf);
70
71      for (NameList *lp = sysIncludeList; lp; lp = lp->next)
72      {
73          strcpy(buf, lp->name);
74          int len = strlen(buf);
75          if ( buf[len-2] == '.' && toupper(buf[len-1]) == 'H' )
76          {
77              buf[len-2] = '\0';
78              char *s = STRBUF();
79              sprintf(s, "\"%s\"", buf);
80              ASM_CODE(_INVOKE, s);
81          }
82      }
83      delName(&sysIncludeList); ASM_OUTP("\n");
84
85      sprintf(buf, "\"pic16e\"");
86      if ( nlist )     // search Name List...
87      {
88          // device RAM name
89          nnp = nlist->search((char*)"__DEVICE", DEFINE);
90          if ( nnp && nnp->np[0] && nnp->np[0]->type == NODE_STR )
91              sprintf(buf, "\"%s\"", nnp->np[0]->str.str);
92
93          // device RAM size
94          nnp = nlist->search((char*)"__SRAM_SIZE", DEFINE);
95          if ( nnp && (nnp->np[0] = prescan.scan(nnp->np[0])) && nnp->np[0]->type == NODE_CON )
96          {
97              ram_size = nnp->np[0]->con.value;
98              sprintf(&buf[strlen(buf)], ", %d", ram_size);
99              stack_addr = 0x2000 + ram_size - 16;
100         }
101
102         // device FLASH size
103         nnp = nlist->search((char*)"__FLASH_SIZE", DEFINE);
104         if ( nnp && (nnp->np[0] = prescan.scan(nnp->np[0])) && nnp->np[0]->type == NODE_CON )
105         {
106             sprintf(&buf[strlen(buf)], ", %d", nnp->np[0]->con.value);
107         }
108
109         nnp = nlist->search((char*)"__STACK_INIT_ADDR", DEFINE);
110         if ( nnp && (nnp->np[0] = prescan.scan(nnp->np[0])) && nnp->np[0]->type == NODE_CON )
111             stack_addr = nnp->np[0]->con.value;
112     }
113     ASM_CODE(_DEVICE, buf);
114     ASM_OUTP("\n");
115
116     if ( stack_addr > 0 )
117     {
118         sprintf(buf, "%s\t0x%X\t; stack init. value\n", _EQU, stack_addr);
119         ASM_LABL((char*)"_$$", true, buf);
120     }
121
122     // allocate memory for fixed address & public data
123     if ( outputData(dataLink) )
124         ASM_OUTP("\n");
125
126     ASM_OUTP("\n");
127     ASM_CODE(_END);
128 }
```


- #62~67: output the header part for the assembly code file.
- #69~80: output ".invoke" lines for system header files in use.
- #84~113: output ".device" line (device name, RAM and ROM sizes, as well as the stack initial address).
- #122~123: output data variables, as data segments, that are assigned at fixed address.

Table-8.7 shows the test result of the section.

Source code (ch8_t1.c)	Assembly code output (ch8_t1.asm)
| ```
#include <pic16f1788.h>

int data1 @ 0x2020;
int data2 @ 0x2022;
``` | ```
;**********************************************************
; Microchip Enhanced PIC16F1xxx C Compiler (CC16E), v0.08.01
; Thu Jun 26 09:47:47 2025
;**********************************************************
        .invoke "F:\p16ecc/include/pic16e"
        .invoke "F:\p16ecc/include/pic16f1788"

        .device "pic16f1788", 2048, 16384

_$$::   .equ  0x27F0 ; stack init. Value

        .segment    BANK (ABS, =8224)
data1:: .rs   2
        .segment    BANK (ABS, =8226)
data2:: .rs   2


        .end
``` |

<div align="right">Table-8.7</div>

8.3 Assembly Code Generation (1)

<u>Project directory</u>: /p16ecc/cc_source_8.2
<u>New source files</u>: pic16e0.cpp, pic16e1.cpp
<u>Modified source files</u>: pic16e.cpp

As shown in the example in the last section, every assembly code line has the general format, described as

>   *label : opcode operand1, operand2 ; comments*

Source file **pic16e0.cpp**, as a part of class `PIC16E`, provides a group of functions that are used to generate the operand strings, from an item in P-code. And it's declared in **pic16e.h**, as illustrated as Code-8.5.

```
28  class PIC16E {
29      public:
30          PIC16E(char *out_file, Nlist *_nlist, Pcoder *_pcoder);
31          ~PIC16E();
        . . .
52      private:
53          void errPrint(const char *);
54          int  outputData(Dlink *dlink);
55          void outputASM0(Pnode *plist);
56
57          // pic16e0.cpp
58          bool useFSR(Item *ip);
59          bool useFSR(Item *ip0, Item *ip1);
60          bool useFSR(Item *ip0, Item *ip1, Item *ip2);
61          char *setFSR(Item *ip, int fsr);
62          char *setFSR0(Item *ip);
63          char *setFSR1(Item *ip);
64          char *itemStr(Item *ip, int offset);
65          char *acceItem(Item *ip, int offset, char *indf = NULL);
66          void fetch(Item *ip, int offset, char *indf);
67          void store(Item *ip, int offset, char *indf);
68          Item *storeToACC(Item *ip, int limit);
69          bool isWREG(Item *ip);
```
`/p16ecc/cc_source_8.2/p16e/pic16e.h`
Code-8.5

Function `run()` in class PIC16E is expanded as in Code-8.6, to scan through the P-codes of `mainPcode` stream:

193

```
 55   void PIC16E :: run(void)
 56   {
 57       time_t t = time(&t);     // current time
 58       char *buf = STRBUF();    // string buffer
 59       Nnode *nnp = NULL;
 60       int ram_size = 0;
 61       int stack_addr = 0;
 62       PreScan prescan(nlist);
 63
 64       sprintf(buf, ";****************************************************\n"
 65                    ";  Microchip Enhanced PIC16F1xxx C Compiler (CC16E), %s\n"
 66                    ";  %s"
 67                    ";****************************************************\n",
 68                    VERSION, ctime(&t));
 69       ASM_OUTP(buf);

122       // allocate memory for fixed address & public data
123       if ( outputData(dataLink) )
124           ASM_OUTP("\n");
125
126       // generate main program ASM code
127       outputASM0(pcoder->mainPcode);
128       ASM_OUTP("\n");
129       ASM_CODE(_END);
130   }
```
/p16ecc/cc_source_8.2/p16e/pic16e.cpp   Code-8.6

- #127: call function `outputASM0()` to scan `mainPcode` stream and generate the assembly code.

`outputASM0()` runs as a dispatching panel, shown in Code-8.7, to call the corresponding functions to generate the assembly code, based on P-code types.

```
181   void PIC16E :: outputASM0(Pnode *plist)
182   {
183       for(; plist; plist = plist->next)
184       {
185           curPnode = plist;
186           int pcode = plist->type;
187           Item *ip0 = plist->items[0];
188           Item *ip1 = plist->items[1];
189           Item *ip2 = plist->items[2];
190
191           switch ( pcode )
192           {
193               case P_FUNC_BEG:     // function begin
194                   curFunc = (Fnode*)ip0->val.p;
195                   if ( curFunc->endLbl > 0 )  // it's func. definition
196                       funcBeg(plist);
197                   else
198                       curFunc = NULL;
199                   break;
200
201               case P_FUNC_END:
202                   curFunc = (Fnode*)ip0->val.p;
203                   if ( curFunc->endLbl > 0 )  // it's func. definition
204                       funcEnd();
205                   curFunc = NULL;
206                   break;
```
/p16ecc/cc_source_8.2/p16e/pic16e.cpp   Code-8.7

- #183: go through `mainPcode` stream.
- #186: get the P-code type.
- #193~199: call `funcBeg()`, to generate assembly code for `P_FUNC_BEG` P-code.
- #201~206: call `funcEnd()`, to generate assembly code for `P_FUNC_END` P-code.

P-code {P_FUNC_BEG, [*fptr*]} contains function head information pointed by *fptr*:
- Function name
- Function return type
- Function input parameters (if any)

When the function being processed has the name "main", which is the entry point of the user application, the following assembly code segments must be generated.
1. Code segment for Reset vector (which starts at address 0x0000 for PIC16Fxxxx processor), as illustrated in Code-8.8.
2. Code segment "wrapper" for the external and static data initialization, as illustrated in Code-8.9.
3. Code segment "wrapper" for ISR (which starts at address 0x0004 for PIC16Fxxxx processor), as illustrated in Code-8.10.

```
26   void PIC16E :: funcBeg(Pnode *pnp)
27   {
28       char *fname = curFunc->name;
29       bool is_isr = ISR_FUNC(curFunc);
30       char func_dname[4096];
31       char *s = STRBUF();
32
33       ASM_OUTP("\n");
34
35       // make local data index name...
36       sprintf(func_dname, "%s_$data$", fname);
37
38       // it's the 'main' function?
39       if ( strcmp(fname, "main") == 0 )
40       {
41           // reset vector init.
42           ASM_CODE(_SEGMENT, "CODE0 (ABS, =0x0000)");
43           ASM_CODE(_NOP);
44           ASM_CODE(_MOVLP, "main >> 8");
45           ASM_CODE(_GOTO, "main");
46           ASM_OUTP("\n");
```
/p16ecc/cc_source_8.2/p16e/pic16e1.cpp   Code-8.8

```
48           // create static data init. link
49           ASM_CODE(_SEGMENT, "CODEi (REL, BEG)");
50           ASM_LABL((char*)"_$init$", true);
51           ASM_CODE(_CLRF, ZERO_LOC);  // ZERO_LOC := b'00000000
52           ASM_CODE(_CLRF, FLAG_LOC);  // bit0: FSR1_SAVED
53           ASM_OUTP("\n");
54
55           ASM_CODE(_SEGMENT, "CODEi (REL, END)");
56           ASM_CODE(_RETURN);
57           ASM_OUTP("\n");
```
/p16ecc/cc_source_8.2/p16e/pic16e1.cpp   Code-8.9

- #49: start of the data initial wrapper segment.
- #51~52: reset locations 0x7F and 0x7E (compiler internal variables).
- #55: end of the data initial wrapper segment.

```
59           ASM_CODE(_SEGMENT, "CODE1 (ABS, =0x0004, BEG)");
60           ASM_CODE(_CLRF, PCLATH);     // set up PCLATH
61           // get stack pointer if needed
62           if ( isrStackSet )
63           {
64               ASM_CODE(_BTFSS, FLAG_LOC, 0);
65               ASM_CODE(_BRA, ".+5");
66               ASM_CODE(_MOVF, BUFFERED_FSR1L, _W_);
67               ASM_CODE(_MOVWF, FSR1L);
68               ASM_CODE(_MOVF, BUFFERED_FSR1H, _W_);
69               ASM_CODE(_MOVWF, FSR1H);
70               ASM_OUTP("\n");
71           }
```

```
72          for (int i = 0; i < accSave; i++)
73          {
74              ASM_CODE(_MOUF, 0x70+i, _W_);
75              ASM_CODE(_MOUWI, "--INDF1");
76          }
77          ASM_OUTP("\n");
78
79          ASM_CODE(_SEGMENT, "CODE1 (REL, END)");
80          for (int i = accSave; i--;)
81          {
82              ASM_CODE(_MOUIW, "INDF1++");
83              ASM_CODE(_MOUWF, 0x70+i);
84          }
85          ASM_CODE(_RETFIE);
86          ASM_OUTP("\n");
87      }
```
`/p16ecc/cc_source_8.2/p16e/pic16e1.cpp`   Code-8.10

- #59: start of the ISR wrapper segment.
- #72~76: save (virtual) accumulator ACC.
- #79: end of the ISR wrapper segment.
- #80~84: restore (virtual) accumulator ACC.
- #85: interrupt return instruction.

The following table, Table-8.8, shows the test result of assembly code generation.

Source code (ch8_t2.c)	Assembly code output (ch8_t2.asm)

`void main()`
`{`
`}`
	`.segment    CODE0 (ABS, =0x0000)`
	`nop`
	`movlp  main >> 8`
	`goto  main`
	
	`.segment    CODEi (REL, BEG)`
	`_$init$::`
	`clrf  127`
	`clrf  126`
	
	`.segment    CODEi (REL, END)`
	`return`
	
	`.segment    CODE1 (ABS, =0x0004, BEG)`
	`clrf  PCLATH`
	`btfss  126, 0`
	`bra   .+5`
	`movf  124, W`
	`movwf  FSR1L`
	`movf  125, W`
	`movwf  FSR1H`
	
	`movf  112, W`
	`movwi  --INDF1`
	`movf  113, W`
	`movwi  --INDF1`
	`movf  114, W`
	`movwi  --INDF1`
	`movf  115, W`
	`movwi  --INDF1`
	
	`.segment    CODE1 (REL, END)`
	`moviw  INDF1++`
	`movwf  115`
	`moviw  INDF1++`
	`movwf  114`
	`moviw  INDF1++`
	`movwf  113`
	`moviw  INDF1++`
	`movwf  112`

```
              retfie

              .segment    CODE2 (REL) main:0
main::
              .psel  main, _$init$
              call   _$init$
              movlw  _$$
              movwf  FSR1L
              movlw  _$$>>8
              movwf  FSR1H
              .psel  _$init$, main
              return
; function(s) called::
              .fcall main, _$init$

              .end
```

Table-8.8

For any function, its internal static data variables should be located in the external dedicated RAM area, as shown in Code-8.11.

```
89          for (int sequence = 0;;)
90          {
91              int offset, depth = sequence++;
92              Dnode *dnp = curFunc->getData(STATIC_DATA, &depth, &offset);
93              if ( dnp == NULL ) break;
94
95              ASM_CODE(_SEGMENT, "BANKi (REL)");
96              sprintf(s, "%s\t%d\n", _RS, dnp->size());
97              ASM_LABL(dnp->nameStr(), false, s);
98          }
```
/p16ecc/cc_source_8.2/p16e/pic16e1.cpp

Code-8.11

- #95~97: locate function static data variables in BANKi segments.

The following is the assembly code output for the function itself.

```
100         Item *list = NULL;
101         int data_size = listData(curFunc, &list, pnp);
102
103         if ( is_isr )
104         {
105             sprintf(s, "CODE1 (REL) %s:%d", fname, data_size);
106             regPCLATH->set(0);
107         }
108         else
109         {
110             sprintf(s, "CODE2 (REL) %s:%d", fname, data_size);
111             regPCLATH->set(fname);
112         }
113         ASM_CODE(_SEGMENT, s);
114
115         // output function local variables...
116         while ( list )
117         {
118             ASM_OUTP(list->val.s);
119             Item *tmp = list->next;
120             delete list; list = tmp;
121         }
122
123         if ( !is_isr )
124             ASM_LABL(fname, !STATIC_FUNC(curFunc));
125
126         regBSR->reset();
127         regWREG->reset();
```

197

```cpp
129         if ( totalParSize > 0 )
130         {   // get function parameter from stack (pointed by FSR1)
131             s = STRBUF();
132             if ( totalParSize <= 2 )     // simple move
133             {
134                 regBSR->load(func_dname);
135                 for (int i = totalParSize; i--;)
136                 {
137                     ASM_CODE(_MOVIW, "INDF1++");
138                     sprintf(s, "%s+%d", func_dname, i);
139                     ASM_CODE(_MOVWF, s);
140                 }
141             }
142             else                         // using system lib routine
143             {
144                 sprintf(s, "%s+%d", func_dname, totalParSize);
145                 ASM_CODE(_MOVLW, s);
146                 ASM_CODE(_MOVWF, FSR0L);
147                 sprintf(s, "(%s+%d)>>8", func_dname, totalParSize);
148                 ASM_CODE(_MOVLW, s);
149                 ASM_CODE(_MOVWF, FSR0H);
150                 ASM_CODE(_MOVLW, totalParSize);
151                 call((char*)"_copyPar");
152             }
153         }
154
155         if ( strcmp(fname, "main") == 0 )
156         {
157             call((char*)"_$init$");
158
159             // init. stack pointer
160             ASM_CODE(_MOVLW, "_$$");     ASM_CODE(_MOVWF, FSR1L);
161             ASM_CODE(_MOVLW, "_$$>>8"); ASM_CODE(_MOVWF, FSR1H);
162         }
163 }
```

`/p16ecc/cc_source_8.2/p16e/pic16e1.cpp`   Code-8.12

- #100~101: calculates the total data size for all the internal variables, including the parameters and temporary variables if any.
- #103~113: output the segment start line, and the function name label line, and initialize the register `PCLATH`.
- #116~121: output all the internal variable macro definitions (offsets to the function data block).
- #129~153: move function parameters from the stack to the function data block (call the system lib function, `_copyPar()`, to move the parameters if the total size > 2 bytes).
- #155~162: if it's `main()` function, call initialization routine, and initialize the stack pointer (FSR1).

The following table, Table-8.9, shows the test result of assembly code generation.

Source code (ch8_t3.c)	Assembly code output (ch8_t3.asm)
| ```c
void func(int a, int b)
{
  int x, y;
}
``` | ```asm
    .segment     CODE2 (REL) func:8
func_$_a:     .equ  func_$data$+0
func_$_b:     .equ  func_$data$+2
func_$1_x:    .equ  func_$data$+4
func_$2_y:    .equ  func_$data$+6
func::
    movlw  func_$data$+4
    movwf  FSR0L
    movlw  (func_$data$+4)>>8
    movwf  FSR0H
    movlb  4
    .psel  func, _copyPar
    call   _copyPar
    .psel  _copyPar, func
    return
; function(s) called::
    .fcall func, _copyPar

    .end
``` |

Table-8.9

The following table, Table-8.10, illustrates how the segment names are used in ".segment" lines.

.segment CODE0	Reset code in ROM
.segment CODE1	Interrupt code in ROM
.segment CODE2	Other running code in ROM
.segment CODEi	Initialization code in ROM
.segment BANK*n*	Located in RAM bank-*n* ($n = 0, 1, 2, …, 31$)
.segment BANKi	Located in any RAM bank
.segment BANKn	Located in RAM for linear data
.segment CONSTi	Constant data in ROM

Table-8.10

All segments with the same segment name will be grouped into a sequence. Additionally, there are a few modifiers that specify how to locate the contents in ROM/RAM.

(ABS, =*N*)	Allocated at the absolute (fixed) address N
(REL, …)	Relocatable (float) address
(… , BEG)	Start segment in a segment sequence
(… , END)	Last segment in a segment sequence

Table-8.11

8.3.2 Assembly code output for P_FUNC_END

Assembly code output for P-code P_FUNC_END is simple, as shown in Code-8.13.

```
165   void PIC16E :: funcEnd(void)
166   {
167       if ( ISR_FUNC(curFunc) )
168           regPCLATH->load(0);
169       else
170       {
171           regPCLATH->load(curFunc->name);
172           ASM_CODE(_RETURN);
173       }
174
175       // list all function called ...
176       if ( curFunc->fcall ) ASM_LABL((char*)"; function(s) called:");
177       for (NameList *fcall = curFunc->fcall; fcall; fcall = fcall->next)
178           ASM_CODE(_FCALL, curFunc->name, fcall->name);
179   }
```
/p16ecc/cc_source_8.2/p16e/pic16e1.cpp
Code-8.13

- #172: output a RETURN instruction at the end of the function if it's not an ISR. (no 'return' instruction generated for an ISR)
- #176~178: output the ".fcall" list that indicates the function calling happened.

8.3.3 Assembly code output for CALL and GOTO

A few P-code assembly code outputs, discussed in this section, are generated within the dispatching panel since the operations are simple, as illustrated in Code-8.14.

```
208              case P_SRC_CODE:     // source code insertion
209                  if ( curFunc )
210                  {
211                      srcCode = ip0->val.s;
212                      ASM_OUTP("; :: "); ASM_OUTP(srcCode); ASM_OUTP("\n");
213                  }
214                  break;
215
216              case LABEL:          // label
217                  if ( curFunc )
218                  {
219                      regPCLATH->load(ip0->val.s);
220                      regBSR->reset();
221                      regWREG->reset();
222                      regFSR0->reset();
223                  }
224                  ASM_LABL(ip0->val.s);
225                  break;
226
227              case CALL:
228                  call(ip0->val.s);
229                  regBSR->reset();
230                  regWREG->reset();
231                  regFSR0->reset();
232                  break;
233
234              case GOTO:
235                  regPCLATH->load(ip0->val.s);
236                  asm16e->code(_GOTO, ip0->val.s);
237                  break;
```
/p16ecc/cc_source_8.2/p16e/pic16e.cpp   Code-8.14

- #208~214: assembly code output for source code P-code {P_SRC_CODE, [*str*]}.
- #216~225: assembly code output for a label P-code {LABEL, [*lbl*]}. Note, the core registers' tracking should be reset.
- #227~232: assembly code output for a (function) call P-code {CALL, [*fname*]}. Note that the core registers' tracking should be reset after the call.
- #234~237: assembly code output for a goto P-code {GOTO, [*lbl*]}.

The following table, Table-8.10, shows the test results for assembly code generation for CALL P-code.

Source code (ch8_t5.c)	Assembly code output (ch8_t5.asm)

<pre>void func();
 	
void foo()
{
    func();
}</pre>
	<pre> .segment CODE2 (REL) foo:0
	foo::
	; :: ch8_t5.c #5: func();
	        .psel foo, func
	        call func
	        .psel func, foo
	        return
	; function(s) called::
	        .fcall foo, func
	
	        .end</pre>

Table-8.10

8.4 Assembly Code Generation (2)

<u>Project directory</u>: /p16ecc/cc_source_8.3
<u>New source files</u>: pic16e2.cpp
<u>Modified source files</u>: pic16e.cpp

This section addresses the assembly code output for P-code that are used most frequently used – assignment:

    {'=', [*X, Y*]}   and
    {P_MOV, [*X, Y*]}

These two P-codes are identical in operation; the only difference is that $\{P_MOV, [X, Y]\}$ will not be affected by optimization.

The assembly code generation for the P-codes is also initiated from `PIC16E::outputASM0()`, and is done in the function of `PIC16E::mov()`.

8.4.1 Assembly code output for constant assignment

Assembly code for constant assignment is done in function `PIC16E::movImmd()`, see Code-8.15~Code-8.16.

```
18   void PIC16E :: mov(Item *ip0, Item *ip1)
19   {
       . . .
23       if ( ip1->type == CON_ITEM || ip1->type == IMMD_ITEM || ip1->type == LBL_ITEM )
24       {
25           movImmd(ip0, ip1);
26           return;
27       }
```
/p16ecc/cc_source_8.3/p16e/pic16e2.cpp   Code-8.15

```
81   void PIC16E :: movImmd(Item *ip0, Item *ip1)
82   {
83       int size0 = ip0->acceSize();
84       char *indf0 = setFSR0(ip0);
85
86       for (int i = 0; i < size0; i++)
87       {
88           if ( ip1->type == CON_ITEM )      // set a solid constant
89           {
90               int n = ip1->val.i >> (i*8);
91               if ( n & 0xff )
92               {
93                   regWREG->load(n & 0xff);
94                   if ( indf0 && (i+1) < size0 ) {
95                       ASM_CODE( _MOVWI, "INDF0++");
96                       regFSR0->inc(1);
97                   }
98                   else if ( !isWREG(ip0) )
99                       ASM_CODE( _MOVWF, acceItem(ip0, i, indf0));
100              }
101              else if ( indf0 && (i+1) < size0 )
102              {
103                  regWREG->load(0);
104                  ASM_CODE( _MOVWI, "INDF0++");
105                  regFSR0->inc(1);
106              }
107              else
108              {
109                  ASM_CODE( _CLRF, acceItem(ip0, i, indf0));
110              }
111          }
112          else if ( i < 2 )                  // set immediate value
113          {
114              ASM_CODE( _MOVLW, acceItem(ip1, i));
115              regWREG->reset();   // reset WREG unconditionally!
116                                  // (don't know real value)
117              if ( isWREG(ip0) )
118                  continue;
119
120              ASM_CODE( _MOVWF, acceItem(ip0, i, indf0));
121          }
122          else                               // fill zero value
123              ASM_CODE( _CLRF, acceItem(ip0, i, indf0));
124      }
125  }
```
/p16ecc/cc_source_8.3/p16e/pic16e2.cpp   Code-8.16

- #88~111: assembly code output for a pure constant assignment.
- #112~123: assembly code output for a label/address assignment.

8.4.2 Assembly code output for indirect assignment

When both X and Y are accessed indirectly, the efficiency of the assembly code will be affected. The assembly code output is implemented in `PIC16E:: movIndir()`, see Code-8.17~Code-8.18.

```
18    void PIC16E :: mov(Item *ip0, Item *ip1)
19    {
      . . .
29        if ( useFSR(ip0, ip1) )
30        {
31            movIndir(ip0, ip1);
32            return;
33        }
```
/p16ecc/cc_source_8.3/p16e/pic16e2.cpp   Code-8.17

```
127   void PIC16E :: movIndir(Item *ip0, Item *ip1)
128   {
129       int size0 = ip0->acceSize();
130       int size1 = ip1->acceSize();
131
132       if ( size0 == 1 || size1 == 1 )
133       {
134           char *indf = setFSR0(ip1);
135           fetch(ip1, 0, indf);
136           ASM_CODE(_MOVWF, ACC0);
137           indf = setFSR0(ip0);
138           ASM_CODE(_MOVF, ACC0, _W_);
139           store(ip0, 0, indf);
140           regWREG->reset();
141
142           for (int i = 1; i < size0; i++)
143           {
144               if ( !SIGNED(ip1) && (i+1) == size0 )
145               {
146                   ASM_CODE(_CLRF, INDF0);
147                   return;
148               }
149               if ( i == 1 )
150               {
151                   regWREG->load(0);
152                   if ( SIGNED(ip1) )
153                   {
154                       ASM_CODE(_BTFSC, ACC0, 7);
155                       ASM_CODE(_MOVLW, 0xff);
156                   }
157               }
158               store(ip0, i, indf);
159           }
160           return;
161       }
162
163       if ( size0 <= 4 && size1 <= 4 )
164       {
165           Item *acc = storeToACC(ip1, size0);
166           mov(ip0, acc);
167           delete acc;
168           return;
169       }
170
171       movBlock(ip0, ip1);
172   }
```
/p16ecc/cc_source_8.3/p16e/pic16e2.cpp   Code-8.18

- #131~161: only move 1 byte, then only FSR0 is used for both X and Y access.
- #163~169: no more than 4 bytes moved, replace it with the operation
        { '=' , [ACC, Y]}
        { '=' , [X, ACC]}
- #171: otherwise, use `PIC16E::movBlock()` for the assembly code generation (see the details in the next section).

8.4.3 Assembly code output for a block data assignment

When the assignment operation size exceeds 4 bytes, it's done using `PIC16E::movBlock()`, illustrated in Code-8.19~Code-8.20, block data assignment or move. In this case, both FSR0 and FSR1 register pairs will be used. The operation is performed in a loop, with ACC used as the counter.

```
18    void PIC16E :: mov(Item *ip0, Item *ip1)
19    {
 . . .
37        if ( size0 > 4 && size1 > 4 )
38        {
39            movBlock(ip0, ip1);
40            return;
41        }
```
/p16ecc/cc_source_8.3/p16e/pic16e2.cpp   Code-8.19

```
174   void PIC16E :: movBlock(Item *ip0, Item *ip1)
175   {
176       int size0 = ip0->acceSize();
177       int n = -size0;
178       if ( size0 == 0 ) return;
179
180       call((char*)"_saveFSR1");
181       regWREG->reset();
182
183       setFSR(ip1, 1);
184       setFSR(ip0, 0);
185       Item *lbl = lblItem(pcoder->getLbl());
186
187       ASM_CODE(_MOVLW, n & 0xff);
188       ASM_CODE(_MOVWF, 0x70);
189       if ( size0 >= 256 )
190       {
191           ASM_CODE(_MOVLW, (n>>8) & 0xff);
192           ASM_CODE(_MOVWF, 0x71);
193       }
194       ASM_LABL(lbl->val.s);
195       ASM_CODE(_MOVIW, "INDF1++");
196       ASM_CODE(_MOVWI, "INDF0++");
197       ASM_CODE(_INCFSZ, 0x70, _F_);
198       if ( size0 >= 256 )
199       {
200           ASM_CODE(_BRA, lbl->val.s);
201           ASM_CODE(_INCFSZ, 0x71, _F_);
202       }
203       ASM_CODE(_BRA, lbl->val.s);
204       delete lbl;
205       call((char*)"_restoreFSR1");
206       regFSR0->reset();
207   }
```
/p16ecc/cc_source_8.3/p16e/pic16e2.cpp   Code-8.20

- #180: save the contents of FSR1.
- #183~184: load FSR0 and FSR1 with the addresses of X and Y.
- #185: generate a label for looping.
- #187~193: load ACC for looping counter.
- #205: restore the contents of FSR1.

8.4.4 Assembly code output for simple assignment

Assembly code output for the simple assignment is shown in Code-8.21. There is one issue that should be addressed when size(X) > size(Y).

```cpp
18    void PIC16E :: mov(Item *ip0, Item *ip1)
19    {
  . . .
43        char *indf0 = setFSR0(ip0);
44        char *indf1 = setFSR0(ip1);
45        for(int i = 0; i < size0; i++)
46        {
47            if ( i < size1 )
48            {
49                if ( overlap(ip0, ip1) )
50                    continue;
51
52                fetch(ip1, i, indf1);
53
54                if ( isWREG(ip0) )       // is WREG target?
55                    return;
56
57                ASM_CODE(_MOVWF, acceItem(ip0, i, indf0));
58            }
59            else if ( !SIGNED(ip1) && (indf0 == NULL || size0 == (size1+1)) )
60                ASM_CODE(_CLRF, acceItem(ip0, i, indf0));
61            else
62            {
63                if ( i == size1 )
64                {
65                    if ( SIGNED(ip1) )
66                    {
67                        if ( overlap(ip0, ip1) )
68                            fetch(ip1, size1-1, indf1);
69
70                        EXTEND_WREG;
71                    }
72                    else
73                        regWREG->load(0);
74                }
75
76                ASM_CODE(_MOVWF, acceItem(ip0, i, indf0));
77            }
78        }
```

`/p16ecc/cc_source_8.3/p16e/pic16e2.cpp` Code-8.21

- #49~50: for the case that X and Y are overlapped.
- #59~77: assembly code output when size(X) > size(Y): fill up the remaining bytes with '0', or the sign of Y.

The following tables, Table-8.11~Table-8.14, show the test results of assembly code generation.

Source code (ch8_t6.c)	Assembly code output (ch8_t6.asm)

<pre>void foo(char *p1, char *p2)
{
  *p1 = *p2;
}</pre>
	<pre>; :: ch8_t6.c #3: *p1 = *p2;
	    .bsel 0, foo_$data$
	    movf foo_$_p2, W
	    movwf FSR0L
	    movf foo_$_p2+1, W
	    movwf FSR0H
	    moviw 0[INDF0]
	    movwf 0x70
	    movf foo_$_p1, W
	    movwf FSR0L
	    movf foo_$_p1+1, W
	    movwf FSR0H
	    movf 0x70, W
	    movwi 0[INDF0]
	    .psel _copyPar, foo
	    return</pre>

Table-8.11

Source code (ch8_t7.c)	Assembly code output (ch8_t7.asm)

<pre>typedef struct {
  int x[4];
  int y[4];</pre>
	<pre> .segment BANKi (REL)
	a:: .rs 16
	    .segment BANKi (REL)</pre>

	b:: .rs 16
| ```
} Data;

Data a, b;

void foo()
{
  a = b;
}
``` | ```
        .segment    CODE2 (REL) foo:0
foo::
; :: ch8_t7.c #10: a = b;
        .psel  foo, _saveFSR1
        call   _saveFSR1
        movlw  (b)
        movwf  FSR1L
        movlw  (b)>>8
        movwf  FSR1H
        movlw  (a)
        movwf  FSR0L
        movlw  (a)>>8
        movwf  FSR0H
        movlw  240
        movwf  112
_$L2:
        moviw  INDF1++
        movwi  INDF0++
        incfsz 112, F
        bra    _$L2
        .psel  _saveFSR1, _restoreFSR1
        call   _restoreFSR1
        .psel  _restoreFSR1, foo
        return
``` |

Table-8.12

Source code (ch8_t8.c)	Assembly code output (ch8_t8.asm)
| ```
void foo()
{
  long x;
  unsigned int y;

  x = y;
}
``` | ```
        .segment    CODE2 (REL) foo:6
foo_$1_x:      .equ  foo_$data$+0
foo_$2_y:      .equ  foo_$data$+4
foo::
; :: ch8_t8.c #6: x = y;
        .bsel  foo_$data$
        movf   foo_$2_y, W
        movwf  foo_$1_x
        movf   foo_$2_y+1, W
        movwf  foo_$1_x+1
        clrf   foo_$1_x+2
        clrf   foo_$1_x+3
        return
``` |

Table-8.13

Source code (ch8_t9.c)	Assembly code output (ch8_t9.asm)
| ```
void foo()
{
  long x;
  int y;

  x = y;
}
``` | ```
        .segment    CODE2 (REL) foo:6
foo_$1_x:      .equ  foo_$data$+0
foo_$2_y:      .equ  foo_$data$+4
foo::
; :: ch8_t9.c #6: x = y;
        .bsel  foo_$data$
        movf   foo_$2_y, W
        movwf  foo_$1_x
        movf   foo_$2_y+1, W
        movwf  foo_$1_x+1
        andlw  128
        btfss  3, 2
        movlw  255
        movwf  foo_$1_x+2
        movwf  foo_$1_x+3
        return
``` |

Table-8.14

8.5 Assembly Code Generation (3)

Project directory: /p16ecc/cc_source_8.4
New source files: pic16e3.cpp
Modified source files: pic16e.cpp

This section addresses the assembly code output for some unary operation P-codes. As the same as that in the last section, the operations are initiated from `PIC16E::outputASM0()`, as shown in Code-8.22.

```
181   void PIC16E :: outputASM0(Pnode *plist)
182   {
183       for(; plist; plist = plist->next)
184       {
185           curPnode = plist;
186           int pcode = plist->type;
187           Item *ip0 = plist->items[0];
188           Item *ip1 = plist->items[1];
189           Item *ip2 = plist->items[2];
190
191           switch ( pcode )
192           {
          .   .   .
243               case INC_OP:    case DEC_OP:
244                   incValue(ip0, (pcode == INC_OP)? ip1->val.i: -ip1->val.i);
245                   break;
246
247               case NEG_OF:
248                   neg(ip0, ip1);
249                   break;
250
251               case '~':
252                   compl1(ip0, ip1);
253                   break;
254
255               case '!':
256                   notop(ip0, ip1);
257                   break;
```
/p16ecc/cc_source_8.4/p16e/pic16e.cpp   Code-8.22

- #243~257: start assembly code output for unary operations.

8.5.1 Assembly code output for INC_OP and DEC_OP

The P-code forms for `INC_OP` and `DEC_OP` are

$$\{INC_OP, [X, N]\} \quad \text{and}$$
$$\{DEC_OP, [X, N]\} \quad \rightarrow \quad \{INC_OP, [X, -N]\}$$

Where N is a constant item (`CON_ITEM`). The assembly code output is illustrated as in Code-8.23.

```
18    void PIC16E :: incValue(Item *ip, int value)
19    {
20        int  size0 = ip->acceSize();
21        char *indf = setFSR0(ip);
22
23        if ( value == 1 && !(ip->type == INDIR_ITEM || ip->type == PID_ITEM) )
24        {
25            for (int i = 0; i < size0; i++)
26            {
27                char *ds = acceItem(ip, i, indf);
28                if ( i ) ASM_CODE(_BTFSC, aSTATUS, 2);
29                ASM_CODE(_INCF, regWREG->reset(ds), _F_);
30            }
31        return;
32    }
```

```cpp
33        bool started = false;
34        for (int i = 0; i < size0; i++, value >>= 8)
35        {
36            int v = value & 0xff;
37            if ( (v == 1 || v == 0xff) && !started && (i+1) == size0 )  // last byte
38            {
39                if ( v == 1 )
40                    ASM_CODE(_INCF, acceItem(ip, i, indf), _F_);
41                else
42                    ASM_CODE(_DECF, acceItem(ip, i, indf), _F_);
43            }
44            else if ( v != 0 || started )                        // skip lower '0' value bytes
45            {
46                regWREG->load(v);
47                if ( started )
48                    ASM_CODE(_ADDWFC, acceItem(ip, i, indf), _F_);
49                else
50                    ASM_CODE(_ADDWF,  acceItem(ip, i, indf), _F_);
51                started = true;
52            }
53        }
54    }
```

`/p16ecc/cc_source_8.4/p16e/pic16e3.cpp`   Code-8.23

- #23~32: when N is '1' and X is a simple item (direct access item), use instruction INCF only.
- #33: boolean flag `started` is for skipping the lower 0x00 bytes of N.
- #37~43: when N has 0x01 or 0xFF byte in the highest byte and 0x00 for the lower bytes.

Table-8.15 shows the test result of assembly code generation for the section.

Source code (ch8_t10.c)	Assembly code output (ch8_t10.asm)

<pre>long *p;
int n, m;
 	
void foo()
{
  *p += 0xff000000;
  n += 1;
  m -= 0xff00;
}</pre>
	<pre> .segment CODE2 (REL) foo:0
	foo::
	; :: ch8_t10.c #6: *p += 0xff000000;
	    .bsel p
	    movf p, W
	    movwf FSR0L
	    .bsel p, p+1
	    movf p+1, W
	    movwf FSR0H
	    addfsr 0, 3
	    decf INDF0, F
	; :: ch8_t10.c #7: n += 1;
	    .bsel p+1, n
	    incf n, F
	    .bsel n, n+1
	    btfsc 3, 2
	    incf n+1, F
	; :: ch8_t10.c #8: m -= 0xff00;
	    .bsel n+1, m+1
	    incf m+1, F
	    return</pre>

Table-8.15

8.5.2 Assembly code output for NEG_OF

NEG_OF P-code has the format

$$\{NEG_OF, [X, Y]\}$$

Mathematically, NEG_OF has the operation, $X = 0 - Y$ (2's complement of Y). The assembly code output is shown in Code-8.24a and Code-8.24b.

```cpp
 96    void PIC16E :: neg(Item *ip0, Item *ip1)
 97    {
 98        if ( same(ip0, ip1) )
 99        {
100            compl2(ip0);
101            return;
102        }
103
104        int size0 = ip0->acceSize();
105        int size1 = ip1->acceSize();
106        char *indf0, *indf1;
107        if ( size0 == 1 )
108        {
109            if ( CONST_ID_ITEM(ip1) )
110            {
111                indf0 = setFSR0(ip0);
112                fetch(ip1, 0, NULL);
113                ASM_CODE(_SUBLW, 0);
114            }
115            else if ( useFSR(ip0, ip1) )
116            {
117                indf1 = setFSR0(ip1);
118                ASM_CODE(_COMF, acceItem(ip1, 0, indf1), _W_);
119                ASM_CODE(_MOVWF, ACC0);
120                indf0 = setFSR0(ip0);
121                ASM_CODE(_INCF, ACC0, _W_);
122            }
123            else
124            {
125                indf1 = setFSR0(ip1);
126                indf0 = setFSR0(ip0);
127                ASM_CODE(_COMF, acceItem(ip1, 0, indf1), _W_);
128                ASM_CODE(_ADDLW, 1);
129            }
130            store(ip0, 0, indf0);
131            regWREG->reset();
132            return;
133        }
134
135        if ( useFSR(ip0, ip1) )
136        {
137            mov(ip0, ip1);
138            compl2(ip0);
139            return;
140        }
```
/p16ecc/cc_source_8.4/p16e/pic16e3.cpp   Code-8.24a

- #98~102: when X and Y are the same, call `PIC16E::compl2()` for self 2's complement.
- #107~133: when X's length is only one byte.
- #135~140: when both X and Y are indirect access items, the operation of assembly code generation becomes
        {'=', [X, Y]}
        {NEG_OF, [X]}

```cpp
142        indf0 = setFSR0(ip0);
143        indf1 = setFSR0(ip1);
144        for(int i = 0; i < size0; i++)
145        {
146            if ( i < size1 )
147            {
148                fetch(ip1, i, indf1);
149                if ( i == 0 )
150                    ASM_CODE(_SUBWF, ZERO_LOC, _W_);
151                else
152                    ASM_CODE(_SUBWFB, ZERO_LOC, _W_);
153            }
154            else if ( i == size1 )
155            {
156                if ( SIGNED(ip1) )
157                    EXTEND_WREG;
158                else
159                {
160                    ASM_CODE(_MOVLW, 0);
161                    ASM_CODE(_SUBWFB, ZERO_LOC, _W_);
162                }
163            }
164            store(ip0, i, indf0);
165            regWREG->reset();
166        }
```
/p16ecc/cc_source_8.4/p16e/pic16e3.cpp   Code-8.24b

- #144~166: Otherwise, the assembly code generation is based on $X = 0 - Y$.

Table-8.16 shows the test result of assembly code generation for the section.

Source code (ch8_t11.c)	Assembly code output (ch8_t11.asm)
<pre>int x, y;

void foo()
{
  x = -x;
  x = -y;
}</pre> | <pre>   .segment   CODE2 (REL) foo:0
foo::
; :: ch8_t11.c #5: x = -x;
        .bsel   x
        comf   x, F
        movlw 1
        addwf x, F
        .bsel   x, x+1
        comf   x+1, F
        movlw 0
        addwfc x+1, F
; :: ch8_t11.c #6: x = -y;
        .bsel   x+1, y
        movf   y, W
        subwf 127, W
        .bsel   y, x
        movwf x
        .bsel   x, y+1
        movf   y+1, W
        subwfb 127, W
        .bsel   y+1, x+1
        movwf x+1
        return</pre> |

Table-8.16

8.5.3 Assembly code output for '~'

The operation of '~' is 1's complement, and it has the P-code format

$$\{ \text{'~'}, [X, Y] \}$$

The assembly code generation of '~' is similar (but simpler) to that of NEG_OF, as shown in Code-8.25.

```
180   void PIC16E :: compl1(Item *ip0, Item *ip1)
181   {
182       if ( same(ip0, ip1) )
183       {
184           compl1(ip0);
185           return;
186       }
187
188       int size0 = ip0->acceSize();
189       int size1 = ip1->acceSize();
190       char *indf0, *indf1;
191       if ( size0 == 1 )
192       {
193           if ( CONST_ID_ITEM(ip1) )
194           {
195               indf0 = setFSR0(ip0);
196               fetch(ip1, 0, NULL);
197               ASM_CODE(_XORLW, 0xff);
198           }
199           else if ( useFSR(ip0, ip1) )
200           {
201               indf1 = setFSR0(ip1);
202               ASM_CODE(_COMF, acceItem(ip1, 0, indf1), _W_);
203               ASM_CODE(_MOVWF, ACC0);
204               indf0 = setFSR0(ip0);
205               ASM_CODE(_MOVF, ACC0, _W_);
206           }
```

209

```
207             else
208             {
209                 indf1 = setFSR0(ip1);
210                 indf0 = setFSR0(ip0);
211                 ASM_CODE(_COMF, acceItem(ip1, 0, indf1), _W_);
212             }
213             store(ip0, 0, indf0);
214             regWREG->reset();
215             return;
216         }
217
218         if ( useFSR(ip0, ip1) )
219         {
220             mov(ip0, ip1);
221             compl1(ip0);
222             return;
223         }
224
225         indf0 = setFSR0(ip0);
226         indf1 = setFSR0(ip1);
227         for(int i = 0; i < size0; i++)
228         {
229             if ( i < size1 )
230             {
231                 if ( CONST_ID_ITEM(ip1) )
232                 {
233                     fetch(ip1, i, NULL);
234                     ASM_CODE(_XORLW, 0xff);
235                 }
236                 else
237                     ASM_CODE(_COMF, acceItem(ip1, i, indf1), _W_);
238             }
239             else if ( i == size1 )
240                 ASM_CODE(_MOVLW, 0xff);
241
242             store(ip0, i, indf0);
243             regWREG->reset();
244         }
245     }
```

`/p16ecc/cc_source_8.4/p16e/pic16e3.cpp`

Code-8.25

Table-8.17 shows the test results for the section.

Source code (ch8_t12.c)	Assembly code output (ch8_t12.asm)

`int x, y;`
 	
`void foo()`
`{`
`  x = ~x;`
`  x = ~y;`
`}`
	`        .segment    CODE2 (REL) foo:0`
	`foo::`
	`; :: ch8_t12.c #5: x = ~x;`
	`        .bsel  x`
	`        comf   x, F`
	`        .bsel  x, x+1`
	`        comf   x+1, F`
	`; :: ch8_t12.c #6: x = ~y;`
	`        .bsel  x+1, y`
	`        comf   y, W`
	`        .bsel  y, x`
	`        movwf  x`
	`        .bsel  x, y+1`
	`        comf   y+1, W`
	`        .bsel  y+1, x+1`
	`        movwf  x+1`
	`        return`

Table-8.17

8.5.4 Assembly code output for '!'

'!' P-code has the format

210

$$\{`!\,', [X, Y]\}$$

Item X will be assigned with '1' or '0', depending on Y is zero or not. In the generation of assembly code, accumulator ACC0 is used to hold the assignment value for X when evaluating Y. Code-8.26 shows the process.

```
247   void PIC16E :: notop(Item *ip0, Item *ip1)
248   {
249       int size1 = ip1->acceSize();
250       char *indf= setFSR0(ip1);
251
252       if ( CONST_ID_ITEM(ip1) )
253       {
254           ASM_CODE(_CLRF, ACC0);
255           ASM_CODE(_CLRF, ACC1);
256           for (int i = 0; i < size1; i++)
257           {
258               fetch(ip1, i, indf);
259               ASM_CODE(_IORWF, ACC1, _F_);
260           }
261           ASM_CODE(_BTFSC, STATUS, 2);
262           ASM_CODE(_INCF, ACC0, _F_);
263       }
264       else
265       {
266           fetch(ip1, 0, indf);
267           for(int i = 1; i < size1; i++)
268               ASM_CODE(_IORWF, acceItem(ip1, i, indf), _W_);
269
270           ASM_CODE(_CLRF, ACC0);
271           ASM_CODE(_ADDLW, 0);
272           ASM_CODE(_BTFSC, STATUS, 2);
273           ASM_CODE(_INCF, ACC0, _F_);
274       }
275
276       attrib *attr = newAttr(CHAR);
277       attr->isUnsigned = 1;
278       Item *acc = accItem(attr);
279       mov(ip0, acc);
280       delete acc;
281       regWREG->reset();
282   }
```
/p16ecc/cc_source_8.4/p16e/pic16e3.cpp   Code-8.26

- #252~263: in the case when Y is a variable stored in ROM, logic OR for all the bytes of Y and store the result in ACC1. Then determine the value in ACC0 depending on the result (#261~262).
- #265~274: otherwise, logic OR for all the bytes of Y and store the result in WREG. Then determine the value in ACC0 depending on the result (#270~273).
- #276~279: assign ACC0 to X.

Table-8.18 shows the test result of assembly code generation for the section.

Source code (ch8_t13.c)	Assembly code output (ch8_t13.asm)

<pre>char x;
int y;
 	
void foo()
{
  x = !y;
}</pre>
	<pre> .segment CODE2 (REL) foo:0
	foo::
	; :: ch8_t13.c #6: x = !y;
	    .bsel y
	    movf y, W
	    .bsel y, y+1
	    iorwf y+1, W
	    clrf 0x70
	    addlw 0
	    btfsc STATUS, 2
	    incf 0x70, F
	    movf 112, W
	    .bsel y+1, x
	    movwf x
	    return</pre>

Table-8.18

8.6 Assembly Code Generation (4)

<u>Project directory</u>: /p16ecc/cc_source_8.5
<u>New source files</u>: pic16e4.cpp
<u>Modified source files</u>: pic16e.cpp

This section presents the assembly code output for both the math and logic compound assignment P-codes. As the same as before, the operation starts in `PIC16E::outputASM0()`, as shown in Code-8.27.

```
181    void PIC16E :: outputASM0(Pnode *plist)
182    {
183        for(; plist; plist = plist->next)
184        {
185            curPnode = plist;
186            int pcode = plist->type;
187            Item *ip0 = plist->items[0];
188            Item *ip1 = plist->items[1];
189            Item *ip2 = plist->items[2];
190
191            switch ( pcode )
192            {
          . . .
259                case ADD_ASSIGN:
260                case SUB_ASSIGN:
261                    addsub(pcode, ip0, ip1);
262                    break;
263
264                case AND_ASSIGN:
265                case OR_ASSIGN:
266                case XOR_ASSIGN:
267                    andorxor(pcode, ip0, ip1);
268                    break;
269            }
270        }
```
/p16ecc/cc_source_8.5/p16e/pic16e.cpp   Code-8.27

8.6.1 Assembly code output for math compound assignments

The P-code has the format

$$\{ADD_ASSIGN, [X, Y]\} \qquad \text{or}$$
$$\{SUB_ASSIGN, [X, Y]\}$$

If Y is a constant, then the operation of assembly code output will be the same as that of `INC_OP` or `DEC_OP`, as shown in Code-8.28.

```
20    void PIC16E :: addsub(int code, Item *ip0, Item *ip1)
21    {
22        if ( ip1->type == CON_ITEM )
23        {
24            int n = (code == ADD_ASSIGN)? ip1->val.i: -ip1->val.i;
25            incValue(ip0, n);
26            return;
27        }
```
/p16ecc/cc_source_8.5/p16e/pic16e4.cpp   Code-8.28

If both X and Y are in indirect access mode, then the operation would become

$$\{\,`='\,, [ACC, Y]\}$$
$$\{ADD_ASSIGN \text{ or } SUB_ASSIGN, [X, ACC]\}$$

Code-8.29 shows the operation.

```
20     void PIC16E :: addsub(int code, Item *ip0, Item *ip1)
21     {
 .  .  .
29         int size0 = ip0->acceSize();
30         int size1 = ip1->acceSize();
31         if ( useFSR(ip0, ip1) )
32         {
33             Item *acc = storeToACC(ip1, size0);
34             addsub(code, ip0, acc);
35             delete acc;
36             return;
37         }
```
/p16ecc/cc_source_8.5/p16e/pic16e4.cpp   Code-8.29

Otherwise, addition or subtraction of X and Y will take place, byte by byte, as illustrated in Code-8.30.

```
20     void PIC16E :: addsub(int code, Item *ip0, Item *ip1)
21     {
 .  .  .
39         char *inst = (code == ADD_ASSIGN)? _ADDWF: _SUBWF;
40         char *indf0 = setFSR0(ip0);
41         char *indf1 = setFSR0(ip1);
42         for(int i = 0; i < size0; i++)
43         {
44             if ( i < size1 )
45                 fetch(ip1, i, indf1);
46             else if ( !ip1->acceSign() )
47                 regWREG->load(0);
48             else if ( i == size1 )
49                 EXTEND_WREG;
50
51             char *ds = acceItem(ip0, i, indf0);
52             ASM_CODE(inst, regWREG->reset(ds), _F_);
53
54             inst = (code == ADD_ASSIGN)? _ADDWFC: _SUBWFB;
55         }
56     }
```
/p16ecc/cc_source_8.5/p16e/pic16e4.cpp   Code-8.30

8.6.2 Assembly code output for logic compound assignments

The logic compound assignment P-codes have the format

{AND_ASSIGN, [X, Y]}   or
{OR_ASSIGN, [X, Y]}   or
{XOR_ASSIGN, [X, Y]}

Again, when Y is a constant, the assembly code generation is done as illustrated in Code-8.31a~Code-8.31b.

```
58     void PIC16E :: andorxor(int code, Item *ip0, int n)
59     {
 .  .  .
```

```
69          int  size0 = ip0->acceSize();
70          char *indf = setFSR0(ip0);
71          for(int i = 0; i < size0; i++, n >>= 8)
72          {
73              char *inst = NULL;
74              char *s = NULL;
75              char *f = NULL;
76              int  b;
77              if ( selectBit(n, &b) && code == OR_ASSIGN )
78              {
79                  inst = _BSF;
80                  s = acceItem(ip0, i, indf);
81                  f = STRBUF(); sprintf(f, "%d", b);
82              }
83              else if ( selectBit(~n, &b) && code == AND_ASSIGN )
84              {
85                  inst = _BCF;
86                  s = acceItem(ip0, i, indf);
87                  f = STRBUF(); sprintf(f, "%d", b);
88              }
```

/p16ecc/cc_source_8.5/p16e/pic16e4.cpp   Code-8.31a

- #77~82: if the constant byte value is 0x01, 0x02, …, 0x80 for OR_ASSIGN operation, then BSF instruction is used.
- #83~88: if the constant byte value is 0xFE, 0xFD, …, 0x7F for AND_ASSIGN operation, then BCF instruction is used.

```
89          else
90          {
91              switch ( n & 0xff )
92              {
93                  case 0x00:
94                      if ( code == AND_ASSIGN )
95                      {
96                          inst = _CLRF;
97                          s = acceItem(ip0, i, indf);
98                      }
99                      break;
100
101                  case 0xff:
102                      if ( code == AND_ASSIGN )
103                          break;
104                      if ( code == XOR_ASSIGN )
105                      {
106                          inst = _COMF;
107                          s = acceItem(ip0, i, indf);
108                          f = _F_;
109                          break;
110                      }
111                      // fall through...
112                  default:
113                      regWREG->load(n);
114                      inst = (code == AND_ASSIGN)? _ANDWF:
115                             (code == XOR_ASSIGN)? _XORWF: _IORWF;
116                      s = acceItem(ip0, i, indf);
117                      f = _F_;
118              }
119          }
120          if ( inst && s )
121          {
122              ASM_CODE(inst, s, f);
123              regWREG->reset(s);
124          }
125      }
```

/p16ecc/cc_source_8.5/p16e/pic16e4.cpp   Code-8.31b

- #93~99: when the constant byte is 0x00,
  - skip it (no assembly code generated) if the P-code is not OR_ASSIGN or XOR_ASSIGN.
  - use CLRF instruction for AND_ASSIGN.
- #101~110: when the constant byte 0xFF,
  - skip it (no assembly code generated) if the P-code is AND_ASSIGN.
  - use COMF instruction for XOR_ASSIGN.
- #112~117: for other cases, use ANDWF, IORWF or XORWF instruction for assembly code output.

When Y is not a constant, assembly code is generated as illustrated in Code-8.32.

```
128    void PIC16E :: andorxor(int code, Item *ip0, Item *ip1)
129    {

136        int size0 = ip0->acceSize();
137        int size1 = ip1->acceSize();
138        if ( useFSR(ip0, ip1) )
139        {
140            Item *acc = storeToACC(ip1, size0);
141            andorxor(code, ip0, acc);
142            delete acc;
143            return;
144        }
145
146        int sign1 = ip1->acceSign();
147        char *indf0 = setFSR0(ip0);
148        char *indf1 = setFSR0(ip1);
149        char *inst = (code == AND_ASSIGN)? _ANDWF:
150                     (code == XOR_ASSIGN)? _XORWF: _IORWF;
151        for(int i = 0; i < size0; i++)
152        {
153            char *ds;
154
155            if ( i >= size1 && !sign1 && code != AND_ASSIGN )
156                return;
157
158            if ( i < size1 )
159                fetch(ip1, i, indf1);
160            else if ( !sign1 )
161            {
162                ds = acceItem(ip0, i, indf0);
163                asm16e->code(_CLRF, regWREG->reset(ds));
164                continue;
165            }
166            else if ( i == size1 )
167            {
168                EXTEND_WREG;
169            }
170
171            ds = acceItem(ip0, i, indf0);
172            asm16e->code(inst, regWREG->reset(ds), _F_);
173        }
```
/p16ecc/cc_source_8.5/p16e/pic16e4.cpp   Code-8.32

- #136~144: when both X and Y are in indirect access mode, then the operation becomes

    {'=', [ACC, Y]}
    {AND_ASSIGN, [X, ACC]}   or

    {'=', [ACC, Y]}
    {OR_ASSIGN, [X, ACC]}   or

    {'=', [ACC, Y]}
    {XOR_ASSIGN, [X, ACC]}

- #146~173: otherwise, regular logic compound assignment will take place, byte by byte.

Table-8.19 shows the test results for the section.

Source code (ch8_t14.c)	Assembly code output (ch8_t14.asm)

`int x, y;`
 	
`void foo()`
`{`
`    x += 0x0100;`
`  x += 0xff00;`
`  x |= 0x0180;`
`  x ^= 0xff00;`
`  x &= y;`
`}`
	`          .segment    CODE2 (REL) foo:0`
	`foo::`
	`;  :: ch8_t14.c #5: x += 0x0100;`
	`          .bsel   x+1`
	`          incf    x+1, F`
	`;  :: ch8_t14.c #6: x += 0xff00;`
	`          decf    x+1, F`
	`;  :: ch8_t14.c #7: x |= 0x0180;`
	`          .bsel   x+1, x`
	`          bsf     x, 7`
	`          .bsel   x, x+1`

```
        bsf    x+1, 0
; :: ch8_t14.c #8: x ^= 0xff00;
        comf   x+1, F
; :: ch8_t14.c #9: x &= y;
        .bsel  x+1, y
        movf   y, W
        .bsel  y, x
        andwf  x, F
        .bsel  x, y+1
        movf   y+1, W
        .bsel  y+1, x+1
        andwf  x+1, F
        return
```

Table-8.19

8.7 Assembly Code Generation (5)

<u>Project directory</u>: /p16ecc/cc_source_8.6
<u>New source files</u>: pic16e5.cpp
<u>Modified source files</u>: pic16e.cpp

The assembly code output for the following P-codes is presented in this part and is started in
`PIC16E::outputASM0()`, as illustrated in Code-8.33.

```cpp
181   void PIC16E :: outputASM0(Pnode *plist)
182   {
183       for(; plist; plist = plist->next)
184       {
185           curPnode = plist;
186           int pcode = plist->type;
187           Item *ip0 = plist->items[0];
188           Item *ip1 = plist->items[1];
189           Item *ip2 = plist->items[2];
190
191           switch ( pcode )
192           {
  .  .  .
270               case P_JZ:  case P_JNZ: // variable test
271                   jzjnz(pcode, ip0, ip1);
272                   break;
273
274               case P_JBZ: case P_JBNZ:// bit test
275                   jbzjbnz(pcode, ip0, ip1, ip2);
276                   break;
277
278               case P_ARG_PASS:
279                   passarg(ip0, ip1, ip2);
280                   break;
281
282               case P_CALL:
283                   pcall(ip0);
284                   regPCLATH->reset();
285                   regWREG->reset();
286                   regBSR->reset();
287                   regFSR0->reset();
288                   break;
289
290               case P_JZ_INC:  case P_JZ_DEC:  case P_JNZ_INC: case P_JNZ_DEC:
291                   jzjnz_incdec(pcode, ip0, ip1, ip2);
292                   break;
293
294               case '+':
295                   add(ip0, ip1, ip2);
296                   regFSR0->reset(ip0);
297                   break;
298               case '-':
299                   sub(ip0, ip1, ip2);
300                   regFSR0->reset(ip0);
301                   break;
```

/p16ecc/cc_source_8.6/p16e/pic16e.cppCode-8.33

216

P_JZ and P_JNZ P-code have the format

$$\{P_JZ, [X, \textit{label}]\} \quad \text{or}$$
$$\{P_JNZ, [X, \textit{label}]\}$$

The evaluation of X (zero or non-zero) determines whether the jump will happen. ACC will be used to hold the evaluation result (OR all the bytes) if X has multiple bytes in length. Code-8.34 shows the operation for the assembly code generation.

```
18   void PIC16E :: jzjnz(int code, Item *ip0, Item *ip1)
19   {
20       int size0 = ip0->acceSize();
21       char *indf0 = setFSR0(ip0);
22       char *inst = (code == P_JZ)? _BTFSC: _BTFSS;
23       for (int i = 0; i < size0; i++)
24       {
25           if ( i == 0 )
26           {
27               fetch(ip0, i, indf0);
28               if ( CONST_ID_ITEM(ip0) && size0 > 1 )
29                   ASM_CODE(_MOVWF, ACC0);
30           }
31           else if ( CONST_ID_ITEM(ip0) )
32           {
33               fetch(ip0, i, indf0);
34               ASM_CODE(_IORWF, ACC0, _F_);
35           }
36           else
37           {
38               ASM_CODE(_IORWF, acceItem(ip0, i, indf0), _W_);
39               regWREG->reset();
40           }
41       }
42
43       if ( CONST_ID_ITEM(ip0) && size0 == 1 )
44           ASM_CODE(_IORLW, 0);     // trigger Z flag
45
46       regPCLATH->load(ip1->val.s);
47       ASM_CODE(inst, aSTATUS, 2);
48       ASM_CODE(_GOTO, ip1->val.s);
49   }
```
/p16ecc/cc_source_8.6/p16e/pic16e5.cpp   Code-8.34

P_JBZ and P_JBNZ P-code have the format

$$\{P_JBZ, [X, \textit{bit}, \textit{label}]\} \quad \text{or}$$
$$\{P_JBNZ, [X, \textit{bit}, \textit{label}]\}$$

It's similar to that in P_JZ and P_JNZ. Instead, the *bit* position value (zero or non-zero) in X determines whether the jump will take place. Code-8.35 shows the operation for assembly code generation.

```cpp
51  void PIC16E :: jbzjbnz(int code, Item *ip0, Item *ip1, Item *ip2)
52  {
53      int size0 = ip0->acceSize();
54      char *indf0 = setFSR0(ip0);
55      char *inst = (code == P_JBZ)? _BTFSS: _BTFSC;
56      int  offset = ip1->val.i/8;
57      char *des;
58
59      if ( ip1->val.i >= (size0*8) ) return;
60
61      if ( CONST_ID_ITEM(ip0) )
62      {
63          fetch(ip0, offset, NULL);
64          des = WREG;
65      }
66      else if ( indf0 && offset > 0 )
67          des = acceItem(ip0, 0, indf0);
68      else
69          des = acceItem(ip0, offset, indf0);
70
71      regPCLATH->load(ip2->val.s);
72      ASM_CODE(inst, des, ip1->val.i%8);
73      ASM_CODE(_GOTO, ip2->val.s);
74  }
```
`/p16ecc/cc_source_8.6/p16e/pic16e5.cpp`   Code-8.35

Table-8.20 shows the test results of assembly code generation for the section.

Source code (ch8_t15.c)	Assembly code output (ch8_t15.asm)

<pre>int x, y;
 	
void foo()
{
  if (x & (1 << 3))
    y++;
}</pre>
	<pre> .segment CODE2 (REL) foo:0
	foo::
	; :: ch8_t15.c #5: if (x & (1 << 3))
	        .bsel x
	        .psel foo, _$L3
	        btfss x, 3
	        goto _$L3
	; :: ch8_t15.c #6: y++;
	        .bsel x, y
	        incf y, F
	        .bsel y, y+1
	        btfsc 3, 2
	        incf y+1, F
	_$L3:
	        .psel _$L3, foo
	        return</pre>

Table-8.20

8.7.3 Assembly code output for P_ARG_PASS

P-code `P_ARG_PASS` is to pass an argument (or parameter) before calling a function, and it has the format

$$\{P_ARG_PASS, [index, X, total_parameters]\}$$

Where
-   *index* is a constant (0, 1, …, *total_parameters* - 1) item, indicating the sequence index in a parameter list for *X*. And the attribute of the item tells that *X* will be scaled to (target size).
-   *total_parameters* tells the total count of the parameters.

The **cc16e** compiler pushes all parameters onto the stack pointed to by **FSR1** before calling a function. In addition, for a parameter with multiple bytes, the stack-pushing operation starts from the LSB to the MSB. The assembly code generation for `P_ARG_PASS` is shown in Code-8.36 below.

```cpp
76    void PIC16E :: passarg(Item *ip0, Item *ip1, Item *ip2)
77    {
78        int size0 = ip0->acceSize();    // arg size
79        int size1 = ip1->acceSize();    // data size
80        char *indf1 = setFSR0(ip1);
81        for (int i = 0; i < size0; i++)
82        {
83            if ( i < size1 )
84                fetch(ip1, i, indf1);
85            else if ( i == size1 )
86            {
87                if ( ip1->acceSign() )
88                    EXTEND_WREG;
89                else
90                    regWREG->load(0);
91            }
92            ASM_CODE(_MOVWI, "--INDF1");
93        }
94    }
```
`/p16ecc/cc_source_8.6/p16e/pic16e5.cpp`   Code-8.36

- #78: size of the parameter to be passed.
- #79: parameter, X, size.

Table-8.21 shows the test results of assembly code generation for the section.

Source code (ch8_t16.c)	Assembly code output (ch8_t16.asm)

<pre>int func(int x, char y);
 	
void foo()
{
  func(1000, 200);
}</pre>
	<pre> .segment CODE2 (REL) foo:0
	foo::
	; :: ch8_t16.c #5: func(1000, 200);
	        movlw 232
	        movwi --INDF1
	        movlw 3
	        movwi --INDF1
	        movlw 200
	        movwi --INDF1
	        .psel foo, func
	        call func
	        .psel func, foo
	        return</pre>

P-code generated

<pre>F:\p16ecc\tests>cc16e ch8_t16.c
cc16e ... ch8_t16.c
P-CODE: P_FUNC_BEG: foo : {void}
P-CODE: SRC ... ch8_t16.c #5: func(1000, 200);
P-CODE: PASS_ARG   0{int}, 1000{int}, 2{int}
P-CODE: PASS_ARG   1{char}, 200{int}, 2{int}
P-CODE: CALL func
P-CODE: ';'
P-CODE: P_FUNC_END: foo</pre>

Table-8.21

8.7.4 Assembly code output for P_CALL

P_CALL is to make a function call using a function pointer (indirect call). In fact, assembly code is generated by calling the library function _pcall(). This way, the returning address will be saved in the hardware return stack. The assembly code generation for P_CALL is shown in Code-8.37.

For PIC16Fxxxx CPU, the indirect function call is made by assigning the pointer's value to PCLATH (higher byte) and PCL (lower byte) registers. The call occurs when PCL is updated. That means, in _pcall(), PCLATH should be assigned first.

```cpp
 96    void PIC16E :: pcall(Item *ip)
 97    {
 98        char *indf = setFSR0(ip);
 99        fetch(ip, 0, indf); // Lo byte of pointer -> stack
100        ASM_CODE(_MOVWI, "--INDF1");
101        fetch(ip, 1, indf); // Hi byte of pointer -> WREG
102        // system lib function call...
103        call((char*)"_pcall");
104    }
```
`/p16ecc/cc_source_8.6/p16e/pic16e5.cpp`

Code-8.37

- #99~100: lower byte of the function pointer (pushed into the stack).
- #101: higher byte of the function pointer (in WREG).

Table-8.22 shows the test result of assembly code generation for the section.

Source code (ch8_t17.c)	Assembly code output (ch8_t17.asm)

<pre>int (*fptr)(int x, char y);
 	
void foo()
{
  fptr(1000, 200);
}</pre>
	<pre> .segment CODE2 (REL) foo:0
	foo::
	; :: ch8_t17.c #5: fptr(1000, 200);
	    movlw 232
	    movwi --INDF1
	    movlw 3
	    movwi --INDF1
	    movlw 200
	    movwi --INDF1
	    .bsel fptr
	    movf fptr, W
	    movwi --INDF1
	    .bsel fptr, fptr+1
	    movf fptr+1, W
	    .psel foo, _pcall
	    call _pcall
	    .psel foo
	    return</pre>

	P-code generated

	<pre>F:\p16ecc\tests>cc16e ch8_t17.c
	cc16e ... ch8_t17.c
	P-CODE: P_FUNC_BEG: foo : {void}
	P-CODE: SRC ... ch8_t17.c #5: fptr(1000, 200);
	P-CODE: PASS_ARG 0{int}, 1000{int}, 2{int}
	P-CODE: PASS_ARG 1{char}, 200{int}, 2{int}
	P-CODE: P_CALL fptr{int}
	P-CODE: ';'
	P-CODE: P_FUNC_END: foo</pre>

Table-8.22

8.7.5 Assembly code output for P_JZ_INC family

It covers four P-codes: `P_JZ_INC`, `P_JNZ_INC`, `P_JZ_DEC` and `P_JNZ_DEC`. Basically, it is the combination of `P_JZ`/`P_JNZ` and `P_INC`/`P_DEC`, and has the format

$$\{\texttt{P_JZ_INC}, [X, N, label]\}$$

In assembly code generation, after evaluating X, it must increment or decrement X by N before making the jump based on the evaluation. Code-8.38 illustrates the process.

220

```cpp
106   void PIC16E :: jzjnz_incdec(int code, Item *ip0, Item *ip1, Item *ip2)
107   {
108       char *indf = setFSR0(ip0);
109       int  size0 = ip0->acceSize();
110       int  n = (code == P_JZ_INC || code == P_JNZ_INC)? ip1->val.i: -ip1->val.i;
111       char *inst = (code == P_JZ_INC || code == P_JZ_DEC)? _BTFSC: _BTFSS;
112
113       if ( size0 == 1 && abs(n) == 1 )
114       {
115           ASM_CODE((n>0)? _INCF: _DECF, acceItem(ip0, 0, indf), _F_);
116           ASM_CODE((n<0)? _INCF: _DECF, acceItem(ip0, 0, indf), _W_);
117       }
118       else
119       {
120           fetch(ip0, 0, indf);
121           for (int i = 1; i < size0; i++)
122           {
123               ASM_CODE(_IORWF, acceItem(ip0, i, indf), _W_);
124           }
125           ASM_CODE(_MOVWF, ACC0);
126           for (int i = 0; i < size0; i++)
127           {
128               regWREG->load(n>>(i*8));
129               ASM_CODE(i? _ADDWFC: _ADDWF, acceItem(ip0, i, indf), _F_);
130           }
131           ASM_CODE(_MOVF, ACC0, _W_);
132       }
133       regWREG->reset();
134       regPCLATH->load(ip2->val.s);
135       ASM_CODE(inst, aSTATUS, 2);
136       ASM_CODE(_GOTO, ip2->val.s);
137   }
```
/p16ecc/cc_source_8.6/p16e/pic16e5.cpp   Code-8.38

- #113~117: in the case that the size of X is one byte and N is 1 or −1.

Table-8.23 shows the test results of assembly code generation for the section.

Source code (ch8_t18.c)	Assembly code output (ch8_t18.asm)

<pre>int x, y;
 	
void foo()
{
  if (x++) y = 0;
}</pre>
	<pre> .segment CODE2 (REL) foo:0
	foo::
	; :: ch8_t18.c #5: if (x++)
	        .bsel x
	        movf x, W
	        .bsel x, x+1
	        iorwf x+1, W
	        movwf 0x70
	        movlw 1
	        .bsel x+1, x
	        addwf x, F
	        movlw 0
	        .bsel x, x+1
	        addwfc x+1, F
	        movf 0x70, W
	        .psel foo, _$L3
	        btfsc 3, 2
	        goto _$L3
	; :: ch8_t18.c #5: y = 0;
	        .bsel x+1, y
	        clrf y
	        .bsel y, y+1
	        clrf y+1
	_$L3:
	        .psel _$L3, foo
	        return</pre>

Table-8.23

'+' P-code has the format

$$\{\,'+'\,,\,[Z, X, Y]\}$$

Which is one of the most common operations. Therefore, the efficiency of the assembly code is essential. Code-8.39 shows how to generate the assembly code for the operation based on the operands X, Y, and Z.

```
139   void PIC16E :: add(Item *ip0, Item *ip1, Item *ip2)
140   {
141       int size0 = ip0->acceSize();
142       int size1 = ip1->acceSize();
143       int size2 = ip2->acceSize();
144
145       if ( overlap(ip0, ip1) )
146       {
147           addsub(ADD_ASSIGN, ip0, ip2);
148           return;
149       }
150       if ( overlap(ip0, ip2) )
151       {
152           addsub(ADD_ASSIGN, ip0, ip1);
153           return;
154       }
155       if ( CONST_ITEM(ip1) && !CONST_ITEM(ip2) )
156       {
157           add(ip0, ip2, ip1);
158           return;
159       }
160       if ( useFSR(ip0, ip1, ip2) )
161       {
162           Item *acc = storeToACC(ip1, size0);
163           addsub(ADD_ASSIGN, acc, ip2);
164           mov(ip0, acc);
165           delete acc;
166           return;
167       }
168       if ( CONST_ID_ITEM(ip1) )
169       {
170           mov(ip0, ip1);
171           addsub(ADD_ASSIGN, ip0, ip2);
172           return;
173       }
174       if ( (size2 < size0 && !CONST_ITEM(ip2) && ip2->acceSign()) ||
175            (size1 < size0 && !CONST_ITEM(ip1) && ip1->acceSign())  )
176       {
177           mov(ip0,                  related(ip0, ip2)? ip2: ip1);
178           addsub(ADD_ASSIGN, ip0, related(ip0, ip2)? ip1: ip2);
179           return;
180       }
181       addsub('+', ip0, ip1, ip2);
182       regFSR0->reset(ip0);
183   }
```

`/p16ecc/cc_source_8.6/p16e/pic16e5.cpp`   Code-8.39

- #145~149: if X and Z overlap (exact data location), the operation can be accomplished as that for $\{\,'+='\,,\,[Z, Y]\}$.
- #150~154: if Y and Z overlap (exact data location), the operation can be accomplished as that for $\{\,'+='\,,\,[Z, X]\}$.
- #155~159: if X is a ROMed data but Y is not, then switch their positions (addition meets exchange rate).
- #160~167: if X, Y, and Z are all in indirect access mode, the operation becomes
$$\{\,'='\,,\,[ACC, X]\}$$
$$\{\,'+='\,,\,[ACC, Y]\}$$
$$\{\,'='\,,\,[Z, ACC]\}$$
- #168~172: if both X and Y are ROMed data, then the operation becomes
$$\{\,'='\,,\,[Z, X]\}$$
$$\{\,'+='\,,\,[Z, Y]\}$$
- #174~180: if Z size bigger than X or Y, and X or Y is a signed data, then the operation also becomes
$$\{\,'='\,,\,[Z, X]\}$$
$$\{\,'+='\,,\,[Z, Y]\}$$

- #181: for the rest of the cases, call `PIC16E::addsub()`, which covers the assembly code generation for both addition and subtraction.

8.7.7 Assembly code output for '-'

'$-$' P-code has the format

$$\{\,'-', [Z, X, Y]\}$$

The assembly code generation for subtraction is similar to that of addition. The significant difference is that operands X and Y are not commutative for subtraction. Code-8.40 shows how to generate the assembly code for the P-code.

```
185   void PIC16E :: sub(Item *ip0, Item *ip1, Item *ip2)
186   {
187       int size0 = ip0->acceSize();
188       int size1 = ip1->acceSize();
189       int size2 = ip2->acceSize();
190
191       if ( same(ip0, ip1) )
192       {
193           addsub(SUB_ASSIGN, ip0, ip2);
194           return;
195       }
196       if ( same(ip0, ip2) )
197       {
198           neg(ip0, ip2);
199           addsub(ADD_ASSIGN, ip0, ip1);
200           return;
201       }
202       if ( useFSR(ip0, ip1, ip2) )
203       {
204           Item *acc = storeToACC(ip1, size0);
205           addsub(SUB_ASSIGN, acc, ip2);
206           mov(ip0, acc);
207           delete acc;
208           return;
209       }
210       if ( CONST_ID_ITEM(ip1) )
211       {
212           mov(ip0, ip1);
213           addsub(SUB_ASSIGN, ip0, ip2);
214           return;
215       }
216       if ( (size2 < size0 && !CONST_ITEM(ip2) && ip2->acceSign()) ||
217            (size1 < size0 && !CONST_ITEM(ip1) && ip1->acceSign()) )
218       {
219           mov(ip0, ip1);
220           addsub(SUB_ASSIGN, ip0, ip2);
221           return;
222       }
223       addsub('-', ip0, ip1, ip2);
224       regFSR0->reset(ip0);
225   }
```
/p16ecc/cc_source_8.6/p16e/pic16e5.cpp   Code-8.40

- #191~195: if X and Z are the same item, the operation can be accomplished as that for $\{\,'-=', [Z, Y]\}$.
- #196~201: if Y and Z are the same item, then the operation becomes
    $$\{NEG_OF, [Z, Y]\}$$
    $$\{\,'+=', [Z, X]\}$$
- 202~209: if X, Y, and Z are all in indirect access mode, the operation becomes
    $$\{\,'=', [ACC, X]\}$$
    $$\{\,'-=', [ACC, Y]\}$$
    $$\{\,'=', [Z, ACC]\}$$
- #210~215: if X is a ROMed data, then the operation becomes
    $$\{\,'=', [Z, X]\}$$
    $$\{\,'-=', [Z, Y]\}$$

- #216~222: if Z size bigger than X or Y, and X or Y is a signed data, then the operation also becomes
$$\{ \,`=', \, [Z, X]\}$$
$$\{ \,`-=', \, [Z, Y]\}$$

Code-8.41 illustrates the function `PIC16E::addsub()` that handles the assembly code generation for both `+` and `-` P-codes.

```
227   void PIC16E :: addsub(int code, Item *ip0, Item *ip1, Item *ip2)
228   {
      . . .
234       int  size0 = ip0->acceSize();
235       int  size1 = ip1->acceSize();
236       int  size2 = ip2->acceSize();
237       char *indf0 = setFSR0(ip0);
238       char *indf1 = useFSR(ip1)? setFSR(ip1, indf0?         1: 0): NULL;
239       char *indf2 = useFSR(ip2)? setFSR(ip2, (indf0||indf1)? 1: 0): NULL;
240       bool started = false;
241       bool ip1_const = CONST_ITEM(ip1);
242       char *inst = (code == '+')? _ADDWF: _SUBWF;
243
244       for (int i = 0; i < size0; i++)
245       {
246           if ( ip2->type == CON_ITEM )
247           {
248               int n = ip2->val.i >> (i*8);
249               if ( (n & 0xff) == 0 && !started )
250               {
251                   if ( indf1 )
252                       ASM_CODE(_MOVIW, INDF_PP(indf1));
253                   else
254                       fetch(ip1, i, NULL);
255
256                   store(ip0, i, indf0);
257                   continue;
258               }
259           }
260
261           if ( ip1_const && started ) {
262               fetch(ip1, i, NULL);
263               if ( i >= size2 ) ASM_CODE(inst, ZERO_LOC, _W_);
264               else              ASM_CODE(inst, acceItem(ip2, i, indf2), _W_);
265               ASM_CODE(_MOVWF, acceItem(ip0, i, indf0));
266           }
267           else {
268               if ( i >= size2 )
269                   ASM_CODE(_CLRW);
270               else
271                   fetch(ip2, i, indf2);
272               if ( ip1_const ) {
273                   inst = (code == '+')? _ADDLW: _SUBLW;
274                   ASM_CODE(inst, acceItem(ip1, i, indf1));
275               }
276               else {
277                   if ( i >= size1 )
278                       ASM_CODE(inst, ZERO_LOC, _W_);
279                   else
280                       ASM_CODE(inst, acceItem(ip1, i, indf1), _W_);
281               }
282               store(ip0, i, indf0);
283           }
284           started = true;
285           regWREG->reset();
286
287           inst = (code == '+')? _ADDWFC: _SUBWFB;
288       }
```
/p16ecc/cc_source_8.6/p16e/pic16e5.cpp

Code-8.41

Table-8.24 shows the test results of assembly code generation for P-code '+' and '−'.

Source code (ch8_t19.c)	Assembly code output (ch8_t19.asm)

<pre>int *p1, *p2;
int x, y;
 	
void foo()
{
  *p1 = *p2 + 1000;
  x = y - x;
}</pre>
	<pre> .segment CODE2 (REL) foo:0
	foo::
	; :: ch8_t19.c #6: *p1 = *p2 + 1000;
	        .psel foo, _saveFSR1
	        call _saveFSR1
	        .bsel p1
	        movf p1, W
	        movwf FSR0L
	        .bsel p1, p1+1
	        movf p1+1, W
	        movwf FSR0H
	        .bsel p1+1, p2
	        movf p2, W
	        movwf FSR1L
	        .bsel p2, p2+1
	        movf p2+1, W
	        movwf FSR1H
	        movlw 232
	        addwf INDF1, W
	        movwi 0[INDF0]
	        movlw 3
	        addfsr 0, 1
	        addwfc INDF1, W
	        movwi 1[INDF0]
	        .psel _saveFSR1, _restoreFSR1
	        call _restoreFSR1
	; :: ch8_t19.c #7: x = y - x;
	        .bsel p2+1, x
	        comf x, F
	        movlw 1
	        addwf x, F
	        .bsel x, x+1
	        comf x+1, F
	        movlw 0
	        addwfc x+1, F
	        .bsel x+1, y
	        movf y, W
	        .bsel y, x
	        addwf x, F
	        .bsel x, y+1
	        movf y+1, W
	        .bsel y+1, x+1
	        addwfc x+1, F
	        .psel _restoreFSR1, foo
	        return</pre>

Table-8.24

8.8 Assembly Code Generation (6)

<u>Project directory</u>: /p16ecc/cc_source_8.7
<u>New source files</u>: pic16e6.cpp
<u>Modified source files</u>: pic16e.cpp

This part talks about the assembly code generation for compare-jump P-codes, which includes all six types of operations

{P_JEQ, [*X, Y, label*]}	if $X == Y$, then jump to *label*
{P_JNE, [*X, Y, label*]}	if $X != Y$, then jump to *label*
{P_JGT, [*X, Y, label*]}	if $X > Y$, then jump to *label*
{P_JGE, [*X, Y, label*]}	if $X >= Y$, then jump to *label*
{P_JLT, [*X, Y, label*]}	if $X < Y$, then jump to *label*
{P_JLE, [*X, Y, label*]}	if $X <= Y$, then jump to *label*

As before, all those operations are initiated from `PIC16E::outputASM0()`, shown in Code-8.42. Here all those six types of P-codes are divided to two groups.

```
181   void PIC16E :: outputASM0(Pnode *plist)
182   {
183       for(; plist; plist = plist->next)
184       {
185           curPnode = plist;
186           int pcode = plist->type;
187           Item *ip0 = plist->items[0];
188           Item *ip1 = plist->items[1];
189           Item *ip2 = plist->items[2];
190
191           switch ( pcode )
192           {
     . . .
303               case P_JEQ: case P_JNE:
304                   jeqjne(pcode, ip0, ip1, ip2);
305                   break;
306
307               case P_JLT: case P_JLE: case P_JGT: case P_JGE:
308                   cmpJump(pcode, ip0, ip1, ip2);
309                   break;
310           }
311       }
```
/p16ecc/cc_source_8.7/p16e/pic16e.cpp   Code-8.42

8.8.1 Assembly code output for P_JEQ and P_JNE

Assembly code generation for P_JEQ and P_JNE is shown in Code-8.43. Basically, the result of $X \wedge Y$, to be stored in ACC, determines if the jump condition is met.

```
18   void PIC16E :: jeqjne(int code, Item *ip0, Item *ip1, Item *ip2)
19   {
20       int size0 = ip0->acceSize();
21       int size1 = ip1->acceSize();
22       int sign1 = ip1->acceSign();
23       int size  = (size0 > size1)? size0: size1;
24       char *inst= (code == P_JEQ)? _BTFSC: _BTFSS;
25
26       if ( useFSR(ip0, ip1) )
27       {
28           Item *acc = storeToACC(ip1, size1);
29           jeqjne(code, ip0, acc, ip2);
30           delete acc;
31           return;
32       }
```

```cpp
63          char *indf0 = setFSR0(ip0);
64          char *indf1 = setFSR0(ip1);
65          for(int i = 0; i < size0; i++)
66          {
67              fetch(ip0, i, indf0);
68              if ( i < size1 )
69              {
70                  ASM_CODE(_XORWF, acceItem(ip1, i, indf1), _W_);
71                  regWREG->reset();
72              }
73              else if ( sign1 )
74              {
75                  ASM_CODE(_BTFSC, acceItem(ip1, size1-1, indf1), 7);
76                  ASM_CODE(_XORLW, 255);
77                  regWREG->reset();
78              }
79
80              if ( i == 0 && size > 1 )
81                  ASM_CODE(_MOVWF, ACC0);
82              else if ( i > 0 )
83                  ASM_CODE(_IORWF, ACC0, _F_);
84          }
85
86          regPCLATH->load(ip2->val.s);
87          ASM_CODE(inst, aSTATUS, 2);
88          ASM_CODE(_GOTO, ip2->val.s);
```

`/p16ecc/cc_source_8.7/p16e/pic16e6.cpp`

Code-8.43

- #26~32: if both *X* and *Y* are in indirect access mode, then the operation becomes

  { '=', [ACC, *X*]}

  {*op-code*, [ACC, *Y*, *label*]}

Table-8.25 shows the test results of assembly code generation for P-code P_JEQ.

Source code (ch8_t20.c)	Assembly code output (ch8_t20.asm)
| ```c
int x, y;

void foo()
{
  if ( x != y )
    x = y;
}
``` | ```
        .segment    CODE2 (REL) foo:0
foo::
; :: ch8_t20.c #5: if ( x != y )
        .bsel  x
        movf   x, W
        .bsel  x, y
        xorwf  y, W
        movwf  0x70
        .bsel  y, x+1
        movf   x+1, W
        .bsel  x+1, y+1
        xorwf  y+1, W
        iorwf  0x70, F
        .psel  foo, _$L3
        btfsc  3, 2
        goto   _$L3
; :: ch8_t20.c #6: x = y;
        .bsel  y+1, y
        movf   y, W
        .bsel  y, x
        movwf  x
        .bsel  x, y+1
        movf   y+1, W
        .bsel  y+1, x+1
        movwf  x+1
_$L3:
        .psel  _$L3, foo
        return
``` |

Table-8.25

Unlike `P_JEQ` and `P_JNE`, the comparison operation in `P_JGT/P_JGE/P_JLT/P_JLE` is based on the result of the subtraction of X and Y. Moreover,

1) The comparing operation can be either 'unsigned comparison' or 'signed comparison'.
2) The PIC16Fxxxx CPU has only Z (zero) and C (carry/borrow) flags, but no overflow flag, in the STATUS register.

This results in the assembly code generation becoming more complicated. The entire process is accomplished in two parts, as shown in Code-8.44 and Code-8.45.

In the first part, it generates assembly code that subtracts Y from X and sets the flags.

```
220   void PIC16E :: cmpJump(int code, Item *ip0, Item *ip1, Item *ip2)
221   {
222        int size0 = ip0->acceSize();
223        int size1 = ip1->acceSize();

        . . .

261        char *indf0 = setFSR0(ip0);
262        char *indf1 = setFSR0(ip1);
263        bool sign0   = ip0->acceSign();
264        bool sign1   = ip1->acceSign();
265        bool signedCmp = sign0 && sign1;
266        char *inst   = _SUBWF;
267
268        if ( indf0 == NULL && CONST_ID_ITEM(ip0) )
269            indf0 = setFSR(ip0, 0);
270
271        for (int i = 0; i < size0; i++, inst = _SUBWFB)
272        {
273            if ( i < size1 )
274                fetch(ip1, i, indf1);
275            else if ( !sign1 )
276                regWREG->load(0);
277            else
278            {
279                regWREG->load(0);
280                ASM_CODE(_BTFSC, acceItem(ip1, size1-1, indf1), 7);
281                ASM_CODE(_MOVLW, 255);
282            }
283            // get sign bit of ip0 ^ ip1 ...
284            if ( (i+1) == size0 && signedCmp )   // MSB of ip0
285            {
286                ASM_CODE(_XORWF, acceItem(ip0, i, indf0), _W_);
287                ASM_CODE(_MOVWI, "--INDF1");
288                ASM_CODE(_XORWF, acceItem(ip0, i, indf0), _W_);
289            }
290
291            ASM_CODE(inst, acceItem(ip0, i, indf0), _W_);
292            regWREG->reset();
293
294            if ( code != P_JLT && code != P_JGE && size0 > 1 )
295            {
296                if ( i ) ASM_CODE(_IORWF, ACC0, _F_);
297                else     ASM_CODE(_MOVWF, ACC0);
298            }
299        }
300        cmpJump(signedCmp, code, ip2);
```

`/p16ecc/cc_source_8.7/p16e/pic16e.cpp`   Code-8.44

- #265: determine if it's a signed comparison.
- #273~282: fetch a byte from Y.
- #283~289: for signed comparison, keep the sign bit (MSbit) of $X \wedge Y$ in the stack (X and Y have the sign or not).
- #291: subtraction of $X - Y$.
- #294~298: set/clear Z flag whether X equals Y (for `P_JLE` and `P_JGT`).

- #300: the second part that determines whether the jump is needed, based on the C and Z flags and other status (see Code-8.45).

In the second part, if either X or Y is an unsigned value, then the C and/or Z flags in the STATUS register will determine whether the jump will happen.

```
408   void PIC16E :: cmpJump(bool signedCmp, int code, Item *ip2)
409   {
410       if ( !signedCmp )
411       {
412           regPCLATH->load(ip2->val.s);
413           switch ( code )
414           {
415               case P_JLE: // jump on (x <= y) : Z = 1 || C = 0
416                   ASM_CODE( _BTFSC, aSTATUS, 2);
417                   ASM_CODE( _BCF, aSTATUS, 0);
418               case P_JLT: // jump on (x < y) : C = 0
419                   ASM_CODE( _BTFSS, aSTATUS, 0);
420                   ASM_CODE( _GOTO, ip2->val.s);
421                   break;
422
423               case P_JGT: // jump on (x > y) : Z = 0 && C = 1
424                   ASM_CODE( _BTFSC, aSTATUS, 2);
425                   ASM_CODE( _BCF, aSTATUS, 0);
426               case P_JGE: // jump on (x >= y) : C = 1
427                   ASM_CODE( _BTFSC, aSTATUS, 0);
428                   ASM_CODE( _GOTO, ip2->val.s);
429                   break;
430           }
431       }
```
`/p16ecc/cc_source_8.7/p16e/pic16e.cpp`   Code-8.45a

- #415~421: for `P_JLE`, jump happens if either Z = 1 or C = 0.
- #418~421: for `P_JLT`, jump happens if C = 0.
- #423~429: for `P_JGT`, jump happens if Z = 0 and C = 1.
- #426~429: for `P_JGE`, jump happens if C = 1.

Note, Z = 1 means $X = Y$; and C = 0 means $X < Y$.

For a signed comparison, things are more complicated. X and Y are in 2's complement format, and both are n-bit in length; then we have $-2^{n-1} \leq X \leq 2^{n-1} - 1$ and $-2^{n-1} \leq Y \leq 2^{n-1} - 1$.

Assume

S_X, S_Y - sign bits (the most significant bit) for X and Y.

$[X]$, $[Y]$ - 2's complement for X and Y.

$$[X] = 2^n - |X| \text{ if } X < 0; \text{ and } [X] = X \text{ if } X \geq 0.$$
$$[Y] = 2^n - |Y| \text{ if } Y < 0; \text{ and } [Y] = Y \text{ if } Y \geq 0.$$

$|X|$, $|Y|$ - absolute values for X and Y.

$$0 < |X| \leq 2^{n-1} \text{ if } X < 0; \; 0 \leq |X| \leq 2^{n-1} - 1 \text{ if } X \geq 0$$
$$0 < |Y| \leq 2^{n-1} \text{ if } Y < 0; \; 0 \leq |Y| \leq 2^{n-1} - 1 \text{ if } Y \geq 0$$

1) If $X \geq 0$ and $Y \geq 0$ ($S_X = 0$ and $S_Y = 0$), then

$$[X] - [Y] = X - Y$$

     $Z = 1$ $\rightarrow X = Y$
     $Z = 0$ and $C = 0$ $\rightarrow X < Y$
     $Z = 0$ and $C = 1$ $\rightarrow X > Y$

2) If $X < 0$ and $Y < 0$ ($S_X = 1$ and $S_Y = 1$), then

$$[X] - [Y] = (2^n - |X|) - (2^n - |Y|) = |Y| - |X|$$

     $Z = 1$ $\rightarrow X = Y$
     $Z = 0$ and $C = 0$ $\rightarrow |Y| < |X| \rightarrow X < Y$

$$Z = 0 \text{ and } C = 1 \qquad \rightarrow |Y| > |X| \rightarrow X > Y$$

3) If $X \geq 0$ and $Y < 0$ ($S_X = 0$ and $S_Y = 1$), then

$$[X] - [Y] = X - (2^n - |Y|) = (|X| + |Y|) - 2^n \qquad Z = 0 \text{ and } C = 0 \qquad \rightarrow X > Y$$

4) If $X < 0$ and $Y \geq 0$ ($S_X = 1$ and $S_Y = 0$), then

$$[X] - [Y] = (2^n - |X|) - |Y| = 2^n - (|X| + |Y|) \qquad Z = 0 \text{ and } C = 1 \qquad \rightarrow X < Y$$

Table-8.26 shows the conclusion about whether a jump condition is met.

P-code		Conditions for jump	
		Native	Equivalent
P_JLE ('<=')	$S_X = S_Y$	$Z = 1$ or $C = 0$	$Z = 1$ or $(S_X \wedge S_Y \wedge C) = 0 \rightarrow (\overline{Z} \mathbin{\&} (S_X \wedge S_Y \wedge C)) = 0$
	$S_X \neq S_Y$	$Z = 1$ or $C = 1$	
P_JGT ('>')	$S_X = S_Y$	$Z = 0$ and $C = 1$	$Z = 0$ and $(S_X \wedge S_Y \wedge C) = 1 \rightarrow (\overline{Z} \mathbin{\&} (S_X \wedge S_Y \wedge C)) = 1$
	$S_X \neq S_Y$	$Z = 0$ and $C = 0$	
P_JLT ('<')	$S_X = S_Y$	$Z = 0$ and $C = 0$	$Z = 0$ and $(S_X \wedge S_Y \wedge C) = 0 \rightarrow (Z \mid (S_X \wedge S_Y \wedge C)) = 0$
	$S_X \neq S_Y$	$Z = 0$ and $C = 1$	
P_JGE ('>=')	$S_X = S_Y$	$Z = 1$ or $C = 1$	$Z = 1$ or $(S_X \wedge S_Y \wedge C) = 1 \rightarrow (Z \mid (S_X \wedge S_Y \wedge C)) = 1$
	$S_X \neq S_Y$	$Z = 1$ or $C = 0$	

Table-8.26

```
432        else
433        {
434            switch ( code )
435            {
436                case P_JLE:
437                    call((char*)"_signedCmpAndZ");   // WREG[0] = ~Z & (Sx ^ Sy ^ C)
438                    regPCLATH->load(ip2->val.s);
439                    ASM_CODE(_BTFSS, aWREG, 0);
440                    ASM_CODE(_GOTO, ip2->val.s);      // jump if WREG[0] = 0
441                    break;
442                case P_JGT:
443                    call((char*)"_signedCmpAndZ");   // WREG[0] = ~Z & (Sx ^ Sy ^ C)
444                    regPCLATH->load(ip2->val.s);
445                    ASM_CODE(_BTFSC, aWREG, 0);
446                    ASM_CODE(_GOTO, ip2->val.s);      // jump if WREG[0] = 1
447                    break;
448                case P_JLT:
449                    call((char*)"_signedCmpOrZ");    // WREG[0] = Z | (Sx ^ Sy ^ C)
450                    regPCLATH->load(ip2->val.s);
451                    ASM_CODE(_BTFSS, aWREG, 0);
452                    ASM_CODE(_GOTO, ip2->val.s);      // jump if WREG[0] = 0
453                    break;
454                case P_JGE:
455                    call((char*)"_signedCmpOrZ");    // WREG[0] = Z | (Sx ^ Sy ^ C)
456                    regPCLATH->load(ip2->val.s);
457                    ASM_CODE(_BTFSC, aWREG, 0);
458                    ASM_CODE(_GOTO, ip2->val.s);      // jump if WREG[0] = 1
459                    break;
460            }
461            regWREG->reset();
462        }
```
`/p16ecc/cc_source_8.7/p16e/pic16e.cpp`
Code-8.45b

- #437 and #443: for P-code P_JLE and P_JGT, a library function _signedCmpAndZ() is called, which does calculation WREG[0] = $(\overline{Z} \mathbin{\&} (S_X \wedge S_Y \wedge C))$.
- #449 and #455: for P-code P_JLT and P_JGE, a library function _signedCmpOrZ() is called, which does calculation WREG[0] = $(Z \mid (S_X \wedge S_Y \wedge C))$.

Table 8.27 shows the test results for assembly code generation of P-code P_JGT.

Source code (ch8_t21.c)	Assembly code output (ch8_t21.asm)
<pre>int x, y;

void foo()
{
   if (x <= y) x = 0;
}</pre> | <pre>   .segment   CODE2 (REL) foo:0
foo::
; :: ch8_t21.c #5: if (x <= y)
        .bsel   y
        movf   y, W
        .bsel   y, x
        subwf   x, W
        movwf   0x70
        .bsel   x, y+1
        movf   y+1, W
        .bsel   y+1, x+1
        xorwf   x+1, W
        movwi   --INDF1
        xorwf   x+1, W
        subwfb x+1, W
        iorwf   0x70, F
        .psel   foo, _signedCmpAndZ
        call   _signedCmpAndZ
        .psel   _signedCmpAndZ, _$L3
        btfsc   9, 0
        goto   _$L3
; :: ch8_t21.c #5: x = 0;
        .bsel   x+1, x
        clrf   x
        .bsel   x, x+1
        clrf   x+1
_$L3:
        .psel   _$L3, foo
        return</pre> |

Table-8.27

8.9 Assembly Code Generation (7)

<u>Project directory</u>: /p16ecc/cc_source_8.8
<u>New source files</u>: pic16e7.cpp
<u>Modified source files</u>: pic16e.cpp

This part illustrates the assembly code output for compound shift P-codes ('`<<=`' and '`>>=`'), which are initiated from `PIC16E::outputASM0()`, as well (see Code-8.46).

```
181   void PIC16E :: outputASM0(Pnode *plist)
182   {
183       for(; plist; plist = plist->next)
184       {
185           curPnode = plist;
186           int pcode = plist->type;
187           Item *ip0 = plist->items[0];
188           Item *ip1 = plist->items[1];
189           Item *ip2 = plist->items[2];
190
191           switch ( pcode )
192           {
        . . . .
```

```
311                         case LEFT_ASSIGN:
312                             leftAssign(ip0, ip1);
313                             break;
314
315                         case RIGHT_ASSIGN:
316                             rightAssign(ip0, ip1, ip0->acceSign());
317                             break;
318                     }
319                 }
/p16ecc/cc_source_8.8/p16e/pic16e.cpp                                    Code-8.46
```

8.9.1 Assembly code output for LEFT_ASSIGN

The P-code LEFT_ASSIGN has the format

$$\{\texttt{LEFT_ASSIGN}, [X, Y]\} \qquad \rightarrow \qquad X <<= Y$$

Which is equivalent to $X *= 2^Y$.

1) If Y is a constant, then the process will be implemented with PIC16E::leftAssignConst(), as shown in Code-8.47.

```
18      void PIC16E :: leftAssign(Item *ip0, Item *ip1)
19      {
20          if ( ip1->type == CON_ITEM )
21          {
22              leftAssignConst(ip0, ip1->val.i);
23              return;
24          }
/p16ecc/cc_source_8.8/p16e/pic16e7.cpp                                   Code-8.47
```

In PIC16E::leftAssignConst(), the shifting operation is accomplished by two stages, byte-shifting and bit-shifting, to improve efficiency. Byte-shifting takes place first, as shown in Code-8.48a.

```
64      void PIC16E :: leftAssignConst(Item *ip0, int n)
65      {
66          int size0 = ip0->acceSize();
67          if ( n > (size0*8) ) n = (size0*8);
68          int byte_shift = n/8;
69          int bit_shift  = n%8;
70          char *indf0 = setFSR0(ip0);
71
72          if ( byte_shift > 0 )
73          {
74              for (int i = size0; i--;)
75              {
76                  if ( i >= byte_shift )
77                  {
78                      fetch(ip0, i-byte_shift, indf0);
79                      store(ip0, i, indf0);
80                      regWREG->reset();
81                  }
82                  else if ( indf0 && byte_shift > 1 )
83                  {
84                      regWREG->load(0);
85                      store(ip0, i, indf0);
86                  }
87                  else
88                      ASM_CODE(_CLRF, acceItem(ip0, i, indf0));
89              }
90          }
/p16ecc/cc_source_8.8/p16e/pic16e7.cpp                                   Code-8.48a
```

And then bit-shifting will be going on, where nibble-shifting might take place, as shown in Code-8.48b.

```
92          if ( bit_shift > 0 )
93          {
94              if ( (size0 - byte_shift) == 1 && bit_shift >= 4 )
95              {
96                  ASM_CODE( _SWAPF, acceItem(ip0, size0-1, indf0), _W_);
97                  ASM_CODE( _ANDLW, 0xf0);
98                  ASM_CODE( _MOVWF, acceItem(ip0, size0-1, indf0));
99                  bit_shift -= 4;
100                 regWREG->reset();
101             }
102
103             Item *startLbl = lblItem(pcoder->getLbl());
104             int  offset = regFSR0->getVarOffset();
105             int  repeat = bit_shift;
106             if ( !indf0 ) regBSR->reset(ip0);
107
108             if ( bit_shift > 1 && (size0 - byte_shift) > 1 )
109             {
110                 regWREG->load(bit_shift);
111                 ASM_LABL(startLbl->val.s);
112                 repeat = 1;
113             }
114
115             while ( repeat-- )
116                 for (int i = byte_shift; i < size0; i++)
117                 {
118                     char *inst = (i == byte_shift)? _LSLF: _RLF;
119                     ASM_CODE(inst, acceItem(ip0, i, indf0), _F_);
120                 }
121
122             if ( bit_shift > 1 && (size0 - byte_shift) > 1 )
123             {
124                 if ( indf0 ) regFSR0->restore(offset);  // reset FSR0 offset
125                 ASM_CODE(_DECFSZ, WREG, _F_);
126                 ASM_CODE(_BRA, startLbl->val.s);
127                 regWREG->set(0);
128             }
129             delete startLbl;
130         }
```
/p16ecc/cc_source_8.8/p16e/pic16e7.cpp   Code-8.48b

- #94~101: nibble-shifting may happen to the highest byte of X.
- #103~128: assembly code generation for bit-shifting.
    - #104: get the current FSR0 offset value.
    - #106: reset BSR (if X is located in the linear space).
    - #110: load WREG for bit-shifting counter.
    - #123: restore FSR0 offset value (for the next loop).

Table-8.28 shows the test result of assembly code generation for P-code LEFT_ASSIGN (when Y is a constant).

Source code (ch8_t22.c)	Assembly code output (ch8_t22.asm)

<pre>int x, y;
 	
void foo()
{
  x <<= 10;
}</pre>
	<pre> .segment CODE2 (REL) foo:0
	foo::
	; :: ch8_t22.c #5: x <<= 10;
	        .bsel x
	        movf x, W
	        .bsel x, x+1
	        movwf x+1
	        .bsel x+1, x
	        clrf x
	        .bsel x, x+1
	        lslf x+1, F
	        lslf x+1, F
	        return</pre>

Table-8.28

2) Otherwise, *Y* will be used as the counter variable for the shifting operation, and it's usually loaded into WREG for loop counting. (To simplify the design, only the lowest byte of *Y* will be used), as shown in Code-8.49.

```
26        char *indf1 = setFSR0(ip1);
27        char *shiftCount = WREG;
28        if ( CONST_ID_ITEM(ip1) )
29        {
30            fetch(ip1, 0, indf1);
31            ASM_CODE(_ADDLW, 1);
32        }
33        else
34            ASM_CODE(_INCF, acceItem(ip1, 0, indf1), _W_);
35
36        regWREG->reset();
37        if ( useFSR(ip0) )
38            ASM_CODE(_MOVWF, shiftCount = ACC0);
39
40        char *indf0 = setFSR0(ip0);
41        int size0 = ip0->acceSize();
42        Item *startLbl = lblItem(pcoder->getLbl());
43        Item *endLbl   = lblItem(pcoder->getLbl());
44        char *shift_inst = _LSLF;
45        int  offset = regFSR0->getVarOffset();
46        if ( !indf0 ) regBSR->reset(ip0);
47
48        ASM_CODE(_BRA, endLbl->val.s);
49        ASM_LABL(startLbl->val.s);
50
51        for (int i = 0; i < size0; i++, shift_inst = _RLF)
52            ASM_CODE(shift_inst, acceItem(ip0, i, indf0), _F_);
53
54        if ( indf0 )
55            regFSR0->restore(offset);    // reset FSR0 offset
56
57        ASM_LABL(endLbl->val.s);
58        ASM_CODE(_DECFSZ, shiftCount, _F_);
59        ASM_CODE(_BRA, startLbl->val.s);
60        delete startLbl;
61        delete endLbl;
```
/p16ecc/cc_source_8.8/p16e/pic16e7.cpp   Code-8.49

- #27~38: shifting counter is loaded into either WREG or ACC0.

Table-8.29 shows the test result of assembly code generation for P-code LEFT_ASSIGN (when *Y* is a variable).

Source code (ch8_t23.c)	Assembly code output (ch8_t23.asm)

`int x, y;`
 	
`void foo()`
`{`
`  x <<= y;`
`}`
	`        .segment      CODE2 (REL) foo:0`
	`foo::`
	`; :: ch8_t23.c #5: x <<= y;`
	`        .bsel   y`
	`        incf   y, W`
	`        bra    _$L3`
	`_$L2:`
	`        .bsel   y, x`
	`        lslf   x, F`
	`        .bsel   x, x+1`
	`        rlf    x+1, F`
	`_$L3:`
	`        decfsz WREG, F`
	`        bra    _$L2`
	`        return`

Table-8.29

The P-code `RIGHT_ASSIGN` has the format

$$\{RIGHT_ASSIGN, [X, Y]\} \qquad \rightarrow \qquad X >>= Y$$

Which is equivalent to $X /= 2^Y$ only when X is a positive value (or an unsigned value).
Assembly code generation for `RIGHT_ASSIGN` is similar to that in `LEFT_ASSIGN`. The difference is that `RIGHT_ASSIGN` needs to extend the sign bit, at the most significant bit positions, if X is a signed value.

1) When Y is a constant, then the process will be handled by `PIC16E::rightAssignConst()`.

```
132   void PIC16E :: rightAssign(Item *ip0, Item *ip1, bool sshift)
133   {
134       if ( ip1->type == CON_ITEM )
135       {
136           rightAssignConst(ip0, ip1->val.i, sshift);
137           return;
138       }
```
/p16ecc/cc_source_8.8/p16e/pic16e7.cpp   Code-8.50

In `PIC16E::rightAssignConst()`, the shifting operation is accomplished by two stages, byte-shifting and bit-shifting, to improve efficiency. Byte-shifting takes place first, as shown in Code-8.51a.

```
183   void PIC16E :: rightAssignConst(Item *ip0, int n, bool sshift)
184   {
185       int size0 = ip0->acceSize();
186       char *indf0 = setFSR0(ip0);
187       if ( n > (size0*8) ) n = (size0*8);
188       int byte_shift = n/8;
189       int bit_shift  = n%8;
190
191       if ( byte_shift > 0 )
192           for (int i = 0; i < size0; i++)
193           {
194               if ( (i+byte_shift) < size0 )
195               {
196                   fetch(ip0, i+byte_shift, indf0);
197                   store(ip0, i, indf0);
198                   regWREG->reset();
199               }
200               else if ( sshift )  // signed shift
201               {
202                   if ( (i+byte_shift) == size0 )
203                   {
204                       if ( byte_shift >= size0 ) fetch(ip0, size0-1, indf0);
205                       EXTEND_WREG;
206                   }
207                   store(ip0, i, indf0);
208               }
209               else
210                   ASM_CODE(_CLRF, acceItem(ip0, i, indf0));
211           }
```
/p16ecc/cc_source_8.8/p16e/pic16e7.cpp   Code-8.51a

- #200~208: when X is a signed value, extend its sign bit and fill in the higher byte(s).
- #209~210: when X is an unsigned value, fill '0' in the higher byte(s).

And then bit-shifting will go on, if needed, as shown in Code-8.51b.

```
213       if ( bit_shift > 0 )
214       {
215           if ( (size0 - byte_shift) == 1 && bit_shift >= 4 && !sshift )
216           {
217               ASM_CODE(_SWAPF, acceItem(ip0, 0, indf0), _W_);
218               ASM_CODE(_ANDLW, 0x0f);
219               ASM_CODE(_MOVWF, acceItem(ip0, 0, indf0));
220               bit_shift -= 4;
221               regWREG->reset();
222           }
223
224           Item *startLbl = lblItem(pcoder->getLbl());
225           int offset = regFSR0->getVarOffset();
226           int repeat = bit_shift;
227           if ( !indf0 ) regBSR->reset(ip0);
228
229           if ( bit_shift > 1 && (size0 - byte_shift) > 1 )
230           {
231               regWREG->load(bit_shift);
232               ASM_LABL(startLbl->val.s);
233               repeat = 1;
234           }
235
236           while ( repeat-- )
237               for (int i = size0 - byte_shift; i--; )
238               {
239                   char *shift_inst = sshift? _ASRF: _LSRF;
240                   if ( (i+1) == (size0-byte_shift) )  // highest byte
241                       ASM_CODE(shift_inst, acceItem(ip0, i, indf0), _F_);
242                   else
243                       ASM_CODE(_RRF,  acceItem(ip0, i, indf0), _F_);
244               }
245
246           if ( bit_shift > 1 && (size0 - byte_shift) > 1 )
247           {
248               if ( indf0 ) regFSR0->restore(offset);  // reset FSR0 offset
249               ASM_CODE(_DECFSZ, WREG, _F_);
250               ASM_CODE(_BRA, startLbl->val.s);
251               regWREG->set(0);
252           }
253           delete startLbl;
254       }
/p16ecc/cc_source_8.8/p16e/pic16e7.cpp                      Code-8.51b
```

- #215~222: nibble-shifting may apply only when X is an unsigned value.

Table-8.30 shows the test result of assembly code generation for P-code `RIGHT_ASSIGN` (when Y is a constant).

Source code (ch8_t24.c)	Assembly code output (ch8_t24.asm)

`int x;`
 	
`void foo()`
`{`
`  x >>= 10;`
`}`
	`        .segment    CODE2 (REL) foo:0`
	`foo::`
	`; :: ch8_t24.c #5: x >>= 10;`
	`        .bsel   x+1`
	`        movf    x+1, W`
	`        .bsel   x+1, x`
	`        movwf   x`
	`        andlw   128`
	`        btfss   3, 2`
	`        movlw   255`
	`        .bsel   x, x+1`
	`        movwf   x+1`
	`        .bsel   x+1, x`
	`        asrf    x, F`
	`        asrf    x, F`
	`        return`

Table-8.30

2) Otherwise, Y is used as the variable (the loop counter) for shifting, and it's usually loaded into WREG (or ACC0). (To simplify the design, only the lowest byte of Y will be used), as shown in Code-8.52.

```
141         char *indf1 = setFSR0(ip1);
142         char *shiftCount = WREG;
143         if ( CONST_ID_ITEM(ip1) )
144         {
145             fetch(ip1, 0, indf1);
146             ASM_CODE(_ADDLW, 1);
147         }
148         else
149             ASM_CODE(_INCF, acceItem(ip1, 0, indf1), _W_);
150
151         regWREG->reset();
152         if ( useFSR(ip0) )
153             ASM_CODE(_MOVWF, shiftCount = ACC0);
154
155         char *indf0 = setFSR0(ip0);
156         int  size0  = ip0->acceSize();
157         int  offset = regFSR0->getVarOffset();
158         Item *startLbl = lblItem(pcoder->getLbl());
159         Item *endLbl   = lblItem(pcoder->getLbl());
160
161         if ( !indf0 ) regBSR->reset(ip0);
162         ASM_CODE(_BRA, endLbl->val.s);
163         ASM_LABL(startLbl->val.s);
164
165         for (int i = size0; i--;)
166         {
167             if ( i == (size0-1) && sshift )
168                 ASM_CODE(_ASRF, acceItem(ip0, i, indf0), _F_);
169             else if ( i == (size0-1) )
170                 ASM_CODE(_LSRF, acceItem(ip0, i, indf0), _F_);
171             else
172                 ASM_CODE(_RRF,  acceItem(ip0, i, indf0), _F_);
173         }
174
175         if ( indf0 ) regFSR0->restore(offset);
176         ASM_LABL(endLbl->val.s);
177         ASM_CODE(_DECFSZ, shiftCount, _F_);
178         ASM_CODE(_BRA, startLbl->val.s);
179         delete startLbl;
180         delete endLbl;
```

`/p16ecc/cc_source_8.8/p16e/pic16e7.cpp`　　　　　　　　Code-8.52

Table-8.31 shows the test result of assembly code generation for P-code `RIGHT_ASSIGN` (when *Y* is a variable).

Source code (ch8_t25.c)	Assembly code output (ch8_t25.asm)
``int x, y;

void foo()
{
  x >>= y;
}``` | ```   .segment   CODE2 (REL) foo:0
foo::
; :: ch8_t25.c #5: x >>= y;
        .bsel   y
        incf   y, W
        bra   _$L3
_$L2:
        .bsel   y, x+1
        asrf   x+1, F
        .bsel   x+1, x
        rrf   x, F
_$L3:
        decfsz WREG, F
        bra   _$L2
        return``` |

Table-8.31

8.10 Assembly Code Generation (8)

<u>Project directory</u>: /p16ecc/cc_source_8.9
<u>New source files</u>: pic16e8.cpp
<u>Modified source files</u>: pic16e.cpp

This part illustrates the assembly code output for LEFT_OP and RIGHT_OP ('<<' and '>>') P-codes,
which are initiated from PIC16E::outputASM0(), as well (see Code-8.53).

Compared to compound shifting operations, the assembly code generation for LEFT_OP and RIGHT_OP
involve two operations, move and shift. If the efficiency is not taken into consideration, the assembly code
generation for the P-codes can be accomplished as

$$\{LEFT_OP, [Z, X, Y]\} \quad \rightarrow \quad \{`=\text{'}, [Z, X]\} \text{ and } \{LEFT_ASSIGN, [Z, Y]\}$$
$$\{RIGHT_OP, [Z, X, Y]\} \quad \rightarrow \quad \{`=\text{'}, [Z, X]\} \text{ and } \{RIGHT_ASSIGN, [Z, Y]\}$$

```
181     void PIC16E :: outputASM0(Pnode *plist)
182     {
183         for(; plist; plist = plist->next)
184         {
185             curPnode = plist;
186             int pcode = plist->type;
187             Item *ip0 = plist->items[0];
188             Item *ip1 = plist->items[1];
189             Item *ip2 = plist->items[2];
190
191             switch ( pcode )
192             {
        . . .
319                 case LEFT_OP:
320                     leftOpr(ip0, ip1, ip2);
321                     break;
322
323                 case RIGHT_OP:
324                     rightOpr(ip0, ip1, ip2);
325                     break;
326             }
327         }
```
/p16ecc/cc_source_8.9/p16e/pic16e.cpp   Code-8.53

8.10.1 Assembly code output for LEFT_OP

Code-8.54a~Code-8.54c illustrate the process of assembly code generation for LEFT_OP.

```
18      void PIC16E :: leftOpr(Item *ip0, Item *ip1, Item *ip2)
19      {
20          if ( same(ip0, ip1) )
21          {
22              leftAssign(ip0, ip2);
23              return;
24          }
25          if ( related(ip0, ip2) || related(ip1, ip2) )
26          {
27              ASM_CODE(_INCF, acceItem(ip2, 0, setFSR0(ip2)), _W_);
28              ASM_CODE(_MOVWI, "--INDF1");
29              leftOprIndf1(ip0, ip1);
30              return;
31          }
32          if ( useFSR(ip0, ip1) || ip2->type != CON_ITEM )
33          {
34              mov(ip0, ip1);
35              leftAssign(ip0, ip2);
36              return;
37          }
```
/p16ecc/cc_source_8.9/p16e/pic16e8.cpp   Code-8.54a

- #20~24: if Z and X are the same item, then the process becomes
    {LEFT_ASSIGN, [Z, Y]}.
- #25~32: if Y is related to either X or Z, then Y (the lowest byte) will be pushed into the stack, and call function
  `PIC16E::leftOprIndf1()`, which does
    {'=', [Z, X]} and
    {LEFT_ASSIGN, [Z, INDF1]}
- #33~38: if both Z and X are in indirect access mode or Y is not a constant, then the process becomes
    {'=', [Z, X]} and
    {LEFT_ASSIGN, [Z, Y]}.

Since Y is a constant, the operation will be performed in two stages: byte moves and bit shifts. The byte move and bit shifting are illustrated in Code-8.54b.

```
39      int  size0 = ip0->acceSize();
40      int  size1 = ip1->acceSize();
41      bool sign1 = ip1->acceSign();
42      int  n = (ip2->val.i >= size0*8)? size0*8: ip2->val.i;
43      int  byte_shift = n/8;
44      int  bit_shift  = n%8;
45      bool shift_move = (bit_shift == 1 && (byte_shift + size1) >= size0);
46      char *indf0 = setFSR0(ip0);
47      char *indf1 = setFSR0(ip1);
48
49      // move/shift at byte level, from ip1 to ip0
50      for (int i = 0; i < size0; i++)
51      {
52          if ( i < byte_shift )
53              ASM_CODE(_CLRF, acceItem(ip0, i, indf0));   // clear lower bytes of ip0
54          else if ( shift_move )
55              break;
56          else if ( (i-byte_shift) < size1 )
57          {
58              fetch(ip1, i-byte_shift, indf1);            // move bytes from ip1 to ip0
59              store(ip0, i, indf0);
60              regWREG->reset();
61          }
62          else if ( !sign1 )
63              ASM_CODE(_CLRF, acceItem(ip0, i, indf0));   // clear higher bytes of ip0
64          else
65          {
66              if ( (i-byte_shift) == size1 )
67              {
68                  fetch(ip1, i-byte_shift, indf1);
69                  EXTEND_WREG;
70                  regWREG->reset();
71              }
72              store(ip0, i, indf0);
73          }
74      }
```
`/p16ecc/cc_source_8.9/p16e/pic16e8.cpp`   Code-8.54b

- #45: `shift_move` sets the condition that combines 'move' and 'shift' operations (from X to Z), to improve the efficiency.
- #50~74: move/shift X to Z, byte by byte, filling lower byte(s).

And then bit-shifting occurs, as shown in Code-8.54c.

```cpp
 76          if ( bit_shift > 0 )
 77          {
 78              if ( bit_shift >= 4 && (size0 - byte_shift) == 1 )
 79              {
 80                  ASM_CODE(_SWAPF, acceItem(ip0, byte_shift, indf0), _W_);
 81                  ASM_CODE(_ANDLW, 0xf0);
 82                  store(ip0, byte_shift, indf0);
 83                  bit_shift -= 4;
 84                  regWREG->reset();
 85              }
 86
 87              Item *startLbl = lblItem(pcoder->getLbl());
 88              int  offset = regFSR0->getVarOffset();
 89              int  repeat = bit_shift;
 90              if ( !indf0 ) regBSR->reset(ip0);
 91
 92              if ( bit_shift > 1 && (size0 - byte_shift) > 1 )
 93              {
 94                  regWREG->load(bit_shift);
 95                  ASM_LABL(startLbl->val.s);
 96                  repeat = 1;
 97              }
 98
 99              while ( repeat-- )
100                  for (int i = byte_shift; i < size0; i++)
101                  {
102                      char *inst = (i == byte_shift)? _LSLF: _RLF;
103                      if ( shift_move )
104                      {
105                          ASM_CODE(inst, acceItem(ip1, i-byte_shift, indf1), _W_);
106                          store(ip0, i, indf0);
107                      }
108                      else
109                          ASM_CODE(inst, acceItem(ip0, i, indf0), _F_);
110                  }
111
112              if ( bit_shift > 1 && (size0 - byte_shift) > 1 )
113              {
114                  if ( indf0 ) regFSR0->restore(offset);  // reset FSR0 offset
115                  ASM_CODE(_DECFSZ, WREG, _F_);
116                  ASM_CODE(_BRA, startLbl->val.s);
117              }
118              delete startLbl;
119              regWREG->reset();
120          }
```
`/p16ecc/cc_source_8.9/p16e/pic16e8.cpp`   Code-8.54c

- #103~107: 'move and shift' combined.

Table 8.32 shows the test results for assembly code generation of P-code `LEFT_OP`.

Source code (ch8_t26.c)	Assembly code output (ch8_t26.asm)

`int x, y;`
`char a, b;`
 	
`void foo()`
`{`
`  a = b << b;`
`  x = y << 9;`
`}`
	`        .segment    CODE2 (REL) foo:0`
	`foo::`
	`; :: ch8_t26.c #6: a = b << b;`
	`        .bsel   b`
	`        incf    b, W`
	`        movwi   --INDF1`
	`        movf    b, W`
	`        .bsel   b, a`
	`        movwf   a`
	`        bra     $L3`
	`$L2:`
	`        lslf    a, F`
	`$L3:`
	`        decfsz  INDF1, F`
	`        bra     $L2`
	`        addfsr  INDF1, 1`
	`; :: ch8_t26.c #7: x = y << 9;`
	`        .bsel   a, x`
	`        clrf    x`
	`        .bsel   x, y`
	`        lslf    y, W`
	`        .bsel   y, x+1`
	`        movwf   x+1`
	`        return`

Table-8.32

In the RIGHT_OP operation, the sign bit in X (if it's a signed value) must be taken into consideration during the shifting. In addition, if the size of X is greater than Z, the shifting operation might need to be done in a temporary holder (ACC) and then copied to the target (Z).

The assembly code generation for RIGHT_OP is shown in Code-8.55a~Code-8.55d.

```
150    void PIC16E :: rightOpr(Item *ip0, Item *ip1, Item *ip2)
151    {
152        int  size0 = ip0->acceSize();
153        int  size1 = ip1->acceSize();
154        bool sign1 = ip1->acceSign();
155
156        if ( same(ip0, ip1) )
157        {
158            rightAssign(ip0, ip2, sign1);
159            return;
160        }
161        if ( related(ip0, ip2) || related(ip1, ip2) )
162        {
163            ASM_CODE(_INCF, acceItem(ip2, 0, setFSR0(ip2)), _W_);
164            ASM_CODE(_MOVWI, "--INDF1");
165            rightOprIndf1(ip0, ip1);
166            return;
167        }
168        if ( useFSR(ip0, ip1) || ip2->type != CON_ITEM )
169        {
170            if ( size0 >= size1 || !sign1 )
171            {
172                mov(ip0, ip1);
173                rightAssign(ip0, ip2, sign1);
174            }
175            else
176            {
177                Item *acc = storeToACC(ip1, size1);
178                rightAssign(acc, ip2, sign1);
179                mov(ip0, acc);
180                delete acc;
181            }
182            return;
183        }
```
/p16ecc/cc_source_8.9/p16e/pic16e8.cpp   Code-8.55a

- #162~167: if X and Z are related or Y and Z are related, then Y will be pushed into the stack (INDF1) and the operation is accomplished by PIC16E::rightOprIndf1(), which does
    - {'=', [Z, X]} and
    - {RIGHT_ASSIGN, [Z, INDF1]}
    - Or
    - {'=', [ACC, X]}, {RIGHT_ASSIGN, [ACC, INDF1]} and
    - {'=', [Z, ACC]}.
- #168~183: if both Z and X are in indirect access mode or Y is not a constant, then the operation is accomplished by
    - {'=', [Z, X]} and
    - {RIGHT_ASSIGN, [Z, Y]}
    - Or
    - {'=', [ACC, X]}, {RIGHT_ASSIGN, [ACC, Y]} and
    - {'=', [Z, ACC]}.

Now, since Y is a constant, both byte_shift and bit_shift can be figured out to determine the further operations.

If the size of X exceeds the limit (size of Z plus `byte_shift`), the operation will become

$$\{\texttt{RIGHT_OP}, [\text{ACC}, X, Y]\},$$
$$\{\texttt{'='}, [Z, \text{ACC}]\}$$

```
185        int  n = ip2->val.i;
186        int  byte_shift = n / 8;
187        int  bit_shift  = n % 8;
188        if ( (size1 > (size0 + byte_shift) && bit_shift > 1) || size1 < size0 )
189        {
190            Item *acc = accItem(newAttr(CHAR + size1 - 1));
191            acc->attr->isUnsigned = sign1? 0: 1;
192            rightOpr(acc, ip1, ip2);
193            mov(ip0, acc);
194            delete acc;
195            return;
196        }
```
/p16ecc/cc_source_8.9/p16e/pic16e8.cpp   Code-8.55b

Otherwise, move and shift, at the byte level, will take place first.

```
198        char *indf0 = setFSR0(ip0);
199        char *indf1 = setFSR0(ip1);
200        if ( byte_shift > size1 ) byte_shift = size1;
201        bool shift_move = (bit_shift == 1 && size1 >= (size0 + byte_shift));
202
203        for (int i = 0; i < size0; i++)
204        {
205            if ( (byte_shift+i) < size1 )
206            {
207                if ( shift_move ) break;
208                fetch(ip1, byte_shift+i, indf1);
209                store(ip0, i, indf0);
210                regWREG->reset();
211            }
212            else if ( sign1 )
213            {
214                if ( (byte_shift+i) == size1 )
215                {
216                    if ( byte_shift >= size1 )
217                        fetch(ip1, size1-1, indf1);
218                    EXTEND_WREG;
219                    regWREG->reset();
220                }
221                store(ip0, i, indf0);
222            }
223            else
224                ASM_CODE(_CLRF, acceItem(ip0, i, indf0));
225        }
```
/p16ecc/cc_source_8.9/p16e/pic16e8.cpp   Code-8.55c

And then, a bit shift takes place if `bit_shift` is not zero.

```
227        if ( bit_shift > 0 && size1 > byte_shift )
228        {
229            int repeat = bit_shift;
230            int offset = regFSR0->getVarOffset();
231            Item *startLbl = lblItem(pcoder->getLbl());
232            if ( !indf0 ) regBSR->reset(ip0);
233            if ( !indf1 && shift_move ) regBSR->reset(ip1);
234
235            if ( bit_shift > 1 )
236            {
237                ASM_CODE(_MOVLW, bit_shift);
238                ASM_LABL(startLbl->val.s);
239                repeat = 1;
240            }
```

```
241
242          while ( repeat-- > 0 )
243              for (int i = size1 - byte_shift; i--;)
244              {
245                  char *inst = _RRF;
246                  if ( (i+byte_shift+1) == size1 )
247                      inst = sign1? _ASRF: _LSRF;
248
249                  if ( shift_move )
250                  {
251                      ASM_CODE(inst, acceItem(ip1, i+byte_shift, indf1), _W_);
252                      if ( i < size0 ) store(ip0, i, indf0);
253                  }
254                  else if ( i < size0 )
255                  {
256                      ASM_CODE(inst, acceItem(ip0, i, indf0), _F_);
257                  }
258
259                  regWREG->reset();
260              }
261
262          if ( bit_shift > 1 )
263          {
264              if ( indf0 || indf1 ) regFSR0->restore(offset);
265              ASM_CODE(_DECFSZ, WREG, _F_);
266              ASM_CODE(_BRA, startLbl->val.s);
267              regWREG->reset();
268          }
269          delete startLbl;
270      }
/p16ecc/cc_source_8.9/p16e/pic16e8.cpp                          Code-8.55d
```

Table 8.33 shows the test results for assembly code generation for P-code `RIGHT_OP`.

Source code (ch8_t27.c)	Assembly code output (ch8_t27.asm)

`int x, y;`
`char a;`
 	
`void foo()`
`{`
`  x = y >> 9;`
`    a = y >> 1;`
`    a = x >> 9;`
`}`
	`        .segment    CODE2 (REL) foo:0`
	`foo::`
	`; :: ch8_t27.c #6: x = y >> 9;`
	`        .bsel   y+1`
	`        movf   y+1, W`
	`        .bsel   y+1, x`
	`        movwf   x`
	`        andlw   128`
	`        btfss   3, 2`
	`        movlw   255`
	`        .bsel   x, x+1`
	`        movwf   x+1`
	`        .bsel   x+1, x`
	`        asrf    x, F`
	`; :: ch8_t27.c #7: a = y >> 1;`
	`        .bsel   x, y+1`
	`        asrf    y+1, W`
	`        .bsel   y+1, y`
	`        rrf    y, W`
	`        .bsel   y, a`
	`        movwf   a`
	`; :: ch8_t27.c #8: a = x >> 9;`
	`        .bsel   a, x+1`
	`        asrf    x+1, W`
	`        .bsel   x+1, a`
	`        movwf   a`
	`        return`

Table-8.33

8.11 Assembly Code Generation (9)

<u>Project directory</u>: /p16ecc/cc_source_8.10
<u>New source files</u>: pic16e9.cpp
<u>Modified source files</u>: pic16e.cpp

This part illustrates the assembly code output for multiplication (`MUL_ASSIGN` and `'*'`) P-codes, which are initiated from `PIC16E::outputASM0()`, as well (see Code-8.56).

```
181   void PIC16E :: outputASM0(Pnode *plist)
182   {
183       for(; plist; plist = plist->next)
184       {
185           curPnode = plist;
186           int pcode = plist->type;
187           Item *ip0 = plist->items[0];
188           Item *ip1 = plist->items[1];
189           Item *ip2 = plist->items[2];
190
191           switch ( pcode )
192           {
                  . . .
327               case MUL_ASSIGN:
328                   mulAssign(ip0, ip1);
329                   break;
330
331               case '*':
332                   mul(ip0, ip1, ip2);
333                   break;
334           }
335       }
```
/p16ecc/cc_source_8.10/p16e/pic16e.cpp   Code-8.56

Actually, most 8-bit CPUs (such as the PIC16Fxxxx series MCU) don't have hardware (instructions) to support multiplication or division. That means they have to be accomplished through software. Moreover, multiplication, division, and so on are performed by calling library functions.

The following lists the library functions that do the multiplications

`_mul8()`	8×8 multiplication (16-bit result)
`_mul16()`	16×16 multiplication (16-bit result)
`_mul24()`	24×24 multiplication (24-bit result)
`_mul32()`	32×32 multiplication (32-bit result)
`_mul16indf1()`	16×16 multiplication (16-bit result)
`_mul24indf1()`	24×24 multiplication (24-bit result)
`_mul32indf1()`	32×32 multiplication (32-bit result)

8.11.1 Assembly code output for MUL_ASSIGN

The P-code of `MUL_ASSIGN` has the format

$$\{MUL_ASSIGN, [X, Y]\}$$

Code-8.57 and Code-8.58 illustrate the assembly code generation process for `MUL_ASSIGN`.

```cpp
18   void PIC16E :: mulAssign(Item *ip0, Item *ip1)
19   {
20       if ( ip0->type == ACC_ITEM )
21       {
22           mul(ip0, ip0, ip1);
23           return;
24       }
25
26       int size0 = ip0->acceSize();
27       if ( size0 == 1 )
28       {
29           mulAssign8(ip0, ip1);
30           return;
31       }
32
33       char *func = (size0 == 2)? (char*)"_mul16irdf":
34                    (size0 == 3)? (char*)"_mul24irdf":
35                                  (char*)"_mul32irdf";
36       pushStack(ip1, size0);   // push y into the stack.
37       setFSR(ip0, 0);          // load FSR0 with the address of x
38       call(func);
39       regFSR0->reset();
40       regWREG->reset();
41   }
```
/p16ecc/cc_source_8.10/p16e/pic16e9.cpp   Code-8.57

- #27~31: if it's an 8-bit multiplication, then call function `PIC16E::mulAssign8()`, (see Code-8.58).
- #33~38: for 16/24/32-bit multiplications, the corresponding library function, `_mul16indf()` or `_mul24indf()` or `_mul32indf()`, will be called. Before calling the library function,
    (1) Push Y into the stack (#38).
    (2) Load FSR0 with the address of X (#37).

In `PIC16E::mulAssign8()`, the library function, `_mul8()`, will be called, which returns a 16-bit result stored in ACC. X and Y are loaded into ACC and WREG individually before calling `_mul8()`.

```cpp
43   void PIC16E :: mulAssign8(Item *ip0, Item *ip1)
44   {
45       Item *acc = storeToACC(ip1, 1);
46       fetch(ip0, 0, setFSR0(ip0));
47
48       call((char*)"_mul8");
49       regFSR0->reset();
50       regWREG->reset();
51
52       mov(ip0, acc);
53       delete acc;
54   }
```
/p16ecc/cc_source_8.10/p16e/pic16e9.cpp   Code-8.58

Table-8.34 shows the test result of assembly code generation for P-code `MUL_ASSIGN`.

Source code (ch8_t28.c)	Assembly code output (ch8_t28.asm)

`int x, y;`
`char a, b;`
 	
`void foo()`
`{`
`  a *= b;`
`  x *= y;`
`}`
	`    .segment    CODE2 (REL) foo:0`
	`foo::`
	`; :: ch8_t28.c #6: a *= b;`
	`    .bsel  b`
	`    movf   b, W`
	`    movwf  112`
	`    .bsel  b, a`
	`    movf   a, W`
	`    .psel  foo, _mul8`
	`    call   _mul8`
	`    movf   112, W`
	`    movwf  a`
	`; :: ch8_t28.c #7: x *= y;`
	`    .bsel  a, y`
	`    movf   y, W`
	`    movwi  --INDF1`
	`    .bsel  y, y+1`
	`    movf   y+1, W`

| | ```
movwi   --INDF1
movlw   (x)
movwf   FSR0L
movlw   (x)>>8
movwf   FSR0H
.psel   _mul8, _mul16indf
call    _mul16indf
.psel   _mul16indf, foo
return
``` |

<div align="right">Table-8.34</div>

8.11.2 Assembly code output for '*'

The P-code of '*' has the format

$$\{\text{'*'}, [Z, X, Y]\}$$

In addition, in some cases, the operation will be done by

$$\{\text{'='}, [Z, X]\} \text{ and}$$
$$\{MUL\_ASSGIN, [Z, Y]\}$$

Assembly code generation for 8-bit multiplication is done as illustrated in Code-8.59 and Code-8.60.

```
56   void PIC16E :: mul(Item *ip0, Item *ip1, Item *ip2)
57   {
58       int size0 = ip0->acceSize();
59       if ( same(ip0, ip2) && ip0->type != ACC_ITEM )
60       {
61           mulAssign(ip0, ip1);
62           return;
63       }
64       if ( size0 == 1 || (ip1->acceSize() == 1 && !ip1->acceSign() &&
65                           ip2->acceSize() == 1 && !ip2->acceSign()  ) )
66       {
67           mul8(ip0, ip1, ip2);
68           return;
69       }
```
`/p16ecc/cc_source_8.10/p16e/pic16e9.cpp`

<div align="right">Code-8.59</div>

- #64~69: Z can be any size if both X and Y are 1-byte unsigned values.

```
86    void PIC16E :: mul8(Item *ip0, Item *ip1, Item *ip2)
87    {
88        Item *acc = NULL;
89        if ( CONST_ITEM(ip1) || CONST_ID_ITEM(ip1) )
90        {
91            acc = storeToACC(ip2, 1);
92            fetch(ip1, 0, NULL);
93        }
94        else if ( CONST_ITEM(ip2) || CONST_ID_ITEM(ip2) )
95        {
96            acc = storeToACC(ip1, 1);
97            fetch(ip2, 0, NULL);
98        }
99        else if ( ip1->type == ACC_ITEM )
100           fetch(ip2, 0, setFSR0(ip2));
101       else if ( ip2->type == ACC_ITEM )
102           fetch(ip1, 0, setFSR0(ip1));
103       else
104       {
105           acc = storeToACC(ip1, 1);
106           fetch(ip2, 0, setFSR0(ip2));
107       }
108
```
```

```
109         call((char*)"_mul8");
110         regFSR0->reset();
111         regWREG->reset();
112
113         if ( acc == NULL )
114             acc = accItem(newAttr(INT));
115
116         acc->attr->type = INT;
117         acc->attr->isUnsigned = 1;
118
119         if ( ip0->type != ACC_ITEM || ip0->acceSize() > 2 )
120             mov(ip0, acc);
121
122         delete acc;
123     }
```
/p16ecc/cc_source_8.10/p16e/pic16e9.cpp   Code-8.60

- #89~107: put both (the lowest byte of) X and Y into ACC and WREG before calling function `_mul8()`.
- #120: move the result in ACC, a 16-bit unsigned value, to Z.

16-bit, 24-bit, and 32-bit multiplications are performed as illustrated in Code-8.61.

```
56  void PIC16E :: mul(Item *ip0, Item *ip1, Item *ip2)
57  {
58      int size0 = ip0->acceSize();

    . . .

71      pushStack((ip2->type == ACC_ITEM)? ip1: ip2, size0);
72      Item *acc  = storeToACC((ip2->type == ACC_ITEM)? ip2: ip1, size0);
73      char *func = (size0 == 2)? (char*)"_mul16":
74                   (size0 == 3)? (char*)"_mul24":
75                                 (char*)"_mul32";
76      call(func);
77      regFSR0->reset();
78      regWREG->reset();
79
80      if ( ip0->type != ACC_ITEM )
81          mov(ip0, acc);
82
83      delete acc;
84  }
```
/p16ecc/cc_source_8.10/p16e/pic16e9.cpp   Code-8.61

- #73~84: if Y (or X) is stored in ACC, X (or Y) will be pushed into the stack (#71). The sizes of both X and Y need to be extended to that of Z (`size0`), before calling the library function (the result returned is in ACC).

Table-8.35 and Table-8.36 show the test results of assembly code generation for P-code '*'.

Source code (ch8_t29.c)	Assembly code output (ch8_t29.asm)

<pre>int x;
int func();
 	
int foo()
{
  return x * func();
}</pre>
	<pre> .segment CODE2 (REL) foo:0
	foo::
	; :: ch8_t29.c #6: return x * func();
	    .psel foo, func
	    call func
	    .bsel x
	    movf x, W
	    movwi --INDF1
	    .bsel x, x+1
	    movf x+1, W
	    movwi --INDF1
	    .psel func, _mul16
	    call _mul16
	    .psel _mul16, foo
	    return</pre>

Table-8.35

Source code (ch8_t30.c)	Assembly code output (ch8_t30.asm)
``int x, y, z;

void foo()
{
  x = y * z;
}``` | ```   .segment   CODE2 (REL) foo:0
foo::
; :: ch8_t30.c #5: x = y * z;
        .bsel   z
        movf   z, W
        movwi   --INDF1
        .bsel   z, z+1
        movf   z+1, W
        movwi   --INDF1
        .bsel   z+1, y
        movf   y, W
        movwf   112
        .bsel   y, y+1
        movf   y+1, W
        movwf   113
        .psel   foo, _mul16
        call   _mul16
        movf   112, W
        .bsel   y+1, x
        movwf   x
        movf   113, W
        .bsel   x, x+1
        movwf   x+1
        .psel   _mul16, foo
        return``` |

Table-8.36

8.12 Assembly Code Generation (10)

<u>Project directory</u>: /p16ecc/cc_source_8.11
<u>New source files</u>: pic16e10.cpp
<u>Modified source files</u>: pic16e.cpp

This part illustrates the assembly code output for division/modulation, and other miscellaneous P-codes. As before, they are added in `PIC16E::outputASM0()` (see Code-8.62).

In fact, division (`'/'` and `'/='`) and modulation (`'%'` and `'%='`) have the same operation. They are treated the same way during the assembly code generation. The only difference is that division yields the quotient, while modulation yields the remainder. As in the last section, the division and modulation are done by calling the library function, listed as following

```
_divmod8()          8-bit division/modulation
_divmod16()         16-bit division/modulation
_divmod24()         24-bit division/modulation
_divmod32()         32-bit division/modulation
```

All those functions will return the result in ACC (quotient or remainder).

```cpp
181   void PIC16E :: outputASM0(Pnode *plist)
182   {
183       for(; plist; plist = plist->next)
184       {
185           curPnode = plist;
186           int pcode = plist->type;
187           Item *ip0 = plist->items[0];
188           Item *ip1 = plist->items[1];
189           Item *ip2 = plist->items[2];
190
191           switch ( pcode )
192           {
        . . .
335               case DIV_ASSIGN:
336               case MOD_ASSIGN:
337                   divmodAssign(pcode, ip0, ip1);
338                   break;
339
340               case '/':
341               case '%':
342                   divmod(pcode, ip0, ip1, ip2);
343                   break;
344
345               case PRAGMA:
346                   pragma(ip0, ip1);
347                   break;
348
349               case P_ASMFUNC:
350                   asmFunc(ip0, ip1, ip2);
351                   regWREG->reset();
352                   regFSR0->reset();
353                   break;
354
355               case AASM:
356                   asm16e->code(ip0->val.s);
357                   regWREG->reset();
358                   regFSR0->reset();
359                   break;
360
361               case P_DJNZ: // dec, jp on NZ
362               case P_IJNZ: // inc, jp on Z
363                   djnz(pcode, ip0, ip1);
364                   break;
365
366               case CASE:
367                   jeqjne(P_JEQ, ip0, ip1, ip2);
368                   break;
369           }
370       }
```

/p16ecc/cc_source_8.11/p16e/pic16e.cpp   Code-8.62

8.12.1 Assembly code output for DIV_ASSIGN and MOD_ASSIGN

The P-code of DIV_ASSIGN and MOD_ASSIGN ('/=' and '%=') have the format

$$\{DIV_ASSIGN, [X, Y]\} \text{ and}$$
$$\{MOD_ASSIGN, [X, Y]\}$$

Code-8.63 illustrates the assembly code generation for DIV_ASSIGN and MOD_ASSIGN.

```cpp
19   void PIC16E :: divmodAssign(int code, Item *ip0, Item *ip1)
20   {
21       int size0 = ip0->acceSize();
22       int size1 = ip1->acceSize();
23       bool sign0 = ip0->acceSign();
24       bool sign1 = ip1->acceSign();
25       int size  = (size0 >= size1)? size0: size1;
26       int op_flag = 0;
27
```

249

```cpp
28        char *func;
29        switch ( size )
30        {
31            case 1: func = (char*)"_divmod8";    break;
32            case 2: func = (char*)"_divmod16";   break;
33            case 3: func = (char*)"_divmod24";   break;
34            case 4: func = (char*)"_divmod32";   break;
35            default: return;
36        }
37
38        if ( sign0 ) op_flag |= 4;                // signed = 1; unsigned = 0
39        if ( sign1 ) op_flag |= 2;                // signed = 1; unsigned = 0
40        if ( code != DIV_ASSIGN ) op_flag |= 1; // division = 0; modulation = 1
41
42        pushStack(ip1, size);
43        Item *acc = storeToACC(ip0, size);
44
45        ASM_CODE(_MOVLW, op_flag);
46        call(func);
47        regFSR0->reset();
48        regWREG->reset();
49
50        if ( (sign0 && sign1) || (ip0->type == CON_ITEM && sign1) ||
51                                 (ip1->type == CON_ITEM && sign0) )
52            acc->attr->isUnsigned = 0;
53        else
54            acc->attr->isUnsigned = 1;
55
56        if ( ip0->type != ACC_ITEM ) mov(ip0, acc);
57        delete acc;
58    }
```
`/p16ecc/cc_source_8.11/p16e/pic16e10.cpp`   Code-8.63

- #25: determine the size for division/modulation operation.
- #29~36: determine the library function to be called.
- #38~40: set the flags for the operation:
- #42: push the divisor into the stack (scale up the size to `size`).
- #43: store the dividend into ACC (scale up the size to `size`).
- #45~46: load the flags and call the library function.
- #50~54: determine the sign property for the result in ACC.

Table 8.37 shows the test results for assembly code generation of P-code `DIV_ASSIGN`.

Source code (ch8_t31.c)	Assembly code output (ch8_t31.asm)

<pre>int x, y;
 	
void foo()
{
  x /= y;
}</pre>
	<pre> .segment CODE2 (REL) foo:0
	foo::
	; :: ch8_t31.c #5: x /= y;
	        .bsel y
	        movf y, W
	        movwi --INDF1
	        .bsel y, y+1
	        movf y+1, W
	        movwi --INDF1
	        .bsel y+1, x
	        movf x, W
	        movwf 112
	        .bsel x, x+1
	        movf x+1, W
	        movwf 113
	        movlw 6
	        .psel foo, _divmod16
	        call _divmod16
	        movf 112, W
	        .bsel x+1, x
	        movwf x
	        movf 113, W
	        .bsel x, x+1
	        movwf x+1</pre>

<table>
<tr><td></td><td>.psel _divmod16, foo
return</td></tr>
</table>

8.12.2 Assembly code output for '/' and '%'

The P-code of `/` and `%` has the format

$$\{ `/', [Z, X, Y]\} \text{ and }$$
$$\{ `\%', [Z, X, Y]\}$$

Code-8.64 illustrates the process of assembly code generation for `/` and `%`.

```
60   void PIC16E :: divmod(int code, Item *ip0, Item *ip1, Item *ip2)
61   {
62       int size0 = ip0->acceSize();
63       int size1 = ip1->acceSize();
64       int size2 = ip2->acceSize();
65       bool sign1 = ip1->acceSign();
66       bool sign2 = ip2->acceSign();
67       int size  = (size1 >= size2)? size1: size2;
68       int op_flag = 0;
69
70       char *func;
71       switch ( size )
72       {
73           case 1: func = (char*)"_divmod8";    break;
74           case 2: func = (char*)"_divmod16";   break;
75           case 3: func = (char*)"_divmod24";   break;
76           case 4: func = (char*)"_divmod32";   break;
77           default: return;
78       }
79       if ( sign1 ) op_flag |= 4;          // signed = 1; unsigned = 0
80       if ( sign2 ) op_flag |= 2;          // signed = 1; unsigned = 0
81       if ( code != '/' ) op_flag |= 1;    // division = 0; modulation = 1
82
83       pushStack(ip2, size);
84       Item *acc = storeToACC(ip1, size);
85
86       ASM_CODE( _MOVLW, op_flag);
87       call(func);
88       regFSR0->reset();
89       regWREG->reset();
90
91       if ( (sign1 && sign2) || (ip1->type == CON_ITEM && sign2) ||
92                                (ip2->type == CON_ITEM && sign1) )
93           acc->attr->isUnsigned = 0;
94       else
95           acc->attr->isUnsigned = 1;
96
97       if ( ip0->type != ACC_ITEM || size0 > size ) mov(ip0, acc);
98       delete acc;
99   }
```
`/p16ecc/cc_source_8.11/p16e/pic16e10.cpp`   Code-8.64

- #67: determine the size for the division/modulation operation.
- #70~78: determine the library function to be called.
- #79~81: set the flags for the operation:
- #83: push the divisor into the stack (scale up the size to `size`).
- #84: store the dividend into ACC (scale up the size to `size`).
- #86~87: load the flags and call the library function
- #91~95: determine the sign property for the result in ACC.

8.12.3 Assembly code output for PRAGMA

As described before, #pragma statements are used to control the compilation and configure the target CPU. In compiler **cc16e**, there are three types of #pragma statement supported:

```
#pragma acc_save      N              - save N bytes of ACC when an interrupt happens.
#pragma isr_no_stack                 - disable stack pointer (FSR1) protection.
#pragme FUSEn         expr           - configure the FUSEn with value expr.
```

Only "#pragma FUSEn ..." will generate the assembly code, which is located in a FUSE segment. Besides, "FUSEn" should be defined in the CPU's specific header file; see the test results in Table-8.38. Code-8.65 illustrates the process of assembly code generation.

```
101   void PIC16E :: pragma(Item *ip0, Item *ip1)
102   {
103       if ( strcmp(ip0->val.s, "acc_save") == 0 )
104       {
105           if ( ip1 && ip1->type == CON_ITEM )
106           {
107               accSave = ip1->val.i;
108               return;
109           }
110       }
111       if ( strcmp(ip0->val.s, "isr_no_stack") == 0 )
112       {
113           isrStackSet = false;
114           return;
115       }
116       if ( memcmp(ip0->val.s, "FUSE", 4) == 0 )
117       {
118           PreScan prescan(nlist);
119           Nnode *nnp = nlist->search(ip0->val.s);
120
121           if ( nnp && (nnp->np[0] = prescan.scan(nnp->np[0])) &&
122                 nnp->np[0]->type == NODE_CON && ip1 && ip1->type == CON_ITEM )
123           {
124               char *s = STRBUF();
125               ASM_OUTP("\n");
126               sprintf(s, "FUSE (ABS, =%ld)", nnp->np[0]->con.value);
127               ASM_CODE(_SEGMENT, s);
128               ASM_CODE(_DW, ip1->val.i);
129               return;
130           }
131       }
132       errPrint("unknown or invalid '#pragma' statement!\n");
133   }
```
/p16ecc/cc_source_8.11/p16e/pic16e10.cpp   Code-8.65

Source code (ch8_t32.c)	Assembly code output (ch8_t32.asm)

`#include <pic12f1840.h>`
 	
`#pragma   FUSE0        _FOSC_INTOSC  &`
`_WDT_DIS`
`#pragma FUSE1  0xffff`
	`.device "pic12f1840", 256, 4096`
	
	`_$$::   .equ  0x20F0 ; stack init. value`
	
	`.segment      FUSE (ABS, =32775)`
	`.dw    16356`
	
	`.segment      FUSE (ABS, =32776)`
	`.dw    65535`
	
	`.end`

Table-8.38

As for "#pragma acc_save ..." and "#pragma isr_no_stack", please check the contents in Section 8.3.1.

8.12.4 Assembly code output for P_ASMFUNC and P_DJNZ/P_IJNZ

The assembly code generation for these P-codes is straightforward and shown in Code-8.66.

```cpp
135    void PIC16E :: asmfunc(Item *ip0, Item *ip1, Item *ip2)
136    {
137        asmCode *afp = (asmCode *)ip0->val.p;
138        int name_len = strlen(afp->name);
139        char *suffix = NULL;
140
141        if ( strcmp(&afp->name[name_len-2], "_W") == 0 ) suffix = _W_;
142        if ( strcmp(&afp->name[name_len-2], "_F") == 0 ) suffix = _F_;
143
144        char *indf = setFSR0(ip1);
145        if ( suffix )
146            ASM_CODE(afp->inst, acceItem(ip1, 0, indf), suffix);
147        else if ( ip2 )
148            ASM_CODE(afp->inst, acceItem(ip1, 0, indf), ip2->val.i);
149        else
150            ASM_CODE(afp->inst, acceItem(ip1, 0, indf));
151    }
152
153    void PIC16E :: djnz(int code, Item *ip0, Item *ip1)
154    {
155        char *indf0 = setFSR0(ip0);
156        char *inst = (code == P_DJNZ)? _DECFSZ: _INCFSZ;
157        regPCLATH->load(ip1->val.s);
158        ASM_CODE(inst, acceItem(ip0, 0, indf0), _F_);
159        ASM_CODE(_GOTO, ip1->val.s);
160    }
```

`/p16ecc/cc_source_8.11/p16e/pic16e1C.cpp`   Code-8.66

8.13 Assembly Code Generation (11)

<u>Project directory</u>: /p16ecc/cc_source_8.12
<u>New source files</u>: pic16e11.cpp
<u>Modified source files</u>: pic16e.cpp

This part illustrates the assembly code output for '&', '|' and '^' P-codes. They have P-code format as

$$\{ \texttt{`\&'}, [Z, X, Y]\}$$
$$\{ \texttt{`|'}, [Z, X, Y]\}$$
$$\{ \texttt{`^'}, [Z, X, Y]\}$$

All those logic P-codes are commutative. So, if eighter X or Y is a constant, then the operation of the assembly code generation can be treated in a different way, to reach higher efficiency of the output. As the same as before, they are added in `PIC16E::outputASM0()`, (see Code-8.67).

In fact, to simplified the design, P-code of logic operations can be transformed as

$$\{ \texttt{`\&'}, [Z, X, Y]\} \quad \rightarrow \quad \{ \texttt{`='}, [Z, X]\} \text{ and } \{\texttt{AND_ASSIGN}, [Z, Y]\}$$
$$\{ \texttt{`|'}, [Z, X, Y]\} \quad \rightarrow \quad \{ \texttt{`='}, [Z, X]\} \text{ and } \{\texttt{OR_ASSIGN}, [Z, Y]\}$$
$$\{ \texttt{`^'}, [Z, X, Y]\} \quad \rightarrow \quad \{ \texttt{`='}, [Z, X]\} \text{ and } \{\texttt{XOR_ASSIGN}, [Z, Y]\}$$

```
181  void PIC16E :: outputASM0(Pnode *plist)
182  {
183      for(; plist; plist = plist->next)
184      {
185          curPnode = plist;
186          int pcode = plist->type;
187          Item *ip0 = plist->items[0];
188          Item *ip1 = plist->items[1];
189          Item *ip2 = plist->items[2];
190
191          switch ( pcode )
192          {

370              case '&':
371              case '|':
372              case '^':
373                  andorxor(pcode, ip0, ip1, ip2);
374                  break;
375
376              default:
377                  if ( pcode != ';' )
378                  {
379                      char *buf = STRBUF();
380                      sprintf(buf, "unknown p-code: '%d'!\n", pcode);
381                      errPrint(buf);
382                  }
383                  break;
384          }
385      }
386  }
```

`/p16ecc/cc_source_8.11/p16e/pic16e.cpp`   Code-8.67

8.13.1 Assembly code output for logic operation

The assembly code generation for P-codes for logic operations can be briefly described as in Code-8.68. In
which neither X nor Y is a constant.

```
29   void PIC16E :: andorxor(int code, Item *ip0, Item *ip1, Item *ip2)
30   {
     . . .
71       int size0   = ip0->acceSize();
72       int size1   = ip1->acceSize();
73       int size2   = ip2->acceSize();
74       char *indf0 = setFSR0(ip0);
75       char *indf1 = setFSR0(ip1);
76       char *indf2 = setFSR0(ip2);
77       char *inst  = (code == '&')? _ANDWF:
78                     (code == '|')? _IORWF: _XORWF;
79       char *inst2 = (code == '&')? _ANDLW:
80                     (code == '|')? _IORLW: _XORLW;
81
82       for(int i = 0; i < size0; i++)
83       {
84           bool zero_fill = false;
85           switch ( code )
86           {
87               case '|':   case '^':
88                   if ( i >= size1 && i >= size2 ) zero_fill = true;
89                   break;
90               case '&':
91                   if ( i >= size1 || i >= size2 ) zero_fill = true;
92                   break;
93           }
94
95           if ( zero_fill )
96           {
97               if ( !indf0 )
98               {
99                   ASM_CODE(_CLRF, acceItem(ip0, i));
100                  continue;
101              }
102              regWREG->load(0);
103          }
```

```
104         else
105         {
106             if ( i < size1 )
107             {
108                 fetch(ip1, i, indf1);
109                 if ( i < size2 )
110                 {
111                     if ( CONST_ITEM(ip2) )
112                         ASM_CODE(inst2, acceItem(ip2, i), _W_);
113                     else
114                         ASM_CODE(inst, acceItem(ip2, i, indf2), _W_);
115                 }
116             }
117             else
118                 fetch(ip2, i, indf2);
119
120             regWREG->reset();
121         }
122
123         store(ip0, i, indf0);
124     }
125 }
```
/p16ecc/cc_source_8.12/p16e/pic16e11.cpp
Code-8.68

- #87~89: for '|' or '^' P-code, if the size of Z is greater than that of both X and Y, then the higher byte(s) of Z needs to be filled with '0's.
- #90~92: for '&' P-code, if the size of Z is greater than that of either X or Y, then the higher byte(s) of Z need to be filled with '0's.
- #106~118: generate the assembly code for the logic operation, or fetch the value of either X or Y (the result is in WREG).
- #123: store the result byte (the content in WREG) to Z.

Table-8.39 shows the test results of assembly code generation for P-code '|'.

Source code (ch8_t33.c)	Assembly code output (ch8_t33.asm)

<pre>char a;
int b;
short c;
 	
void foo()
{
  c = a | b;
}</pre>
	<pre> .segment CODE2 (REL) foo:0
	foo::
	; :: ch8_t33.c #7: c = a | b;
	        .bsel a
	        movf a, W
	        .bsel a, b
	        iorwf b, W
	        .bsel b, c
	        movwf c
	        .bsel c, b+1
	        movf b+1, W
	        .bsel b+1, c+1
	        movwf c+1
	        .bsel c+1, c+2
	        clrf c+2
	        return</pre>

Table-8.39

8.13.2 Assembly code output for logic operation with a constant

The assembly code generation for P-codes for logic operations can be briefly described as in Code-8.69. In which Y is a constant.

In logic operations, we have the following rules (or equivalents):

$$a = b \mid 1 \quad \rightarrow \quad a = 1$$
$$a = b \mid 0 \quad \rightarrow \quad a = b$$

255

$$a = b \ \char94 \ 1 \quad \rightarrow \quad a = \overline{b}$$
$$a = b \ \char94 \ 0 \quad \rightarrow \quad a = b$$
$$a = b \ \& \ 1 \quad \rightarrow \quad a = b$$
$$a = b \ \& \ 0 \quad \rightarrow \quad a = 0$$

That means the assembly code could be optimized if the constant byte has an exceptional value (0xFF or 0x00).

```cpp
127   void PIC16E :: andorxor(int code, Item *ip0, Item *ip1, int n)
128   {
129       int size0 = ip0->acceSize();
130       int size1 = ip1->acceSize();
131       char *indf0 = setFSR0(ip0);
132       char *indf1 = setFSR0(ip1);
133
134       for (int i = 0; i < size0; i++, n >>= 8)
135       {
136           int  num = n & 0xff;
137           bool load_constant = false;
138
139           switch ( num )
140           {
141               case 0x00:  // constant byte = 0x00
142                   if ( i < size1 && code != '&' )
143                       fetch(ip1, i, indf1);
144                   else
145                       load_constant = true;
146                   break;
147               case 0xff:  // constant byte = 0xFF
148                   if ( i < size1 && code == '&' )
149                       fetch(ip1, i, indf1);
150                   else if ( i < size1 && code == '^' )
151                       ASM_CODE(_COMF, acceItem(ip1, i, indf1), _W_);
152                   else
153                       load_constant = true;
154                   break;
155               default:
156                   if ( i < size1 )
157                   {
158                       char *inst = (code == '|')? _IORLW:
159                                    (code == '^')? _XORLW: _ANDLW;
160                       fetch(ip1, i, indf1);
161                       ASM_CODE(inst, num);
162                   }
163                   else
164                       load_constant = true;
165           }
166
167           if ( load_constant )
168           {
169               if ( num || indf0 )
170                   regWREG->load(num);
171               else
172               {
173                   ASM_CODE(_CLRF, acceItem(ip0, i, indf0));
174                   continue;
175               }
176           }
177           else
178               regWREG->reset();
179
180           store(ip0, i, indf0);
181       }
182   }
```


Code-8.69

- #141~146: assembly code generation when the value of a constant byte = 0x00.
- #147~154: assembly code generation when the value of a constant byte = 0xFF.

Table-8.40 shows the test results for assembly code generation in this section.

Source code (ch8_t34.c)	Assembly code output (ch8_t34.asm)
| ```c
short x, y;

void foo()
{
   x = y | 0x01ff00;
   x = y & 0x01ff00;
   x = y ^ 0x00ff01;
}
``` | ```
        .segment     CODE2 (REL) foo:0
foo::
; :: ch8_t34.c #5: x = y | 0x01ff00;
        .bsel  y
        movf   y, W
        .bsel  y, x
        movwf  x
        movlw  255
        .bsel  x, x+1
        movwf  x+1
        .bsel  x+1, y+2
        movf   y+2, W
        iorlw  1
        .bsel  y+2, x+2
        movwf  x+2
; :: ch8_t34.c #6: x = y & 0x01ff00;
        .bsel  x+2, x
        clrf   x
        .bsel  x, y+1
        movf   y+1, W
        .bsel  y+1, x+1
        movwf  x+1
        .bsel  x+1, y+2
        movf   y+2, W
        andlw  1
        .bsel  y+2, x+2
        movwf  x+2
; :: ch8_t34.c #7: x = y ^ 0x00ff01;
        .bsel  x+2, y
        movf   y, W
        xorlw  1
        .bsel  y, x
        movwf  x
        .bsel  x, y+1
        comf   y+1, W
        .bsel  y+1, x+1
        movwf  x+1
        .bsel  x+1, y+2
        movf   y+2, W
        .bsel  y+2, x+2
        movwf  x+2
        return
``` |

Table-8.40

8.14 Assembly Code Generation (12)

<u>Project directory</u>: /p16ecc/cc_source_8.13
<u>New source file</u>: pic16e12.cpp

This part presents the assembly code generation for bit-field data (or, bit-field in short) operations/manipulations. Bit-field data appear as data members defined in **struct/union** data structures, and they are assigned the offsets within the byte during the parsing (in **cc.y**). Bit-field data sizes are limited to 1-bit~8-bit, and are functioning as unsigned integers. Adjacent bit-field data members are packed into a single byte 'carrier' for 8-bit CPUs (such as the PIC16F CPU family).

Here is an example of bit-field data, STATUS register, defined in pic16e.h.

```
typedef struct {
      unsigned char C   : 1;
      unsigned char DC : 1;
      unsigned char Z   : 1;
      unsigned char PD : 1;
      unsigned char TO : 1;
} STATUS_t;
#define STATUSbits        (STATUS_t*)(&STATUS)
```

Bit fields are helpful and suitable for CPU registers access (configurations, controlling bits checking, …). In theory, bit-field data works the same way as regular integers. In this book, only the following operations are covered for bit-field data:
- Assign a value to a bit-field.
- Read the value from a bit-field.
- Compare the value of a bit-field with a constant.

Note,
1) Readers who have an interest can expand the capabilities to cover more operations needed.
2) Using bit-field data will save RAM data space, but may cost more ROM code space and slow down the application execution.

Code-8.70 lists the service routines used to manipulate bit-field data, which are defined in pic16e.h.

```
28    class PIC16E {
29        public:
30            PIC16E(char *out_file, Nlist *_nlist, Pcoder *_pcoder);
31            ~PIC16E();
32            void run(void);

          . . .

150           // pic16e12.cpp
151           #define TO_STACK      false
152           void getBF(Item *ip, int left_shift=0, int req_size=8, bool in_acc=true);
153           void movFromBF(Item *ip0, Item *ip1);
154           void movToBF(Item *ip0, Item *ip1);
155           void movToBF(Item *ip0, int n);
156           void movToBF(Item *ip0, char *s);
```
/p16ecc/cc_source_8.13/p16e/pic16e.h   Code-8.70

The following sections will expand the existing routines to handle bit-field data/variables.

8.14.1 Assembly code output for bit-field assignments

Source file updated: /p16e/pic16e2.cpp

The Code-8.71 shows the expansion in PIC16E::mov() that supports bit-field assignments (P-code '=').

```
18    void PIC16E :: mov(Item *ip0, Item *ip1)
19    {
20        if ( same(ip0, ip1) )
21            return;
22
23        if ( ip0->isBF() )
24        {
25            movToBF(ip0, ip1);
26            return;
27        }
28        if ( ip1->isBF() )
29        {
30            movFromBF(ip0, ip1);
31            return;
32        }
```
/p16ecc/cc_source_8.13/p16e/pic16e2.cpp   Code-8.71

- #23~32: the expanded part for bit-field data
- #23~27 - assigning a value to a bit-field.
- #28~32 - assigning a value from a bit-field.

Table-8.41 shows the test results for assembly code generation for assignments.

Source code (ch8_t35.c)	Assembly code output (ch8_t35.asm)
``#include <pic12f1840.h>

void foo()
{
   char a;
   STATUSbits->Z = a;
   OSCCONbits->IRCF = 2;

}``` | ```   .segment   CODE2 (REL) foo:1
foo_$1_a: .equ   foo_$data$+0
foo::
; :: ch8_t35.c #6: STATUSbits->Z = a;
          .bsel foo_$data$
          rlf   foo_$1_a, W
          rlf   WREG, F
          xorwf 3, W
          andlw 4
          xorwf 3, F
; :: ch8_t35.c #7: OSCCONbits->IRCF = 2;
          .bsel foo_$data$, 153
          movf   25, W
          andlw -61
          iorlw 8
          .bsel 153, 153
          movwf 25
          return``` |

Table-8.41

8.14.2 Assembly code output for bit-field increment

Source file updated: /p16e/pic16e3.cpp

The Code-8.72 shows the expansion in PIC16E::incValue() that supports bit-field increment (P-code INC_OP) that does

$$b \mathrel{+}= N$$

Where N is a constant.

```
18   void PIC16E :: incValue(Item *ip, int value)
19   {
20       int  size0 = ip->acceSize();
21       char *indf = setFSR0(ip);
22
23       if ( ip->isBF() )
24       {
25           size0 = BF_SIZE(ip->attr);
26           int shift = BF_OFFSET(ip->attr);
27           uint8_t mask = ((1 << size0) - 1) << shift;
28           int n = (value << shift) & mask;
29
30           if ( (size0 + shift) >= 8 )
31           {
32               regWREG->load(n);
33               ASM_CODE(_ADDWF, acceItem(ip, 0, indf), _F_);
34           }
35           else if ( n )
36           {
37               regWREG->load(n);
38               if ( size0 > 1 )
39               {
40                   ASM_CODE(_ADDWF, acceItem(ip, 0, indf), _W_);
41                   ASM_CODE(_XORWF, acceItem(ip, 0, indf), _W_);
42                   ASM_CODE(_ANDLW, mask);
43                   regWREG->reset();
44               }
45               ASM_CODE(_XORWF, acceItem(ip, 0, indf), _F_);
46           }
47       return;
48   }
```
/p16ecc/cc_source_8.13/p16e/pic16e3.cpp

Code-8.72

- #23~48: the expanded part for bit-field data
  - #30~34: if the bit-field is at the high end of the byte.

8.14.3 Assembly code output for bit-field logic compound assignments

```
Source file updated: /p16e/pic16e4.cpp
```

The operation is limited only to the situations

$$\{\texttt{AND_ASSIGN}, [b, N]\} \quad \rightarrow \quad b \mathbin{\&}= N$$
$$\{\texttt{OR_ASSIGN}, [b, N]\} \quad \rightarrow \quad b \mathbin{|}= N$$
$$\{\texttt{XOR_ASSIGN}, [b, N]\} \quad \rightarrow \quad b \mathbin{\char`\^}= N$$

Where b is a bit-field, and N is a constant.

Code 8.73 shows the expansion in `PIC16E::andorxor()` that supports compound bit-field logic operations.

```
 58   void PIC16E :: andorxor(int code, Item *ip0, int n)
 59   {

 . . .

 69       int  size0 = ip0->acceSize();
 70       char *indf = setFSR0(ip0);
 71       if ( ip0->isBF() )
 72       {
 73           size0    = BF_SIZE(ip0->attr);
 74           int shift = BF_OFFSET(ip0->attr);
 75           uint8_t mask = ((1 << size0) - 1) << shift;
 76           uint8_t num  = (n << shift) & mask;
 77           int b;
 78
 79           switch ( code )
 80           {
 81               case OR_ASSIGN:
 82                   if ( selectBit(num, &b) )
 83                       ASM_CODE(_BSF, acceItem(ip0, 0, indf), b);
 84                   else if ( num )
 85                   {
 86                       regWREG->load(num);
 87                       ASM_CODE(_IORWF, acceItem(ip0, 0, indf), _F_);
 88                   }
 89                   break;
 90               case XOR_ASSIGN:
 91                   if ( num )
 92                   {
 93                       regWREG->load(num);
 94                       ASM_CODE(_XORWF, acceItem(ip0, 0, indf), _F_);
 95                   }
 96                   break;
 97               case AND_ASSIGN:
 98                   if ( selectBit(~(~mask | num), &b) )
 99                       ASM_CODE(_BCF, acceItem(ip0, 0, indf), b);
100                   else if ( num != mask )
101                   {
102                       regWREG->load(~mask | num);
103                       ASM_CODE(_ANDWF, acceItem(ip0, 0, indf), _F_);
104                   }
105                   break;
106           }
107       return;
108   }
```
```
/p16ecc/cc_source_8.13/p16e/pic16e4.cpp
```
Code-8.73

- #71~108: the expanded part for bit-field data
    - #81~89: assembly code generation for OR_ASSIGN.
    - #90~96: assembly code generation for XOR_ASSIGN.
    - #97~105: assembly code generation for AND_ASSGN.

Table-8.42 shows the test results for assembly code generation of bit-field data logic compound assignments.

Source code (ch8_t36.c)	Assembly code output (ch8_t36.asm)

<pre>typedef struct {
  char a: 3;
  char b: 2;
  char c: 3;
} Data;
 	
Data data;
 	
void foo()
{
  data.a |= 5;
  data.b &= 1;
  data.c ^= 6;
}</pre>
	<pre> .segment CODE2 (REL) foo:0
	foo::
	; :: ch8_t36.c #11: data.a |= 5;
	    movlw 5
	    .bsel data
	    iorwf data, F
	; :: ch8_t36.c #12: data.b &= 1;
	    bcf data, 4
	; :: ch8_t36.c #13: data.c ^= 6;
	    movlw 192
	    xorwf data, F
	    return</pre>

Table-8.42

8.14.4 Assembly code output for bit-field P_JZ and P_JNZ

Source file updated: /p16e/pic16e5.cpp

Code-8.74 shows the expansion in PIC16E::jzjnz() that supports bit-field data in generating the assembly code for P-code P_JZ and P_JNZ,

    {P_JZ, [b, label]}   or
    {P_JNZ, [b, label]}

```
18   void PIC16E :: jzjnz(int code, Item *ip0, Item *ip1)
19   {
20       if ( ip0->isBF() )
21       {
22           int size0 = BF_SIZE(ip0->attr);
23           int shift = BF_OFFSET(ip0->attr);
24           char *indf0 = setFSR0(ip0);
25
26           if ( size0 == 1 )
27           {
28               char *inst = (code == P_JZ)? _BTFSS: _BTFSC;
29               regPCLATH->load(ip1->val.s);
30               ASM_CODE(inst, acceItem(ip0, 0, indf0), shift);
31           }
32           else
33           {
34               uint8_t mask = ((1 << size0) - 1) << shift;
35               char *inst = (code == P_JZ)? _BTFSC: _BTFSS;
36               if ( size0 < 8 )
37               {
38                   ASM_CODE(_MOVLW, mask);
39                   ASM_CODE(_ANDWF, acceItem(ip0, 0, indf0), _W_);
40               }
41               else
42                   ASM_CODE(_MOVF, acceItem(ip0, 0, indf0), _W_);
43
44               regPCLATH->load(ip1->val.s);
45               ASM_CODE(inst, STATUS, 2);
46               regWREG->reset();
47           }
48           ASM_CODE(_GOTO, ip1->val.s);
49           return;
50       }
```
/p16ecc/cc_source_8.13/p16e/pic16e5.cpp

Code-8.74

- #20~50: the expanded part for bit-field data
    - #26~31: when the bit-field is 1-bit in size.
    - #32~47: when the bit-field is greater than 1-bit.

Table-8.43 shows the test results for assembly code generation for bit-field data P_JZ.

Source code (ch8_t37.c)	Assembly code output (ch8_t37.asm)
``#include <pic12f1840.h>

void foo()
{
  if (PORTAbits->b0)
    LATAbits->b1 ^= 1;
}``` | ```   .segment   CODE2 (REL) foo:0
foo::
; :: ch8_t36.c #5: if (PORTAbits->b0)
          .psel foo, _$L3
          movlb 0
          btfss 12, 0
          goto   _$L3
; :: ch8_t36.c #6: LATAbits->b1 ^= 1;
          movlw 2
          .bsel 12, 268
          xorwf 12, F
_$L3:
          .psel _$L3, foo
          return``` |

Table-8.43

8.14.5 Assembly code output for bit-field P_JZ_INC/P_JZ_DEC

Source file updated: /p16e/pic16e5.cpp

The Code-8.75 shows the expansion in PIC16E::jzjnz_incdec() that supports bit-field data in generating the assembly code for P-code P_JZ_INC, P_JZ_DEC, P_JNZ_INC, and P_JNZ_DEC.

```
138   void PIC16E :: jzjnz_incdec(int code, Item *ip0, Item *ip1, Item *ip2)
139   {
140       char *indf = setFSR0(ip0);
141       int  size0 = ip0->acceSize();
142       int  n = (code == P_JZ_INC || code == P_JNZ_INC)? ip1->val.i: -ip1->val.i;
143       char *inst = (code == P_JZ_INC || code == P_JZ_DEC)? _BTFSC: _BTFSS;
144
145       if ( ip0->isBF() )
146       {
147           int size0 = BF_SIZE(ip0->attr);
148           int shift = BF_OFFSET(ip0->attr);
149           getBF(ip0, shift, size0, TO_STACK);
150           incValue(ip0, ip1->val.i);
151           ASM_CODE(_MOVIW, "INDF1++");
152       }
```
/p16ecc/cc_source_8.13/p16e/pic16e5.cpp Code-8.75

- #145~152: the expansion for bit-field.

Table-8.44 shows the test results for assembly code generation for bit-field data P_JZ.

Source code (ch8_t38.c)	Assembly code output (ch8_t38.asm)
``typedef struct {
  char a: 3;
  char b: 2;
  char c: 3;
} Data;

Data data;

void foo()
{
  if (data.b++)
    data.c++;
}``` | ```   .segment   CODE2 (REL) foo:0
foo::
; :: ch8_t38.c #11: if (data.b++)
          .bsel data
          movf   data, W
          andlw 24
          movwi --INDF1
          movlw 8
          addwf data, W
          xorwf data, W
          andlw 24
          xorwf data, F
          moviw INDF1++
          .psel foo, _$L3
          btfsc 3, 2
          goto   _$L3
; :: ch8_t38.c #12: data.c++;
          movlw 32
          addwf data, F
_$L3:
          .psel _$L3, foo
          return``` |

Table-8.44

```
Source file updated: /p16e/pic16e6.cpp
```
The Code-8.76 shows the expansion in PIC16E:jeqjne() that supports bit-field data in generating the assembly code for P-code P_JEQ and P_JNE,

```
18    void PIC16E :: jeqjne(int code, Item *ip0, Item *ip1, Item *ip2)
19    {
20        int size0 = ip0->acceSize();
21        int size1 = ip1->acceSize();
22        int sign1 = ip1->acceSign();
23        int size  = (size0 > size1)? size0: size1;
24        char *inst= (code == P_JEQ)? _BTFSC: _BTFSS;
25
26        if ( ip0->isBF() )
27        {
28            if ( ip1->type != CON_ITEM )
29                errPrint("SORRY - not support this compare!");
30            else
31            {
32                size0       = BF_SIZE(ip0->attr);
33                int shift = BF_OFFSET(ip0->attr);
34                uint8_t mask = ((1 << size0) - 1) << shift;
35                int n = (ip1->val.i << shift) & mask;
36
37                if ( ip1->val.i >= (1 << size0) )    // out of range
38                {
39                    if ( code == P_JNE )
40                    {
41                        regPCLATH->load(ip2->val.s);
42                        ASM_CODE(_GOTO, ip2->val.s);
43                    }
44                    return;
45                }
46
47                if ( size0 == 1 )                    // 1-bit compare
48                {
49                    if ( code == P_JEQ )
50                        jzjnz(n? P_JNZ: P_JZ, ip0, ip2);
51                    else
52                        jzjnz(n? P_JZ: P_JNZ, ip0, ip2);
53                }
54                else
55                {
56                    getBF(ip0, shift, size0);
57                    ASM_CODE(_XORLW, n);
58                    regWREG->reset();
59
60                    regPCLATH->load(ip2->val.s);
61                    ASM_CODE(inst, STATUS, 2);
62                    ASM_CODE(_GOTO, ip2->val.s);
63                }
64            }
65            return;
66        }
67        if ( ip1->isBF() )
68        {
69            jeqjne(code, ip1, ip0, ip2);
70            return;
71        }
```
```
/p16ecc/cc_source_8.13/p16e/pic16e6.cpp                          Code-8.76
```

Table-8.45 shows the test results for assembly code generation for bit-field data `P_JZ`.

Source code (ch8_t39.c)	Assembly code output (ch8_t39.asm)

<pre>typedef struct {
  char a: 3;
  char b: 2;
  char c: 3;
} Data;
 	
Data data;
char x;
 	
void foo ()
{
  if (data.c == 7)
    x++;
}</pre>
	<pre> .segment CODE2 (REL) foo:0
	foo::
	; :: ch8_t39.c #12: if (data.c == 7)
	        .bsel data
	        movf data, W
	        andlw 224
	        xorlw 224
	        .psel foo, _$L3
	        btfss STATUS, 2
	        goto _$L3
	; :: ch8_t39.c #13: x++;
	        .bsel data, x
	        incf x, F
	_$L3:
	        .psel _$L3, foo
	        return</pre>

Table-8.45

8.15 Assembly Code Generation (13)

```
Project directory: /p16ecc/cc source 8.14
New source file: pic16e13.cpp
Modified source files: pic16e.cpp
```

As described in Section 8.2, there are three P-code streams generated during P-code generation:

```
mainPcode        Program running code
initPcode        External and static variable initialization code
constPcode       ROMed constants (initialization data images)
```

In addition, constant strings (grouped in `constGroup`), collected from `NODE_STR` type of nodes, shall also need to be added to the assembly code output.

The function `PIC16E::outputASM0()` only generates the assembly code for `mainPcode` stream. The following sections will address the rest, as the accomplishment for the assembly code output.

The process initiation is illustrated in `PIC16E::run()`, shown in Code-8.77.

```
54   ////////////////////////////////////////////////////////////////////////////////
55   void PIC16E :: run(void)
56   {
57       time_t t = time(&t);      // current time
58       char *buf = STRBUF();     // string buffer
59       Nnode *nnp = NULL;
60       int ram_size = 0;
61       int stack_addr = 0;
62       PreScan prescan(nlist);
     . . .
```

```
126        // generate main program ASM code
127        outputASM0(pcoder->mainPcode);
128
129        // generate init code
130        outputInit(pcoder->initPcode);
131
132        // generate constant code
133        outputConst(pcoder->constPcode);
134
135        // output constant strings
136        outputString(pcoder->constGroup->list);
137
138        ASM_OUTP("\n");
139        ASM_CODE(_END);
140    }
```

/p16ecc/cc_source_8.14/p16e/pic16e.cppCode-8.77

8.15.1 Assembly code output for data initialization assignments

The P-code of initialization assignments are collected in `initPcode`. The following example shows the typical initialization of variables.

```
int x = 1000;
char array[3] = {'a','b','c'};
...
void foo()
{
    static int seed = 0;
}
```

And all those initializations are happening outside the function, before the code runs. The assembly codes of initialization are generated and put in `CODEi` segments, as show in Code-8.78. There are two types of P-codes that are used for initialization: `P_COPY` and `'='`. The former is used for array or struct/union initialization, and it copies the constant image (block) stored in ROM to the target.

```cpp
18   void PIC16E :: outputInit(Pnode *plist)
19   {
20       if ( plist == NULL ) return;
21
22       ASM_OUTP("\n");
23       ASM_CODE(_SEGMENT, "CODEi (REL)");
24       regWREG->reset();
25       regBSR->reset();
26       for(int seq = 0; plist; plist = plist->next)
27       {
28           Item *ip0 = plist->items[0];
29           Item *ip1 = plist->items[1];
30           Item *ip2 = plist->items[2];
31           int size;
32           Item *lbl;
33           if ( seq++ ) ASM_OUTP("\n");
34           switch ( plist->type )
35           {
36               case P_COPY:
37                   if ( !(ip0 && ip1 && ip2 && ip2->val.i > 0) ) break;
38                   ASM_CODE(_MOVLW, acceItem(ip1, 0)); ASM_CODE(_MOVWF, FSR1L);
39                   ASM_CODE(_MOVLW, acceItem(ip1, 1)); ASM_CODE(_MOVWF, FSR1H);
40                   ASM_CODE(_MOVLW, acceItem(ip0, 0)); ASM_CODE(_MOVWF, FSR0L);
41                   ASM_CODE(_MOVLW, acceItem(ip0, 1)); ASM_CODE(_MOVWF, FSR0H);
42
43                   size = -ip2->val.i;
44                   ASM_CODE(_MOVLW, size & 0xff);
45                   ASM_CODE(_MOVWF, ACC0);
46                   if ( ip2->val.i >= 256 )
47                   {
48                       ASM_CODE(_MOVLW, (size >> 8) & 0xff);
49                       ASM_CODE(_MOVWF, ACC1);
50                   }
51                   lbl = lblItem(pcoder->getLbl());
52                   ASM_LABL(lbl->val.s);
53                   ASM_CODE(_MOVIW, "INDF1++");
54                   ASM_CODE(_MOVWI, "INDF0++");
55                   regWREG->reset();
56
57                   ASM_CODE(_INCFSZ, ACC0, _F_);
58                   if ( ip2->val.i >= 256 )
59                   {
60                       ASM_CODE(_BRA, lbl->val.s);
61                       ASM_CODE(_INCFSZ, ACC1, _F_);
62                   }
63                   ASM_CODE(_BRA, lbl->val.s);
64                   delete lbl;
65                   break;
66
67               case '=':
68                   mov(ip0, ip1);
69                   break;
70           }
71       }
72   }
```

/p16ecc/cc_source_8.14/p16e/pic16e13.cpp   Code-8.78

- #36~65: assembly code generation for P_COPY, in which both FSR0 and FSR1 are used for initialization data image copy.

8.15.2 Assembly code output for data initialization values

Code-8.79 shows the assembly code generated for constPcode, which holds the constant images used to initialize variables. There are two types of P-code in the stream: one for segment type and the other for defining constants.

```cpp
74   void PIC16E :: outputConst(Pnode *pcode)
75   {
76       for (; pcode; pcode = pcode->next)
77       {
78           Item *ip0 = pcode->items[0];
79           Item *ip1 = pcode->items[1];
80           char *buf = STRBUF();
81           bool is_public;
82           switch ( pcode->type )
83           {
84               case P_SEGMENT:
85                   is_public = !(strchr(ip0->val.s, '$') || ip0->attr->isStatic);
86                   ASM_OUTP("\n");
87                   if ( ip1 && ip1->val.i > 0 )
88                       sprintf(buf, "CONST0 (ABS =0x%04X)", ip1->val.i);
89                   else
90                       sprintf(buf, "CONSTi (REL)");
91
92                   if ( is_public && ip0->attr->dimVect )
93                       sprintf(&buf[strlen(buf)], " %s", ip0->val.s);
94
95                   ASM_CODE(_SEGMENT, buf);
96                   ASM_LABL(ip0->val.s, is_public);
97                   break;
98
99               case P_FILL:
100                  for (int i = 0; i < ip1->val.i; i++)
101                      ASM_CODE(_RETLW, acceItem(ip0, i));
102                  break;
103          }
104      }
105  }
```
`/p16ecc/cc_source_8.14/p16e/pic16e13.cpp`   Code-8.79

- #84~97: output the assembly code for the segment type.
- #99~102: output the assembly code for constants (constant bytes take lower byte `RETLW` instructions).

8.15.3 Assembly code output for constant strings

All the constant strings are collected and kept in `constGroup` of class `Pcoder`. Every character in a string takes the lower byte of `RETLW` instruction, as illustrated in Code-8.80.

```cpp
107  void PIC16E :: outputString(Const_t *list)
108  {
109      for (; list; list = list->next)
110      {
111          ASM_OUTP("\n");
112          ASM_CODE(_SEGMENT, "CONSTi (REL)");
113          ASM_LABL(list->strName());
114          for(int i = 0; i < (int)(strlen(list->str)+1); i++)
115              ASM_CODE(_RETLW, list->str[i]);
116      }
117  }
```
`/p16ecc/cc_source_8.14/p16e/pic16e13.cpp`   Code-8.80

Table-8.46 shows the test results for assembly code generation across all the sections above.

Source code (ch8_t40.c)	Assembly code output (ch8_t40.asm)
| ```c
typedef struct {
  int a, b, c;
} Data;

Data data = {1000, 2000,
3000};

int n = 0x0fff;
char *s = "Hello, world";
``` | ```
        .segment    BANKi (REL)
data:: .rs    6
        .segment    BANKi (REL)
n::    .rs    2
        .segment    BANKi (REL)
s::    .rs    2

        .segment    CODEi (REL)
        movlw  (data$init$)
        movwf  FSR1L
        movlw  (data$init$)>>8
        movwf  FSR1H
        movlw  (data)
        movwf  FSR0L
        movlw  (data)>>8
        movwf  FSR0H
        movlw  250
        movwf  0x70
_$L1:
        moviw  INDF1++
        movwi  INDF0++
        incfsz 0x70, F
        bra    _$L1

        movlw  255
        .bsel  n
        movwf  n
        movlw  15
        .bsel  n, n+1
        movwf  n+1

        movlw  (_$CS1)
        .bsel  n+1, s
        movwf  s
        movlw  (_$CS1)>>8
        .bsel  s, s+1
        movwf  s+1

        .segment    CONSTi (REL)
data$init$:
        retlw  232
        retlw  3
        retlw  208
        retlw  7
        retlw  184
        retlw  11

        .segment    CONSTi (REL)
_$CS1:
        retlw  72
        retlw  101
        retlw  108
        retlw  108
        retlw  111
        retlw  44
        retlw  32
        retlw  119
        retlw  111
        retlw  114
        retlw  108
        retlw  100
        retlw  0
``` |

Table-8.46

Chapter-9

Finalizing and Enhancing the cc16e Compiler

As the final chapter wrapping up the design of the cc16e compiler, a few enhancements and features are added. Beyond that, floating-point data type support will be addressed as a research topic.

9.1 Optimization of Assembly Code Output

```
Project directory: /p16ecc/cc_source_9.1
New source file: pic16e_asm_opt.h, pic16e_asm_opt.cpp,
                 pic16e_asm_opt1.cpp, pic16e_asm_opt2.cpp
Modified source files: pic16e_asm.h, pic16e_asm.cpp
```

The P-code optimization described in Chapter 7 occurs at a higher (or abstraction) level and is independent of the target machine (MCU). In contrast, the optimization applied to assembly code output presented here applies only to the specific target instructions.

9.1.1 The class of assembly code optimization

A new class, `P16E_asmOPT`, is created to support the assembly code optimization shown in Code-9.1 and Code-9.2 below. And it becomes the assembly code output pipeline. Naturally, assembly code optimization is optional during compilation.

269

```cpp
 6    typedef struct {
 7        bool insertTAB;
 8        std::string inst, opr1, opr2;
 9        bool nullLine(void) {
10            return (inst.c_str()[0] == '\n' ||
11                           inst.c_str()[0] == ';'   );
12        }
13        bool isInst(char *_inst, char *des=NULL)
14        {
15            if ( inst != _inst ) return false;
16            return (des == NULL || opr2 == des);
17        }
18        bool isInst1(char *_inst, const char *opr)
19        {
20            if ( inst != _inst ) return false;
21            return (opr && opr1 == opr);
22        }
23    } AsmLine;
24
25    #define ASM_BUFFER_SIZE        32
26    #define NEXT_INDEX(pt)         ((pt + 1) & (ASM_BUFFER_SIZE-1))
27    #define LAST_INDEX(pt)         ((pt - 1) & (ASM_BUFFER_SIZE-1))
28    #define SKIP_BSEL              true
29
30    class P16E_asmOPT {
31
32        public:
33            P16E_asmOPT(FILE *fileout);
34            ~P16E_asmOPT() { flush(); }
35            void output(char *s);
36            void label(char *lbl, bool c=false, char *opr=NULL);
37            void code(char *inst);
38            void code(char *inst, char *opr1, char *opr2=NULL);
39            void code(char *inst, int opr1, char *opr2=NULL);
40            void code(char *inst, char *opr1, int opr2);
41            void code(char *inst, int opr1, int opr2);
42
43        private:
44            FILE *fout;
45            AsmLine outputBuffer[ASM_BUFFER_SIZE];
46            int outputHead;
47            int outputTail;
48            bool append_str;

          . . .

70            int optimize(void);
71            int case1(void);
72            int case2(void);
73            int case3(void);
74            int case4(void);
75            int case5(void);
76            int case6(void);
77            int case7(void);
78            int case8(void);
79            int case9(void);
80    };
```

/p16ecc/cc_source_9.1/p16e/pic16e_asm_opt.h   Code-9.1

- #6~23: data structure that is used to construct the loop buffer, which will hold output assembly code lines.
- #35~41: member functions that are called from class `P16E_ASM` to output assembly code.
- #45~47: assembly code line buffer, and its indexes.
- #70: start optimization.
- #71~79: optimization cases.

```cpp
16    P16E_asmOPT :: P16E_asmOPT(FILE *fileout)
17    {
18        fout = fileout;
19        outputHead = 0;
20        outputTail = 0;
21        append_str = false;
22    }
23
24    void P16E_asmOPT :: flush(void)
25    {
26        while ( outputHead != outputTail )
27            flush(1);
28    }
29
30    void P16E_asmOPT :: flush(int flush_lines)
31    {
32        while ( outputHead != outputTail && flush_lines > 0 )
33        {
34            int n = optimize();      // run optimization before flushing a line...
35            if ( n > 0 )
36            {
37                flush_lines -= n;
38                if ( flush_lines < 0 ) flush_lines = 0;
39                continue;
40            }
41
42            flush_lines--;
43            AsmLine *p = &outputBuffer[outputHead];
44            outputHead = NEXT_INDEX(outputHead);
45
46            if ( fout )
47            {
48                if ( p->insertTAB )
49                    fprintf(fout, "\t");
50
51                fprintf(fout, "%s", p->inst.c_str());
52                if ( p->opr1 != "" )
53                {
54                    fprintf(fout, "\t%s", p->opr1.c_str());
55                    if ( p->opr2 != "" )
56                        fprintf(fout, ", %s", p->opr2.c_str());
57                }
58
59                if ( !(strchr(p->inst.c_str(), '\n') ||
60                       strchr(p->opr1.c_str(), '\n') ||
61                       strchr(p->opr2.c_str(), '\n')) )
62                    fprintf(fout, "\n");         // terminate the line.
63            }
64        }
65    }
```

```
/p16ecc/cc_source_9.1/p16e/pic16e_asm_opt.cpp                              Code-9.2
```

- #24~28: in the destructor of the class, it needs to flush out all the assembly code in the buffer.
- #30~65: the member function `P16E_asmOPT::flush()` drives the optimization (#34) and assembly code output (#46~63).

9.1.2 Embed the assembly code optimization

Code-9.3 shows how the optimization class is added to the assembly code output class `P16E_ASM`.

```
 7    P16E_ASM :: P16E_ASM(FILE *fileout)
 8    {
 9        fout = fileout;
10        asmOpt = new P16E_asmOPT(fileout);
11    }
12
13    P16E_ASM :: ~P16E_ASM()
14    {
15        delete asmOpt;
16    }
17
18    void P16E_ASM :: output(char *s)
19    {
20        if ( s )
21            asmOpt->output(s);
22    }
23
24    void P16E_ASM :: label(char *lbl, bool c, char *opr)
25    {
26        asmOpt->label(lbl, c, opr);
27    }
28
29    void P16E_ASM :: code(char *inst)
30    {
31        asmOpt->code(inst);
32    }
33
34    void P16E_ASM :: code(char *inst, char *opr1, char *opr2)
35    {
36        asmOpt->code(inst, opr1, opr2);
37    }
38
39    void P16E_ASM :: code(char *inst, int opr1, char *opr2)
40    {
41        asmOpt->code(inst, opr1, opr2);
42    }
43
44    void P16E_ASM :: code(char *inst, char *opr1, int opr2)
45    {
46        asmOpt->code(inst, opr1, opr2);
47    }
48
49    void P16E_ASM :: code(char *inst, int opr1, int opr2)
50    {
51        asmOpt->code(inst, opr1, opr2);
52    }
```

`/p16ecc/cc_source_9.1/p16e/pic16e_asm.cpp`   Code-9.3

- #7~11: instantiate the optimization class – `P16E_asmOPT`.
- #13~16: delete the optimization class – `P16E_asmOPT`, thus flushing out the contents in its butter.
- #18~52: all the output operations are led to `P16E_asmOPT`, directly.

9.1.3 Assembly code optimization operations

As mentioned above, in class `P16E_asmOPT`, the output of assembly code (lines) is fed in through the routines described in Code-9.3, to the optimizer. And these routines will put the output lines into the (loop) buffer – `outputBuffer`, and start scanning the lines for optimization when the buffer is full, as illustrated in Code-9.4 (only presented few of the routines).

```cpp
54    void P16E_asmOPT :: code(char *inst)
55    {
56        if ( bufferFull() ) flush(1);
57
58        AsmLine *p = &outputBuffer[outputTail];
59        outputTail = NEXT_INDEX(outputTail);
60
61        p->insertTAB = true;
62        p->inst = inst;
63        p->opr1 = "";
64        p->opr2 = "";
65        AsmLine();
66        append_str = false;
67    }
68
69    void P16E_asmOPT :: code(char *inst, char *opr1, char *opr2)
70    {
71        if ( bufferFull() ) flush(1);
72
73        AsmLine *p = &outputBuffer[outputTail];
74        outputTail = NEXT_INDEX(outputTail);
75
76        p->insertTAB = true;
77        p->inst = inst;
78        p->opr1 = opr1;
79        p->opr2 = opr2? opr2: "";
80        append_str = false;
81    }
82
83    void P16E_asmOPT :: code(char *inst, int opr1, char *opr2)
84    {
85        if ( bufferFull() ) flush(1);
86
87        AsmLine *p = &outputBuffer[outputTail];
88        outputTail = NEXT_INDEX(outputTail);
89
90        char buff[16];
91        p->insertTAB = true;
92        p->inst = inst;
93        sprintf(buff, "%d", opr1);   p->opr1 = buff;
94        p->opr2 = opr2? opr2: "";
95        append_str = false;
96    }
97
98    void P16E_asmOPT :: code(char *inst, char *opr1, int opr2)
99    {
100       if ( bufferFull() ) flush(1);
101
102       AsmLine *p = &outputBuffer[outputTail];
103       outputTail = NEXT_INDEX(outputTail);
104
105       char buff[16];
106       p->insertTAB = true;
107       p->inst = inst;
108       p->opr1 = opr1;
109       sprintf(buff, "%d", opr2);   p->opr2 = buff;
110       append_str = false;
111   }
```
/p16ecc/cc_source_9.1/p16e/pic16e_asm_opt1.cpp   Code-9.4

- #56, #71, #85, #100: when the buffer is full, start optimization. And then add the input line into the buffer.

In fact, the actual optimization operation is activated in the function P16E::flush(), as described in Section 9.1.1, by calling function P16E::optimize(), in which it goes through all case inspecting and matching routines, case1()~case8(), to remove redundant lines, as illustrated in Code-9.5.

```cpp
16     int P16E_asmOPT :: optimize(void)
17     {
18         int optimized = 0, n;
19         while ( outputHead != outputTail )
20         {
21             if ( (n = case1()) ) { optimized += n; continue; }
22             if ( (n = case2()) ) { optimized += n; continue; }
23             if ( (n = case3()) ) { optimized += n; continue; }
24             if ( (n = case4()) ) { optimized += n; continue; }
25             if ( (n = case5()) ) { optimized += n; continue; }
26             if ( (n = case6()) ) { optimized += n; continue; }
27             if ( (n = case7()) ) { optimized += n; continue; }
28             if ( (n = case8()) ) { optimized += n; continue; }
29             break;
30         }
31
32         return optimized;
33     }
```
`/p16ecc/cc_source_9.1/p16e/pic16e_asm_opt2.cpp`   Code-9.5

P16E_asmOPT::case1() is shown in Code-9.6.

```cpp
35     int P16E_asmOPT :: case1(void)
36     {
37         if ( bufferDepth() >= 2 )
38         {
39             int idx1, idx2;
40             AsmLine *p0 = &outputBuffer[outputHead];
41             AsmLine *p1 = nextLine(outputHead, &idx1);
42
43             if ( p0->isInst(_MOVWF) && p1->isInst(_MOVWF) &&
44                  sameOperand(p0, p1)                        )
45             {
46                 removeLine(idx1);
47                 return 1;
48             }
49             if ( p0->isInst(_MOVF, _W_) &&
50                  p1->isInst(_MOVWF) && sameOperand(p0, p1) )
51             {
52                 removeLine(idx1);
53                 return 1;
54             }
55             if ( isUpdate_WandZ(p0) && bufferDepth() >= 3 )
56             {
57                 AsmLine *p2 = nextLine(idx1, &idx2);
58                 if ( p1->isInst(_MOVWF) && p2->isInst(_MOVF, _W_) &&
59                      sameOperand(p1, p2)                            )
60                 {
61                     removeLine(idx2);
62                     return 1;
63                 }
64             }
65         }
66         return 0;
67     }
```
`/p16ecc/cc_source_9.1/p16e/pic16e_asm_opt2.cpp`   Code-9.6

- #43~48: remove the duplication:

Before	After
MOVWF *x* MOVWF *x*	MOVWF *x*

- #49~54: remove redundant write operation:

Before	After
MOVF *x*, W MOVWF *x*	MOVF *x*, W

- #55~64: remove redundant read operation:

Before	After

ADDWF *?*, W
 MOVWF *x*
 MOVF *x*, W
	ADDWF *?*, W
	 MOVWF *x*

`P16E_asmOPT::case2()` is shown in Code-9.7.

```
69    int P16E_asmOPT :: case2(void)
70    {
71        if ( bufferDepth() >= 1 )
72        {
73            AsmLine *p0 = &outputBuffer[outputHead];
74
75            if ( (p0->isInst(_BSEL) || p0->isInst(_PSEL)) &&
76                  p0->opr1 == p0->opr2                      )
77            {
78                removeLine(outputHead);
79                return 1;
80            }
81        }
82        return 0;
83    }
```
/p16ecc/cc_source_9.1/p16e/pic16e_asm_opt2.cpp   Code-9.7

- #75~80: remove the lines like ".bsel *label*, *label*" or ".psel *label*, *label*".

`P16E_asmOPT::case3()` is shown in Code-9.8.

```
85     int P16E_asmOPT :: case3(void)
86     {
87         if ( bufferDepth() >= 2 )
88         {
89             int idx, n, m;
90             AsmLine *p0 = &outputBuffer[outputHead];
91             AsmLine *p1 = nextLine(outputHead, &idx);
92
93             if ( p0->isInst(_ADDFSR) && isINDF0(p0->opr1) &&
94                   p1->isInst(_ADDFSR) && isINDF0(p1->opr1) &&
95                   isConst(p0->opr2, &n) && isConst(p1->opr2, &m) )
96             {
97                 if ( (n + m) >= -32 && (n + m) <= 31 )
98                 {
99                     int c = 1;
100                    char buff[16];
101                    sprintf(buff, "%d", n + m);
102                    p0->opr2 = buff;
103                    removeLine(idx);
104
105                    if ( (n + m) == 0 )
106                    {
107                        removeLine(outputHead);
108                        c++;
109                    }
110                    return c;
111                }
112            }
113
114            if ( p0->isInst(_ADDFSR) && isINDF1(p0->opr1) &&
115                  p1->isInst(_ADDFSR) && isINDF1(p1->opr1) &&
116                  isConst(p0->opr2, &n) && isConst(p1->opr2, &m) )
117            {
118                if ( (n + m) >= -32 && (n + m) <= 31 )
119                {
120                    int c = 1;
121                    char buff[16];
122                    sprintf(buff, "%d", n + m);
123                    p0->opr2 = buff;
124                    removeLine(idx);
125
126                    if ( (n + m) == 0 )
127                    {
128                        removeLine(outputHead);
129                        c++;
130                    }
131                    return c;
132                }
133            }
134        }
135        return 0;
136    }
```
/p16ecc/cc_source_9.1/p16e/pic16e_asm_opt2.cpp   Code-9.8

- #93~112: combine consecutive lines of "addfsr 0, *N*".
- #114~133: combine consecutive lines of "addfsr 1, *N*".

`P16E_asmOPT::case4()` is shown in Code-9.9.

```
138   int P16E_asmOPT :: case4(void)
139   {
140       if ( bufferDepth() >= 4 )
141       {
142           int idx1, idx2, idx3;
143           AsmLine *p0 = &outputBuffer[outputHead];
144           AsmLine *p1 = nextLine(outputHead, &idx1);
145                         nextLine(idx1, &idx2);
146           AsmLine *p3 = nextLine(idx2, &idx3);
147
148           if ( (p0->isInst(_BTFSS) || p0->isInst(_BTFSC)) &&
149               p1->isInst(_GOTO)                           )
150           {
151               std::string lbl = p1->opr1 + ":";
152               if ( lbl == p3->inst )
153               {
154                   p0->inst = p0->isInst(_BTFSS)? _BTFSC: _BTFSS;
155                   removeLine(idx1);
156                   return 1;
157               }
158           }
159       }
160       return 0;
161   }
```
/p16ecc/cc_source_9.1/p16e/pic16e_asm_opt2.cpp   Code-9.9

- #148~158: remove redundant GOTO line. Such as

Before	After
BTFSC … GOTO label *any instruction* label:	BTFSS … *any instruction* label:

`P16E_asmOPT::case5()` is shown in Code-9.10.

```
163   int P16E_asmOPT :: case5(void)
164   {
165       if ( bufferDepth() >= 3 )
166       {
167           int idx1, idx2, n;
168           AsmLine *p0 = &outputBuffer[outputHead];
169           AsmLine *p1 = nextLine(outputHead, &idx1);
170           AsmLine *p2 = nextLine(idx1, &idx2);
171
172           if ( (p0->isInst(_MOVLW) && isConst(p0->opr1, &n) && n == 0) ||
173               p0->isInst(_CLRW)                                        )
174           {
175               if ( p1->isInst(_MOVWF) && renewWREG(p2) )
176               {
177                   p1->inst = _CLRF;
178                   removeLine(outputHead);
179                   return 1;
180               }
181           }
182       }
183       return 0;
184   }
```
/p16ecc/cc_source_9.1/p16e/pic16e_asm_opt2.cpp   Code-9.10

- #172~181: remove redundant MOVLW line. Such as

Before	After
MOVLW 0 MOVWF *x* MOVF *y*, W	CLRF *x* MOVF *y*, W

`P16E_asmOPT::case6()` is shown in Code-9.11a~Code-9.11c.

```cpp
186    int P16E_asmOPT :: case6(void)
187    {
188        if ( bufferDepth() >= 8 )
189        {
190            int idx1, idx2, idx3, idx4, idx5, idx6, idx7;
191            AsmLine *p0 = &outputBuffer[outputHead];
192            AsmLine *p1 = nextLine(outputHead, &idx1, SKIP_BSEL);
193            AsmLine *p2 = nextLine(idx1, &idx2, SKIP_BSEL);
194            AsmLine *p3 = nextLine(idx2, &idx3, SKIP_BSEL);
195            AsmLine *p4 = nextLine(idx3, &idx4, SKIP_BSEL);
196            AsmLine *p5 = nextLine(idx4, &idx5, SKIP_BSEL);
197            AsmLine *p6 = nextLine(idx5, &idx6, SKIP_BSEL);
198            AsmLine *p7 = nextLine(idx6, &idx7, SKIP_BSEL);
199
200            if ( p0->isInst(_MOVF, _W_)      &&
201                 p1->isInst(_ADDWF, _F_)     &&
202                 p2->isInst(_MOVF, _W_)      &&
203                 p3->isInst(_ADDWFC, _F_)    &&
204                 p4->isInst(_MOVF, _W_)      &&
205                 p5->isInst(_MOVWF) && p5->opr1 == FSR0L &&
206                 p6->isInst(_MOVF, _W_)      &&
207                 p7->isInst(_MOVWF) && p7->opr1 == FSR0H )
208            {
209                if ( isVarPair(p0->opr1, p2->opr1) &&
210                     isTempPair(p1->opr1, p3->opr1) &&
211                     sameOperand(p1, p4) && sameOperand(p3, p6) )
212                {
213                    p1->opr2 = _W_;
214                    p3->opr2 = _W_;
215
216                    AsmLine t = *p5;
217                    *p5 = *p7;
218                    *p4 = *p3;
219                    *p3 = *p2;
220                    *p2 = t;
221                    removeLine(idx7);
222                    removeLine(idx6);
223                    return 2;
224                }
```

          Code-9.11a

- #200~225: remove redundant temporary variable read-back:

Before	After
MOVF *x*, W	
ADDWF *temp*, F	MOVF *x*, W
MOVF *x*+1, W	ADDWF *temp*, W
ADDWFC *temp*+1, F	MOVWF FSR0L
MOVF *temp*, W	MOVF *x*+1, W
MOVWF FSR0L	ADDWFC *temp*+1, W
MOVF *temp*+1, W	MOVWF FSR0H
MOVWF FSR0H	

```cpp
227            if ( p0->isInst(_MOVLW )         &&
228                 p1->isInst(_ADDWF, _F_)     &&
229                 p2->isInst(_MOVLW)          &&
230                 p3->isInst(_ADDWFC, _F_)    &&
231                 p4->isInst(_MOVF, _W_)      &&
232                 p5->isInst(_MOVWF)          &&
233                 p6->isInst(_MOVF, _W_)      &&
234                 p7->isInst(_MOVWF)          )
235            {
236                if ( isAddrPair(p0->opr1, p2->opr1) &&
237                     isTempPair(p1->opr1, p3->opr1) &&
238                     sameOperand(p1, p4) && sameOperand(p3, p6) &&
239                     p5->opr1 == FSR0L && p7->opr1 == FSR0H )
```

```
240             {
241                 p1->opr2 = _W_;
242                 p3->opr2 = _W_;
243
244                 AsmLine t = *p5;
245                 *p5 = *p7;
246                 *p4 = *p3;
247                 *p3 = *p2;
248                 *p2 = t;
249                 removeLine(idx7);
250                 removeLine(idx6);
251                 return 2;
252             }
253         }
```
/p16ecc/cc_source_9.1/p16e/pic16e_asm_opt2.cpp   Code-9.11b

- #227~252: remove redundant temporary variable read-back, for setting FSR0:

Before	After
MOVLW *x*	
ADDWF *temp*, F	MOVLW *x*
MOVLW *x* >> 8	ADDWF *temp*, W
ADDWFC *temp*+1, F	MOVWF FSR0L
MOVF *temp*, W	MOVLW *x* >> 8
MOVWF FSR0L	ADDWFC *temp*+1, W
MOVF *temp*+1, W	MOVWF FSR0H
MOVWF FSR0H	

```
255         if ( p0->isInst(_MOVLW)          &&
256              p1->isInst(_ADDWF, _F_)     &&
257              p2->isInst(_MOVLW)          &&
258              p3->isInst(_ADDWFC, _F_)    &&
259              p4->isInst(_MOVF, _W_)      &&
260              p5->isInst(_MOVWI) && p5->opr1 == "--INDF1" &&
261              p6->isInst(_MOVF, _W_)      &&
262              p7->isInst(_MOVWI) && p7->opr1 == "--INDF1" )
263         {
264             if ( isAddrPair(p0->opr1, p2->opr1) &&
265                  isTempPair(p1->opr1, p3->opr1) &&
266                  sameOperand(p1, p4) && sameOperand(p3, p6) )
267             {
268                 p1->opr2 = _W_;
269                 p3->opr2 = _W_;
270
271                 AsmLine t = *p5;
272                 *p5 = *p7;
273                 *p4 = *p3;
274                 *p3 = *p2;
275                 *p2 = t;
276                 removeLine(idx7);
277                 removeLine(idx6);
278                 return 2;
279             }
280         }
281     }
282     return 0;
283 }
```
/p16ecc/cc_source_9.1/p16e/pic16e_asm_opt2.cpp   Code-9.11c

- #255~280: remove redundant temporary variable read-back, for stack pushing operation:

Before	After
MOVLW *x*	
ADDWF *temp*, F	MOVLW *x*
MOVLW *x* >> 8	ADDWF *temp*, W
ADDWFC *temp*+1, F	MOVWI --INDF1
MOVF *temp*, W	MOVLW *x* >> 8
MOVWI --INDF1	ADDWFC *temp*+1, W
MOVF *temp*+1, W	MOVWI --INDF1
MOVWI --INDF1	

P16E_asmOPT::case7() is shown in Code-9.12.

```
285    int P16E_asmOPT :: case7(void)
286    {
287        if ( bufferDepth() >= 10 )
288        {
289            int idx1, idx2, idx3, idx4, idx5, idx6, idx7, idx8, idx9;
290            AsmLine *p0 = &outputBuffer[outputHead];
291            AsmLine *p1 = nextLine(outputHead, &idx1, SKIP_BSEL);
292            AsmLine *p2 = nextLine(idx1, &idx2, SKIP_BSEL);
293            AsmLine *p3 = nextLine(idx2, &idx3, SKIP_BSEL);
294            AsmLine *p4 = nextLine(idx3, &idx4, SKIP_BSEL);
295            AsmLine *p5 = nextLine(idx4, &idx5, SKIP_BSEL);
296            AsmLine *p6 = nextLine(idx5, &idx6, SKIP_BSEL);
297            AsmLine *p7 = nextLine(idx6, &idx7, SKIP_BSEL);
298            AsmLine *p8 = nextLine(idx7, &idx8, SKIP_BSEL);
299            AsmLine *p9 = nextLine(idx8, &idx9, SKIP_BSEL);

301            if ( p0->isInst(_MOVLW)          &&
302                 p1->isInst(_ADDWF, _W_)     &&
303                 p2->isInst(_MOVWF)          &&
304                 p3->isInst(_MOVLW)          &&
305                 p4->isInst(_ADDWFC, _W_)    &&
306                 p5->isInst(_MOVWF)          &&
307                 p6->isInst(_MOVF, _W_)      &&
308                 p7->isInst(_MOVWF) && p7->opr1 == FSR0L &&
309                 p8->isInst(_MOVF, _W_)      &&
310                 p9->isInst(_MOVWF) && p9->opr1 == FSR0H      )
311            {
312                if ( isAddrPair(p0->opr1, p3->opr1) &&
313                     isTempPair(p2->opr1, p5->opr1) &&
314                     sameOperand(p2, p6) && sameOperand(p5, p8) )
315                {
316                    *p2 = *p7;
317                    *p5 = *p9;
318                    removeLine(idx9);
319                    removeLine(idx8);
320                    removeLine(idx7);
321                    removeLine(idx6);
322                    return 4;
323                }
324            }
325        }
326        return 0;
327    }
```

/p16ecc/cc_source_9.1/p16e/pic16e_asm_opt2.cpp   Code-9.12

- #301~325: remove redundant read/write of temporary variable:

Before	After
MOVLW x	
ADDWF y, W	
MOVWF $temp$	MOVLW x
MOVLW $x \gg 8$	ADDWF y, W
ADDWFC y+1, W	MOVWF FSR0L
MOVWF $temp$+1	MOVLW $x \gg 8$
MOVF $temp$, W	ADDWFC y+1, W
MOVWF FSR0L	MOVWF FSR0H
MOVF $temp$+1, W	
MOVWF FSR0H	

P16E_asmOPT::case8() is shown in Code-9.13a ~ Code-9.13e.

```
329    int P16E_asmOPT :: case8(void)
330    {
331        if ( bufferDepth() >= 2 )
332        {
333            int idx1, m;
334            AsmLine *p0 = &outputBuffer[outputHead];
335            AsmLine *p1 = nextLine(outputHead, &idx1, SKIP_BSEL);
336
337            if ( p0->isInst(_ADDFSR) && isINDF0(p0->opr1) &&
338                 isConst(p0->opr2, &m) && (m == 1 || m == -1) )
339            {
340                if ( p1->isInst(_MOVF, _W_) && isINDF0(p1->opr1) )
341                {
342                    p1->inst = _MOVIW;
343                    p1->opr1 = (m > 0) ? "++INDF0" : "--INDF0";
344                    p1->opr2 = "";
345                    removeLine(outputHead);
346                    return 1;
347                }
348            }
```
/p16ecc/cc_source_9.1/p16e/pic16e_asm_opt2.cpp   Code-9.13a

- #337~348: combine the lines:

<table>
<tr><td>Before</td><td>After</td></tr>
<tr><td>ADDFSR 0, 1
MOVF INDF0, W</td><td>MOVIW ++INDF0</td></tr>
</table>

```
350            if ( p0->isInst(_ADDFSR) && isINDF1(p0->opr1) &&
351                 isConst(p0->opr2, &m) && (m == 1 || m == -1) )
352            {
353                if ( p1->isInst(_MOVF, _W_) && isINDF1(p1->opr1) )
354                {
355                    p1->inst = _MOVIW;
356                    p1->opr1 = (m > 0)? "++INDF1" : "--INDF1";
357                    p1->opr2 = "";
358                    removeLine(outputHead);
359                    return 1;
360                }
361            }
```
/p16ecc/cc_source_9.1/p16e/pic16e_asm_opt2.cpp   Code-9.13b

- #350~361: combine the lines:

<table>
<tr><td>Before</td><td>After</td></tr>
<tr><td>ADDFSR 1, 1
MOVF INDF1, W</td><td>MOVIW ++INDF1</td></tr>
</table>

```
363            if ( p0->isInst(_ADDFSR) && isINDF0(p0->opr1) &&
364                 isConst(p0->opr2, &m) && (m == 1 || m == -1) )
365            {
366                if ( p1->isInst(_MOVWF) && isINDF0(p1->opr1) )
367                {
368                    p1->inst = _MOVWI;
369                    p1->opr1 = (m > 0) ? "++INDF0" : "--INDF0";
370                    p1->opr2 = "";
371                    removeLine(outputHead);
372                    return 1;
373                }
374            }
```
/p16ecc/cc_source_9.1/p16e/pic16e_asm_opt2.cpp   Code-9.13c

- #363~374: combine the lines:

<table>
<tr><td>Before</td><td>After</td></tr>
<tr><td>ADDFSR 0, 1
MOVWF INDF0</td><td>MOVWI ++INDF0</td></tr>
</table>

```
376     if ( p0->isInst(_ADDFSR) && isINDF1(p0->opr1) &&
377         isConst(p0->opr2, &m) && (m == 1 || m == -1) )
378     {
379         if ( p1->isInst(_MOVWF) && isINDF1(p1->opr1) )
380         {
381             p1->inst = _MOVWI;
382             p1->opr1 = (m > 0) ? "++INDF1" : "--INDF1";
383             p1->opr2 = "";
384             removeLine(outputHead);
385             return 1;
386         }
387     }
```
/p16ecc/cc_source_9.1/p16e/pic16e_asm_opt2.cpp   Code-9.13d

- #363~374: combine the lines:

Before	After
ADDFSR 1, 1 MOVWF INDF1	MOVWI ++INDF1

```
389     if ( p0->isInst1(_MOVWI, "--INDF1") && p1->isInst1(_MOVIW, "INDF1++") )
390     {
391         p0->inst = _IORLW;
392         p0->opr1 = "0";
393         removeLine(idx1);
394         return 1;
395     }
396     }
397     return 0;
398 }
```
/p16ecc/cc_source_9.1/p16e/pic16e_asm_opt2.cpp   Code-9.13e

- #389~395: remove redundant stack pointer operation:

Before	After
MOVWI --INDF1 MOVIW INDF1++	IORLW 0

9.2 Options for Compiling Command

<u>Project directory</u>: /p16ecc/cc_source_9.2
<u>New source file</u>: option.h, option.cpp
<u>Modified source files</u>: main.cpp, pic16e_asm.cpp

Compiling options, or flags, are commands and settings that users provide to a compiler to control how it transforms source code into an executable, affecting performance, debugging, and the final code's behavior. For example, GNU GCC compiler has the compiling option

```
gcc -Os file.c
```

This sets the optimization level to the maximum during compilation. The following table, Table-9.1, lists the options that cc16e supports.

Option	Usage	Example
-? or -help	Help prompt for cc16e command	
-O0	Disable P-code optimization	
-a	Disable assembly code optimization	
-d or -debug	Enable debug output (print out P-codes)	
-D*id*=*value*	Macro definition	-DMAX_VALUE=100
-M=*cpu_model*	Target CPU model selection	-M="pic16f1788"

Table-9.1

9.2.1 The class of compiling option

The following, Code-9.14 and Code-9.15, are the class definition that handles the compiling option control.

```
1    #ifndef _OPTION_H
2    #define _OPTION_H
3
4       #include "nlist.h"
5
6    class Option {
7        public:
8            Option();
9            ~Option();
10
11           bool   add(char *s);
12           Nnode *get(int type, int index, char *id = NULL);
13
14       public:
15           int    level;
16           bool   debug;
17           bool   asmopt;
18           char  *mcuFile;
19
20       private:
21           Nnode *nnpList;
22           node  *makeNode(char *s);
23           void   addNode(Nnode *np);
24    };
25
26    extern Option *option;
27
28    #define DEBUG_ON        (option->debug)
29    #define PCODE_OPT_ON    (option->level == 's')
30    #define ASM_OPT_ON      (option->asmopt)
31
32    #endif
```
/p16ecc/cc_source_9.2/option.h
Code-9.14

- #11: add an option (flag).
- #12: get an option item.
- #15~18: current option items.

```cpp
1    #include <stdio.h>
2    #include <stdlib.h>
3    #include <string.h>
4    #include <ctype.h>
5    #include <string>
6    #include "common.h"
7    extern "C" {
8    #include "cc.h"
9    int aConstant(char *s, int type);
10   }
11   #include "option.h"
12
13   #define HELP_PROMPT "\n'cc16e' Command Line Format:\n\n\tcc16e [option(s)] file1.c file2.c ...\n\n"
14   #define TEMP_FILE   (char*)"~mcu.c"
15   #define MCU_FILE    TEMP_FILE"_"
16
17   Option *option;
18
19   Option :: Option()
20   {
21       nnpList = NULL;
22       level   = 's';  // -Os
23       debug   = false;
24       asmopt  = true;
25       mcuFile = NULL;
26   }
27
28   Option :: ~Option()
29   {
30       while ( nnpList )
31       {
32           Nnode *next = nnpList->next;
33           delete nnpList;
34           nnpList = next;
35       }
36
37       if ( mcuFile )
38           remove(mcuFile);
39   }
40
```

```cpp
41   bool Option :: add(char *str)
42   {
43       char *p;
44       const char *prompt = "unknown/unsupported option";
45       int length = strlen(str);
46
47       switch ( str[0] )
48       {
49           case '?': case 'h':
50               printf(HELP_PROMPT);
51               return false;
52
53           case 'O':    // optimization level
54               if ( length == 2 )
55               {
56                   level = str[1];
57                   return true;
58               }
59               break;
60
61           case 'd':
62               return (debug = true);
63
64           case 'D':    // #define ...
65               if ( str[1] != '=' && length > 2 )
66               {
67                   p = strchr(str, '=');
68                   if ( p && p[1] )     // search start of ID
69                   {
70                       int  id_len = strlen(str) - strlen(p);  // ID length
71                       char id_buf[id_len];
72                       memcpy(id_buf, &str[1], id_len); id_buf[id_len] = 0;
73
74                       if ( !get('D', 0, id_buf) )             // re-defined?
75                       {
76                           addNode(new Nnode('D', id_buf, makeNode(p + 1)));
77                           return true;
78                       }
79                       prompt = "re-defined";
80                   }
81                   if ( p == NULL && !get('D', 0, &str[1]) )
82                   {
83                       addNode(new Nnode('D', &str[1]));
84                       return true;
85                   }
86               }
87               break;
88
89           case 'a':
90               asmopt = false;
91               return true;
92
93           case 'M':    // mcu model (indlude <xxx.h>)
94               if ( str[1] == '=' && length > 2 && !get('M', 0) )
95               {
96                   FILE *mcu_file = fopen(TEMP_FILE, "w");
97                   if ( mcu_file )
98                   {
99                       fprintf(mcu_file, "#include <%s.h>\n", &str[2]);
100                      fclose(mcu_file);
101
102                      std::string mcu_f = "cpp1 "; mcu_f += TEMP_FILE;
103                      int rtcode = system(mcu_f.c_str());
104                      remove(TEMP_FILE);
105
106                      if ( rtcode != 0 )
107                          return false;
108
109                      addNode(new Nnode('M', &str[2]));
110                      mcuFile = MCU_FILE;
111                      return true;
112                  }
113              }
114              break;
115      }
116      printf("%s - %s\n", prompt, str);
117      return false;
118  }
```

/p16ecc/cc_source_9.2/option.cpp

- #93~113: for option −M, a temporary file, named "~mcu.c_", is generated (after being processed by cpp1).

9.2.2 Embed the option control

Option class is instantiated in the root function `main()` at the beginning, as illustrated in Code-9.16.

```cpp
16  int main(int argc, char *argv[])
17  {
18      option = new Option();
19      for (int i = 1; i < argc; i++)
20      {
21          char *p = argv[i];
22          int l = strlen(p);
23
24          if ( p[0] == '-' )  // compiling option...
25          {
26              if ( !option->add(&p[1]) ) break;
27              continue;
28          }
29
30          if ( !(l > 2 && p[l-2] == '.' && toupper(p[l-1]) == 'C') )
31              continue;
32
33          printf("parse ... %s\n", p);
34          std::string str = "cpp1 ";
35          str += p;
37          if ( system(str.c_str()) == 0 )
38          {
39              str = argv[i]; str += "_";
40
41              if ( option->mcuFile )  // merge files
42              {
43                  char buf[l + 32];
44                  sprintf(buf, "cat %s %s > ~.c_", option->mcuFile, str.c_str());
45                  system(buf);
46                  remove(str.c_str());
47                  rename("~.c_", str.c_str());
48              }
49
50              int rtcode = _main((char*)str.c_str());
51              remove(str.c_str());
52
53              if ( rtcode == 0 )
54              {
55                  Nlist nlist;
56                  for (int i = 0;;)
57                  {
58                      Nnode *np = option->get('D', i++);
59                      if ( np == NULL ) break;
60                      nlist.add(np->name, DEFINE, cloneNode(np->np[0]));
61                  }
62
63                  PreScan preScan(&nlist);
64                  progUnit = preScan.scan(progUnit);
65
66                  Pcoder pcoder;              // P-code generation
67                  pcoder.run(progUnit);       // run it now.
68
69                  if ( pcoder.errorCount == 0 )
70                  {
71                      if ( PCODE_OPT_ON )
72                      {
73                          Optimizer opt(&pcoder);
74                          opt.run();
75                      }
76
77                      if ( DEBUG_ON )
78                          display(pcoder.mainPcode);
79
80                      str.replace(str.length()-3, 3, ".asm");
81                      PIC16E asm_gen((char*)str.c_str(), &nlist, &pcoder);
82                      asm_gen.run();
83                  }
84              }
85          }
86      }
87      delete option;
88      return 0;
89  }
```

/p16ecc/cc_source_9.2/main.cpp

- #18: instantiate `Option` class.
- #24~28: if the item of the command line starts with '-', it will be added to `Option` class.
- #33~35: generate the command line for pre-processor "cpp1 *file*.c".
- #37: start pre-processor, generate *file*.c.
- #41~48: if the target CPU model option has been added, using -M option, then its head file, ~mcu.c_, will be merged in.
- $55~61: if any external macro has been defined, using -D option, then it needs to be added into the name list `nlist`, before pre-scan operation.
- #69~75: option -O control.
- #77~78: option -d control.

Code-9.17 shows how option -a controls the assembly code optimization in class `P16E_ASM`.

```
8    P16E_ASM :: P16E_ASM(FILE *fileout)
9    {
10       fout = fileout;
11       asmOpt = new P16E_asmOPT(fileout);
12   }
13
14   P16E_ASM :: ~P16E_ASM()
15   {
16       delete asmOpt;
17   }
18
19   void P16E_ASM :: output(char *s)
20   {
21       if ( s )
22       {
23           if ( ASM_OPT_ON )
24               asmOpt->output(s);
25           else
26               fprintf(fout, "%s", s);
27       }
28   }
29
30   void P16E_ASM :: label(char *lbl, bool c, char *opr)
31   {
32       if ( ASM_OPT_ON )
33           asmOpt->label(lbl, c, opr);
34       else
35       {
36           fprintf(fout, "%s:", lbl);
37           if ( c ) fprintf(fout, ":");
38           if ( opr ) fprintf(fout, "\t%s", opr);
39           fprintf(fout, "\n");
40       }
41   }
42
43   void P16E_ASM :: code(char *inst)
44   {
45       if ( ASM_OPT_ON )
46           asmOpt->code(inst);
47       else
48           fprintf(fout, "\t%s\n", inst);
49   }
     . . .
```
/p16ecc/cc_source_9.2/p16e/pic16e_asm.cpp   Code-9.17

9.3 Basic Library Functions

<u>New source file</u>: /p16ecc/lib/crt0.c, /p16ecc/lib/crt1.c

Library functions are used to simplify and improve the efficiency of the compiler's operation. A compiler tool set usually supplies a group of library functions, as a package, that are transparent to the users. The library functions are typically written in C and come with the corresponding header files.

In this section, the document addresses the library functions inherent to the compiling operation – basic library functions. That means the library functions are automatically included during linking, without being specified with header files. In addition, those function codes are highly optimized to improve efficiency.

9.3.1 Basic library functions for integer operations

As described in previous chapters, cc16e relies on the support from the library functions, which are wrapped in a file name crt0.c, located in the dedicated folder /lib. The following, Table-9.2, lists all the functions in it.

Library function	Purpose
_pcall	Indirect function call
_copyPar	Move function parameters
_mul8	8-bit multiplication
_mul16	16-bit multiplication
_mul24	24-bit multiplication
_mul32	32-bit multiplication
_mul16indf	16-bit multiplication
_mul24indf	24-bit multiplication
_mul32indf	32-bit multiplication
_divmod8	8-bit division/modulation
_divmod16	16-bit division/modulation
_divmod24	24-bit division/modulation
_divmod32	32-bit division/modulation
_signedCmpAndZ	Adjust flags for comparison
_signedCmpOrZ	Adjust flags for comparison
_saveFSR1	Save FSR1
_restoreFSR1	Restore FSR1

Table-9.2

Code-9.18 lists few library functions in crt0.c:

```
21   void _pcall(void)
22   {
23       PCLATH = WREG;
24       asm("moviw INDF1++");
25       PCL = WREG;
26   }
27
28   /*******************************
29       Copy function parameters,
30       from stack pointed by FSR1 to the RAM pointed by FSR0
31   *******************************/
32   void _copyPar(void)
33   {
34       ACC0 = WREG;
35       do {
36           asm("moviw   INDF1++");
37           asm("movwi   --INDF0");
38       } while ( --ACC0 );
39   }
```

/p16ecc/lib/crt0.cCode-9.18

9.3.2 Basic library functions for float operations

The following table, Table-9.3, gives the library functions that support float value operations. Note, this group of routines does not cover the math float functions (such as `sin()`, `cos()`, `sqrt()`, …). It's just a case study about how to support floating-point operations in an 8-bit CPU architecture/environment, which will be discussed in the next section.

Library function	Purpose
_uint4ToFloatACC	Unsigned integer to float (store in ACC)
_int4ToFloatACC	Signed integer to float (store in ACC)
_uint4ToFloatStack	Unsigned integer to float (store in stack)
_int4ToFloatStack	Signed integer to float (store in stack)
_floatToInt4ACC	Float value to integer (store in ACC)
_floatEqu	Compare float value (equal or not)
_floatAdd	Addition for float values
_floatSub	Subtraction for float values
_floatMul	Multiplication for float values
_floatDiv	Division for float values
_floatCmpJLT	Float compare (jump on '<')
_floatCmpJLE	Float compare (jump on '<=')

Table-9.3

9.4 Study of Floating-point

```
Project directory: /p16ecc/cc_source_10.1
New source file: fformat.h, fformat.cpp, /p16e/pic16e14.cpp
Modified source files: cc.l, cc.h, common.c, common.h, …
```

As a new type of data/structure, support for floating-point will affect most parts of the compiler design.

9.4.1 IEEE-754 floating-point format

A float-point value is formatted in 32 bits (4 bytes), 1 bit sign (s), 8 bits exponent ($e_7 \sim e_0$), and 23 bits fraction or mantissa ($f_{22} \sim f_0$), as shown below in IEEE-754 format.

Byte-1 (MSB)								Byte-2								Byte-3								Byte-4 (LSB)							
s	e_7	e_6	e_5	e_4	e_3	e_2	e_1	e_0	f_{22}	f_{21}	f_{20}	f_{19}	f_{18}	f_{17}	f_{16}	f_{15}	f_{14}	f_{13}	f_{12}	f_{11}	f_{10}	f_9	f_8	f_7	f_6	f_5	f_4	f_3	f_2	f_1	f_0

s: 0 = positive value; 1 = negative value.

$e_7 \sim e_0$: excessive-127 (or shift-127); that means: $e = [e_7 \sim e_0]$ - 127

$f_{22} \sim f_0$: fraction in original format.

A float value x can be described as:

$$x = (-1)^s \times 1.f \times 2^e$$

The following table (Table 9.4) shows the tests generated with gcc, along with the binary format of float values.

C source code

```c
#include <stdio.h>

void print(float f)
{
   int *p = (int*)&f;
   printf("%f = %08X\n", f, (unsigned int)*p);
}

void main()
{
   print(0.0);
   print(-0.0);
   print(1.0);
   print(-1.0);
   print(0.5);
   print(1.5);
   print(-67.125);
}
```

Running result

```
F:\p16ecc>a
0.000000 = 00000000
-0.000000 = 80000000
1.000000 = 3F800000
-1.000000 = BF800000
0.500000 = 3F000000
1.500000 = 3FC00000
-67.125000 = C2864000
```

Table-9.4

9.4.2 Expansion of lexer and parer

Code-9.19a~Code-9.19c shows the expansion of the lexer, cc.l, to support floating-point.

```
29   %}
30
31   letter                  [A-Za-z]
32   digit                   [0-9]
33   underscore              "_"
34   following_character     ({letter}|{digit}|{underscore})
35   identifier              ({letter}|{underscore}){following_character}*
36
37   oct_digit               [0-7]
38   hex_digit               [0-9a-fA-F]
39   integer_suffix          [uU]|[lL]|([uU][lL])|([lL][uU])
40   hex_constant            (0[xX]{hex_digit}+)
41   bin_constant            (0[bB][01]+)
42   dec_constant            ([0-9]|([1-9][0-9]+))
43   oct_constant            (0{oct_digit}+)
44
45   flt_constant1           {dec_constant}[.]([0-9]*)?
46   flt_constant2           [.][0-9]+
47   flt_exponent            [eE][+-]?{dec_constant}
/p16ecc/cc_source_10.1/cc.l                                    Code-9.19a
```

- #45~47: macros for floating-point value.

```
127  {dec_constant}[fF]            |
128  {dec_constant}{flt_exponent}[fF]?    |
129  {flt_constant1}{flt_exponent}?[fF]?  |
130  {flt_constant2}{flt_exponent}?[fF]? { addSrcCode();
131                                         yylval.f = atof(yytext);
132                                         return F_CONSTANT;
133                                       }
/p16ecc/cc_source_10.1/cc.l                                    Code-9.19b
```

- #127~133: rules for identifying a float value (constant).

```
142  "unsigned"{SP}+"char"   { addSrcCode(); return UCHAR; }
143  "unsigned"{SP}+"int"    { addSrcCode(); return UINT; }
144  "unsigned"{SP}+"short"  { addSrcCode(); return USHORT; }
145  "unsigned"{SP}+"long"   { addSrcCode(); return ULONG; }
146  ("signed"{SP}+)?"char"  { addSrcCode(); return CHAR; }
147  ("signed"{SP}+)?"short" { addSrcCode(); return SHORT; }
148  ("signed"{SP}+)?"int"   { addSrcCode(); return INT; }
149  ("signed"{SP}+)?"long"  { addSrcCode(); return LONG; }
150  "float"                 { addSrcCode(); return FLOAT; }
151  "void"                  { addSrcCode(); return VOID; }
/p16ecc/cc_source_10.1/cc.l                                    Code-9.19c
```

- #150: rule for identifying the keyword "float".

Code-9.20a~Code-9.20c shows the expansion of the parser, cc.y, to support floating-point.

```
22    %expect 1
23
24    %union {
25        char   *s;  // string value
26        int     i;  // integer value
27        float   f;  // float value
28        node   *n;
29        attrib *a;
30        int    *p;
31    }
32
33    %token CHAR INT SHORT LONG UCHAR UINT USHORT ULONG VOID FLOAT
34    %token FUNC_HEAD FUNC_DECL DATA_DECL
35    %token _IF _ELSE _ENDIF _IFDEF _IFNDEF DEFINE UNDEF PRAGMA
36    %token BREAK CASE CONST CONTINUE DEFAULT DO ELSE ENUM EXTERN
37    %token FOR GOTO IF REGISTER RETURN SIZEOF STATIC SWITCH
38    %token VOLATILE WHILE INTERRUPT UNION STRUCT TYPEDEF ELIPSIS
39    %token LINEAR
40    %token ADD_ASSIGN SUB_ASSIGN MUL_ASSIGN DIV_ASSIGN MOD_ASSIGN
41    %token LEFT_ASSIGN RIGHT_ASSIGN AND_ASSIGN XOR_ASSIGN OR_ASSIGN
42    %token EQ_OP NE_OP LE_OP GE_OP LEFT_OP RIGHT_OP INC_OP DEC_OP
43    %token OR_OP AND_OP PTR_OP FPTR CAST CALL
44    %token POST_INC POST_DEC PRE_INC PRE_DEC POS_OF ADDR_OF NEG_OF
45    %token EOL LABEL
46
47    %token <i> CONSTANT L_CONSTANT U_CONSTANT UL_CONSTANT C_CONSTANT MEM_BANK
48    %token <i> RELATIONAL_OP EQUALITY_OP SHIFT_OP ASSIGN_OP
49    %token <s> IDENTIFIER IDENTIFIER_ STRING AASM
50    %token <a> TYPEDEF_NAME
51    %token <f> F_CONSTANT
/p16ecc/cc_source_10.1/cc.y                                    Code-9.20a
```

- #27: value item for float type.
- #33: new token for keyword, FLOAT.
- #51: new type constant (float constant).

```
226   type_specifier
227       : VOID                        { $$ = newAttr(VOID); }
228       | CHAR                        { $$ = newAttr(CHAR); }
229       | SHORT                       { $$ = newAttr(SHORT); }
230       | INT                         { $$ = newAttr(INT); }
231       | LONG                        { $$ = newAttr(LONG); }
232       | UCHAR                       { $$ = newAttr(CHAR);  $$->isUnsigned = 1; }
233       | USHORT                      { $$ = newAttr(SHORT); $$->isUnsigned = 1; }
234       | UINT                        { $$ = newAttr(INT);   $$->isUnsigned = 1; }
235       | ULONG                       { $$ = newAttr(LONG);  $$->isUnsigned = 1; }
236       | FLOAT                       { $$ = newAttr(FLOAT); }
237       | enum_specifier              { $$ = $1; }
238       | struct_or_union_specifier   { $$ = $1; }
239       | TYPEDEF_NAME                { $$ = $1; }
240       ;
/p16ecc/cc_source_10.1/cc.y                                    Code-9.20b
```

- #236: new type added for float.

```
435   primary_expr
436       : identifier                  { $$ = $1; }
437       | CONSTANT                    { $$ = conNode($1, INT); }
438       | C_CONSTANT                  { $$ = conNode($1, CHAR); }
439       | U_CONSTANT                  { $$ = conNode($1, INT); $$->con.attr->isUnsigned = 1; }
440       | L_CONSTANT                  { $$ = conNode($1, LONG); }
441       | UL_CONSTANT                 { $$ = conNode($1, LONG); $$->con.attr->isUnsigned = 1; }
442       | STRING                      { $$ = strNode($1); free($1); }
443       | F_CONSTANT                  { $$ = fconNode($1); }
444       | '(' expr ')'                { $$ = $2; }
445       ;
/p16ecc/cc_source_10.1/cc.y                                    Code-9.20c
```

- #443: new rule and action for parsing float constant.

9.4.3 Expansion of data structure

common.h and common.c added the expansion for supporting floating-point, as illustrated in Code-9.21.

```
41     /* constants */
42     typedef struct {
43         int          type;          /* type: NODE_CON */
44         src_t        *src;
45         attrib       *attr;
46         union {
47             long     value;
48             float    fvalue;
49         };
50     } conNode_t;
       .  .  .
143    // routines for 'node'
144    node *idNode  (char *s);
145    node *conNode (int v, int type);
146    node *fconNode(float v);
147    node *listNode(int length);
148    node *strNode (char *s);
149    node *lblNode (char *s);
150    node *oprNode (int type, int cnt, ...);
```
/p16ecc/cc_source_10.1/common.h   Code-9.21

- #48: added float value field in conNode_t.
- #146: added new routine to create float constant node.

9.4.4 Expansion in Pre-scan/Sizer/Item classes

The expansion in Pre-scan class only happens by merging the operations for constants, shown as in Code-9.22.

```
456          switch ( op->oper )
457          {
458              case '+': case '-': case '*': case '/': case '%':
459              case '&': case '|': case '^':
460              case LEFT_OP: case RIGHT_OP:
461              case EQ_OP: case NE_OP: case AND_OP: case OR_OP:
462              case '>': case '<': case LE_OP: case GE_OP:
463                  if ( op->op[0]->type == NODE_CON &&
464                       op->op[1]->type == NODE_CON )
465                  {
466                      if ( op->op[0]->con.attr->type == FLOAT ||
467                           op->op[1]->con.attr->type == FLOAT )
468                      {
469                          int err;
470                          node *np = mergeFloatCon(op->op[0], op->op[1], op->oper, &err);
471                          CLR_COMMENT(op); delNode((node*)op);
472                          if ( err ) errPrint("illegal float operation!");
473                          return np;
474                      }
475
476                      int n = mergeCon(op->op[0]->con.value,
477                                       op->op[1]->con.value, op->oper);
478                      CLR_COMMENT(op); delNode((node*)op);
479                      return conNode(n, INT);
480                  }
481                  break;
482
483              case '~':
484              case '!':
485              case NEG_OF:
486                  if ( op->op[0]->type == NODE_CON )
487                  {
488                      np = op->op[0]; op->op[0] = NULL;
489                      int  n = np->con.value;
490                      bool is_fp = (np->con.attr->type == FLOAT);
491
492                      if ( op->oper == '~' && is_fp )
493                          errPrint("illegal operation!");
494                      else if ( op->oper == NEG_OF && is_fp )
495                          np->con.fvalue = -np->con.fvalue;
496                      else
497                      {
498                          n = (op->oper == '~')? ~n:
499                              (op->oper == '!')? (n? 0:1):
500                                                 -n;
501                          np->con.value = n;
502                          if ( op->oper == NEG_OF ) np->con.attr->isNeg ^= 1;
503                          if ( is_fp ) np->con.attr->type = INT;
504                      }
505                      CLR_COMMENT(op); delNode((node*)op);
506                      return np;
507                  }
508                  break;
509          }
/p16ecc/cc_source_10.1/prescan.cpp                                      Code-9.22
```

- #466~474: float binary operations, they will be merged to a single float value. Note: float values cannot support shifting or logic operations.
- #492~495: float unary operations.

In `Sizer` class, it needs to report the float value size as 4 bytes, shown as Code-9.23.

```
41          switch ( ap->type )
42          {
43              case CHAR: n = (BF_OFFSET(ap) && opt == TOTAL_SIZE)? 0: 1; break;
44              case INT:  n = 2;  break;
45              case SHORT: n = 3;  break;
46              case LONG: n = 4;  break;
47              case FLOAT: n = 4;  break;
      . . .
/p16ecc/cc_source_10.1/sizer.cpp                                        Code-9.23
```

In `Item` class, it adds a function that checks if the item has the float value property (see Code-9.24).

```
30    class Item {
31
32        public:
33            ITEM_TYPE     type;     // item type
34            value         val;
35            int           bias;     // bias for indirect access
36            attrib        *attr;    // attributes
37            void          *home;    // owner of id (ID_ITEM only)
38            Item          *next;    // linking chain
39
40        public:
41            Item(ITEM_TYPE t);
42            ~Item();
43
44            Item *clone (void);
45            void updateAttr (attrib *ap, int data_bank = -1);
46            void updatePtr  (int *p);
47            void updateName (char *new_name);
48            bool isWritable (void);
49            bool isOperable (void);
50            bool isMonoVal  (void);
51            bool isAccePtr  (void);
52            bool isBitVal   (void);
53            attrib *acceAttr(void);
54            int   stepSize  (void);
55            int   acceSize  (void);
56            int   acceSign  (void);
57            int   storSize  (void);
58            bool isBF       (void);
59            bool acceFloat  (void);
60    };
```
/p16ecc/cc_source_10.1/item.h   Code-9.24

- #59: the member function tells if the item has the float-point property.

9.4.5 Expansion in P-code generation

As a case study, it makes the expansion in the following places, in P-code generation. (Interested Readers can accomplish the expansions when needed)

The following code, shown in Code-9.25, converts integers to a floating-point value format for the ROMed constants. For example,

```
float array[4] = {0, 1, 2, 3};
```

It shows that initial values for `array[4]` are in integer format. They should be converted to the floating-point format and stored in ROM. Table-9.5 is the test result.

```
 1    #include <stdio.h>
 2    #include <stdlib.h>
 3    #include <string.h>
 4    #include "common.h"
 5    extern "C" {
 6    #include "cc.h"
 7    }
 8    #include "nlist.h"
 9    #include "item.h"
10    #include "pnode.h"
11    #include "pcoder.h"
12    #include "dlink.h"
13    #include "flink.h"
14    #include "sizer.h"
15    #include "display.h"
16    #include "fformat.h"
     . . . .
```

```
20   /////////////// --- init rom data --- ///////////////
21   void Pcoder :: romDataInit(char *dname, attrib *attr, node *init_data, Dnode *dp, int addr)
22   {
23       if ( dimDepth(attr) > 0 || (attr->ptrVect == NULL && attr->type == STRUCT) )
24       {
25           const char *err = arrayDimVerify(attr, init_data, dp? true: false);
26           if ( err != NULL ) {
27               errPrint(err, dname);
28               return;
29           }
30       }
31

       .  .  .

135          bool init_ok  = false;
136          Pnode *backup = mainPcode;
137          mainPcode = NULL;
138          run(init_data);
139          if ( DEPTH() == (depth+1) )
140          {
141              Item *ip = POP();
142
143              if ( FFormat.reform(ip, attr) != 0 )
144                  errPrint("improper form for float!", ip->val.s);
145
146              ip = parseInitItem(ip);
```

/p16ecc/cc_source_10.1/pcoder0.cpp   Code-9.25

- #143: format conversion for float-point constants.

Source code (ch9_t1.c)	Assembly code output (ch9_t1.asm)
`float array[4] = {0, 1, 2, 3};`	```.segment CONSTi (REL)` `array$init$:` ` retlw 0` ` retlw 0` ` retlw 0` ` retlw 0` ` retlw 0` ` retlw 0` ` retlw 128` ` retlw 63` ` retlw 0` ` retlw 0` ` retlw 0` ` retlw 64` ` retlw 0` ` retlw 0` ` retlw 64` ` retlw 64```

Table-9.5

The following code, Code-9.26, which generates the P-codes for compound assignments, will validate the data type. Only `'*='` and `'/='` can apply to floating-point data.

```
135    case AND_ASSIGN:    case OR_ASSIGN:    case XOR_ASSIGN:
136    case MUL_ASSIGN:    case DIV_ASSIGN:    case MOD_ASSIGN:
137    case LEFT_ASSIGN:    case RIGHT_ASSIGN:
138        run(OPR_NODE(op, 0));
139        run(OPR_NODE(op, 1));
140        if ( DEPTH() == (depth+2) )
141        {
142            ip1 = POP();
143            ip0 = POP();
144
145            if ( !(code == MUL_ASSIGN  || code == DIV_ASSIGN) )
146            {
147                if ( ip0->acceFloat() || ip1->acceFloat() )
148                {
149                    delete ip0;
150                    delete ip1;
151                    errPrint("invalid operation for float-point!");
152                    return;
153                }
154            }
```

/p16ecc/cc_source_10.1/pcoder5.cpp   Code-9.26

- #145~154: P-code type validation for floating-point data.

The following code, Code-9.27, which generates the P-code for '?', will unify the data type to floating-point if either operand is a floating-point.

```
151    case '?':
152        true_lbl  = getLbl();
153        false_lbl = getLbl();
154        end_lbl   = getLbl();
155        logicBranch(op->op[0], true_lbl, false_lbl, true_lbl);
156        ip0 = ip1 = NULL;

    . . .

181        if ( ip0 && ip1 && ip0->isOperable() && ip1->isOperable() )
182        {
183            Item *ip;
184            if ( cmpAttr(ip0->attr, ip1->attr) == 0 && ip0->acceSize() && ip1->acceSize() )
185                ip = makeTemp(ip0->attr);
186            else
187            {
188                int size0 = ip0->acceSize();
189                int size1 = ip1->acceSize();
190                if ( size0 < size1 ) size0 = size1;
191                size0 = (size0 > 0)? size0-1: 0;
192                ip = makeTemp();
193                ip->attr = newAttr(size0+CHAR);
194            }
195
196            if ( ip0->acceFloat() || ip1->acceFloat() )
197                ip->attr->type = FLOAT;
198
199            pp0->items[0] = ip->clone();
200            pp1->items[0] = ip->clone();
201            PUSH(ip);
202        }
```

/p16ecc/cc_source_10.1/pcoder8.cpp   Code-9.27

- #196~197: unify the type for temporary item as FLOAT, if either ip0 or ip1 is a floating-point data.

Besides the library file **crt1.c**, a new source file, **pic16e14.cpp**, is added to support floating-point assembly code generation, and it includes the functions listed below (shown in Code-9.28).

```
28    class PIC16E {
29        public:
30            PIC16E(char *out_file, Nlist *_nlist, Pcoder *_pcoder);
31            ~PIC16E();
32            void run(void);

      .  .  .

163           // pic16e14.cpp
164           void moveToFP(Item *ip0, Item *ip1);     // float <- ip1
165           void moveFromFP(Item *ip0, Item *ip1);   // ip0 <- float
166           void pushToStackAsFloat(Item *ip);
167           void storeToAccAsFloat(Item *ip);
168           void addFP(Item *ip0, Item *ip1);
169           void subFP(Item *ip0, Item *ip1);
170           void mulFP(Item *ip0, Item *ip1);
171           void divFP(Item *ip0, Item *ip1);
172           void cmpFP(int code, Item *ip0, Item *ip1, Item *p2);
173           void jeqjneFP(int code, Item *ip0, Item *ip1, Item *p2);
174    };
```
/p16ecc/cc_source_10.1/p16e/pic16e.h   Code-9.28

1)   Assembly code generation for P-code '='

The following, Code-9.29, is the expansion part that fulfills the assignment for floating-point, in which PIC16E::moveToFP() or PIC16E::moveFromFP() is called. Table-9.6 shows the test results.

```
18    void PIC16E :: mov(Item *ip0, Item *ip1)
19    {
20        if ( same(ip0, ip1) )
21            return;
22
23        if ( ip0->acceFloat() && !ip1->acceFloat() )
24        {
25            moveToFP(ip0, ip1);
26            return;
27        }
28        if ( ip1->acceFloat() && !ip0->acceFloat() )
29        {
30            moveFromFP(ip0, ip1);
31            return;
32        }

      .  .  .
```
/p16ecc/cc_source_10.1/p16e/pic16e2.cpp   Code-9.29

Source code (ch9_t2.c)	Assembly code output (ch9_t2.asm)

``float x;
int n;
 	
void foo()
{
  x = 10;
  n = x;
}```
	`` .segment CODE2 (REL) foo:0
	foo::
	; :: ch9_t2.c #6: x = 10;
	        .bsel x
	        clrf x
	        .bsel x, x+1
	        clrf x+1
	        movlw 32
	        .bsel x+1, x+2
	        movwf x+2
	        movlw 65
	        .bsel x+2, x+3
	        movwf x+3
	; :: ch9_t2.c #7: n = x;
	        .bsel x+3, x
	        movf x, W
	        movwf 112
	        .bsel x, x+1```

```
                              movf   x+1, W
                              movwf  113
                              .bsel  x+1, x+2
                              movf   x+2, W
                              movwf  114
                              .bsel  x+2, x+3
                              movf   x+3, W
                              movwf  115
                              .psel  foo, _floatToIntACC
                              call   _floatToIntACC
                              movf   112, W
                              .bsel  x+3, n
                              movwf  n
                              movf   113, W
                              .bsel  n, n+1
                              movwf  n+1
                              .psel  _floatToIntACC, foo
                              return
```

Table-9.6

2) Assembly code generation for P-code `+=` and `-=`

The following, Code-9.30, is the expansion part that fulfills the `+=` and `-=` for floating-point, in which
`PIC16E::addFP()` or `PIC16E::subFP()` is called. Table-9.7 shows the test results.

```
20    void PIC16E :: addsub(int code, Item *ip0, Item *ip1)
21    {
22        if ( ANY_FLOAT(ip0, ip1) )
23        {
24            if ( code == ADD_ASSIGN )
25                addFP(ip0, ip1);
26            else
27                subFP(ip0, ip1);
28            Item *acc = accItem(newAttr(FLOAT));
29            mov(ip0, acc);
30            delete acc;
31            return;
32        }

      . . .
/p16ecc/cc_source_10.1/p16e/pic16e4.cpp
```

Code-9.30

* #22~32: the expansion of `+=`/`-=` for floating-point.

Source code (ch9_t3.c)	Assembly code output (ch9_t3.asm)

<pre>float x;
int n;
 	
void foo()
{
  x += n;
  n -= x;
}</pre>
	<pre> .segment CODE2 (REL) foo:0
	foo::
	; :: ch9_t3.c #6: x += n;
	      .bsel x
	      movf x, W
	      movwi --INDF1
	      .bsel x, x+1
	      movf x+1, W
	      movwi --INDF1
	      .bsel x+1, x+2
	      movf x+2, W
	      movwi --INDF1
	      .bsel x+2, x+3
	      movf x+3, W
	      movwi --INDF1
	      .bsel x+3, n
	      movf n, W
	      movwf 112
	      .bsel n, n+1
	      movf n+1, W
	      movwf 113
	      movlw 0</pre>

298

<table>
<tr><td></td><td>

```
        btfsc  113, 7
        movlw  255
        movwf  114
        movwf  115
        .psel  foo, _int4ToFloatACC
        call   _int4ToFloatACC
        .psel  _int4ToFloatACC, _floatAdd
        call   _floatAdd
        movf   112, W
        .bsel  n+1, x
        movwf  x
        movf   113, W
        .bsel  x, x+1
        movwf  x+1
        movf   114, W
        .bsel  x+1, x+2
        movwf  x+2
        movf   115, W
        .bsel  x+2, x+3
        movwf  x+3
;  :: ch9_t3.c #7: n -= x;
        .bsel  x+3, n
        movf   n, W
        movwi  --INDF1
        .bsel  n, n+1
        movf   n+1, W
        movwi  --INDF1
        andlw  128
        btfss  3, 2
        movlw  255
        movwi  --INDF1
        movwi  --INDF1
        .psel  _floatAdd, _int4ToFloatStack
        call   _int4ToFloatStack
        .bsel  n+1, x
        movf   x, W
        movwf  112
        .bsel  x, x+1
        movf   x+1, W
        movwf  113
        .bsel  x+1, x+2
        movf   x+2, W
        movwf  114
        .bsel  x+2, x+3
        movf   x+3, W
        movwf  115
        .psel  _int4ToFloatStack, _floatSub
        call   _floatSub
        .psel  _floatSub, _floatToIntACC
        call   _floatToIntACC
        movf   112, W
        .bsel  x+3, n
        movwf  n
        movf   113, W
        .bsel  n, n+1
        movwf  n+1
        .psel  _floatToIntACC, foo
        return
```

</td></tr>
</table>

Table-9.7

299

3) Assembly code generation for P-code '+' and '-'

The expansion of '+' and '-' for floating-point are almost identical, as shown in Code-9.31 and Code-9.32.

```
179   void PIC16E :: add(Item *ip0, Item *ip1, Item *ip2)
180   {
181       int size0 = ip0->acceSize();
182       int size1 = ip1->acceSize();
183       int size2 = ip2->acceSize();
184
185       if ( ANY_FLOAT(ip1, ip2) )
186       {
187           addFP(ip1, ip2);
188           Item *acc = accItem(newAttr(FLOAT));
189           mov(ip0, acc);
190           delete acc;
191           return;
192       }
193       if ( ip0->acceFloat() )
194       {
195           Item *acc = accItem(newAttr(LONG));
196           acc->attr->isUnsigned = SIGNED_RESULT(ip1, ip2)? 0: 1;
197           add(acc, ip1, ip2);
198           mov(ip0, acc);
199           delete acc;
200           return;
201       }
```
/p16ecc/cc_source_10.1/p16e/pic16e5.cpp   Code-9.31

- #185~201: the expansion of '+' for floating-point.

```
242   void PIC16E :: sub(Item *ip0, Item *ip1, Item *ip2)
243   {
244       int size0 = ip0->acceSize();
245       int size1 = ip1->acceSize();
246       int size2 = ip2->acceSize();
247
248       if ( ANY_FLOAT(ip1, ip2) )
249       {
250           subFP(ip1, ip2);
251           Item *acc = accItem(newAttr(FLOAT));
252           acc->attr->isUnsigned = SIGNED_RESULT(ip1, ip2)? 0: 1;
253           mov(ip0, acc);
254           delete acc;
255           return;
256       }
257       if ( ip0->acceFloat() )
258       {
259           Item *acc = accItem(newAttr(LONG));
260           sub(acc, ip1, ip2);
261           mov(ip0, acc);
262           delete acc;
263           return;
264       }
```
/p16ecc/cc_source_10.1/p16e/pic16e5.cpp   Code-9.32

- #248~264: the expansion of '-' for floating-point.

4) Assembly code generation for conditional jump P-codes

The conditional jump here includes two groups of P-codes:

        P_JEQ/P_JNE   and
        P_JLT/P_JLE/P_JGT/P_JGE

The expansions for them are shown in Code-9.33 and Code-9.34 below.

```
18     void PIC16E :: jeqjne(int code, Item *ip0, Item *ip1, Item *ip2)
19     {
20         int size0 = ip0->acceSize();
21         int size1 = ip1->acceSize();
22         int sign1 = ip1->acceSign();
23         int size  = (size0 > size1)? size0: size1;
24         char *inst= (code == P_JEQ)? _BTFSC: _BTFSS;
25
26         if ( ANY_FLOAT(ip0, ip1) )
27         {
28             jeqjneFP(code, ip0, ip1, ip2);
29             return;
30         }
```
/p16ecc/cc_source_10.1/p16e/pic16e6.cpp   Code-9.33

- #26~30: the expansion of P_JEQ/P_JNE for floating-point.

```
272    void PIC16E :: cmpJump(int code, Item *ip0, Item *ip1, Item *ip2)
273    {
274        int size0 = ip0->acceSize();
275        int size1 = ip1->acceSize();
276        int code2 = convertCompare(code);
277
278        if ( ANY_FLOAT(ip0, ip1) )
279        {
280            cmpFP(code, ip0, ip1, ip2);
281            return;
282        }
```
/p16ecc/cc_source_10.1/p16e/pic16e6.cpp   Code-9.34

- #278~282: the expansion of P_JLT/P_JLE/P_JGT/P_JGE for floating-point.

5) Assembly code generation for multiplication

The expansion of multiplication here includes '*=' and '*'. And they are pretty the same way to treat floating-point data as that above. Code-9.35 and Code-9.36 show the operations.

```
18     void PIC16E :: mulAssign(Item *ip0, Item *ip1)
19     {
20         if ( ANY_FLOAT(ip0, ip1) )
21         {
22             mulFP(ip0, ip1);
23             Item *acc = accItem(newAttr(FLOAT));
24             mov(ip0, acc);
25             delete acc;
26             return;
27         }
       . . . .
```
/p16ecc/cc_source_10.1/p16e/pic16e9.cpp   Code-9.35

- #20~27: the expansion of '*=' for floating-point.

```cpp
64     void PIC16E :: mul(Item *ip0, Item *ip1, Item *ip2)
65     {
66         if ( ANY_FLOAT(ip1, ip2) )
67         {
68             mulFP(ip1, ip2);
69             Item *acc = accItem(newAttr(FLOAT));
70             mov(ip0, acc);
71             delete acc;
72             return;
73         }
74         if ( ip0->acceFloat() )
75         {
76             Item *acc = accItem(newAttr(LONG));
77             acc->attr->isUnsigned = SIGNED_RESULT(ip1, ip2)? 0: 1;
78             mul(acc, ip1, ip2);
79             mov(ip0, acc);
80             delete acc;
81             return;
82         }
```

- #66~82: the expansion of `*` for floating-point.

6)   Assembly code generation for division

The expansion of multiplication here includes `/=` and `/`. Code-9.37 and Code-9.38 show the operations.

```
19    void PIC16E :: divmodAssign(int code, Item *ip0, Item *ip1)
20    {
21        int size0 = ip0->acceSize();
22        int size1 = ip1->acceSize();
23        bool sign0 = ip0->acceSign();
24        bool sign1 = ip1->acceSign();
25        int size  = (size0 >= size1)? size0: size1;
26        int op_flag = 0;
27
28        if ( ANY_FLOAT(ip0, ip1) && code == DIV_ASSIGN )
29        {
30            divFP(ip0, ip1);
31            Item *acc = accItem(newAttr(FLOAT));
32            mov(ip0, acc);
33            delete acc;
34            return;
35        }
```
/p16ecc/cc_source_10.1/p16e/pic16e10.cpp   Code-9.37

- #28~35: the expansion of `/=` for floating-point.

```
69    void PIC16E :: divmod(int code, Item *ip0, Item *ip1, Item *ip2)
70    {
71        int size0 = ip0->acceSize();
72        int size1 = ip1->acceSize();
73        int size2 = ip2->acceSize();
74        bool sign1 = ip1->acceSign();
75        bool sign2 = ip2->acceSign();
76        int size  = (size1 >= size2)? size1: size2;
77        int op_flag = 0;
78
79        if ( code == '/' )
80        {
81            if ( ANY_FLOAT(ip1, ip2) )
82            {
83                Item *acc = accItem(newAttr(FLOAT));
84                divFP(ip1, ip2);
85                mov(ip0, acc);
86                delete acc;
87                return;
88            }
89            if ( ip0->acceFloat() )
90            {
91                Item *acc = accItem(newAttr(LONG));
92                acc->attr->isUnsigned = SIGNED_RESULT(ip1, ip2)? 0: 1;
93                divmod(code, acc, ip1, ip2);
94                mov(ip0, acc);
95                delete acc;
96                return;
97            }
98        }
```
/p16ecc/cc_source_10.1/p16e/pic16e10.cpp   Code-9.38

- #79~98: the expansion of `/` for floating-point.

Table-9.8 is the test result of '/' for floating-point.

Source code (ch9_t4.c)	Assembly code output (ch9_t4.asm)

<pre>float x, y;
 	
void foo()
{
  x = y / 10.0;
}</pre>
	<pre> .segment CODE2 (REL) foo:0
	foo::
	; :: ch9_t4.c #5: x = y / 10.0;
	        .bsel y
	        movf y, W
	        movwi --INDF1
	        .bsel y, y+1
	        movf y+1, W
	        movwi --INDF1
	        .bsel y+1, y+2
	        movf y+2, W
	        movwi --INDF1
	        .bsel y+2, y+3
	        movf y+3, W
	        movwi --INDF1
	        movlw 0
	        movwf 112
	        movwf 113
	        movlw 32
	        movwf 114
	        movlw 65
	        movwf 115
	        .psel foo, _floatDiv
	        call _floatDiv
	        movf 112, W
	        .bsel y+3, x
	        movwf x
	        movf 113, W
	        .bsel x, x+1
	        movwf x+1
	        movf 114, W
	        .bsel x+1, x+2
	        movwf x+2
	        movf 115, W
	        .bsel x+2, x+3
	        movwf x+3
	        .psel _floatDiv, foo
	        return</pre>

Table-9.8

PART-III:
PIC16F Assembler (as16e.exe)

The assembler, as16e, plays an intermediate role in the compiler toolset. It takes the assembly files, with the extension ".asm", generated by cc16e, and converts them into PIC16F binary code as the output files, with extension ".obj". Note that the outputs generated are relocatable or floating, ready for the final process – linking.

In general, an assembly (source) code line in an .asm file has the format as

      [label]   **mnemonic**   **[operands]**   **[; comment]**

The fields in the square brackets are optional. The 'operands' field can contain a register name or expressions, and a comma ',' is inserted between operands as the separator.

As in design **cc16e**, the command tools **flex** and **bison** are used during the design of the assembler **as16e**. On the other hand, the grammar or language structure of assembly code is a lot simpler compared with that in the C language. So, the data structures and parsing rules are simpler and easier to understand.

Project directory: /p16ecc/as_source_1
Source files: main. cpp, asm. l, asm.y, common.h, common.c

10.1 Basic Data Structures

common.h, as shown in Code-10.1a~Code-10.1b, defines the data structure for holding the assembly lines (including all the fields of each line).

```
4    enum {
5        TYPE_VALUE=1,
6        TYPE_STRING,
7        TYPE_SYMBOL
8    };
9
10   /* /////////////////////////////////////////////////////////////////// */
11   typedef union data_ {
12       int     val;
13       char *str;
14   } data_t;
15
16   /* /////////////////////////////////////////////////////////////////// */
17   typedef struct item_ {
18       int         type;
19       data_t      data;
20       struct item_ *left;
21       struct item_ *right;
22       struct item_ *next;
23   } item_t;
```
/p16ecc/as_source_1/common.h   Code-10.1a

- #4~8: operand value type.
- #11~14: operand value.
- #17~23: operand holder (it supports math/logic expressions).

Figure-10.1 shows an example of an expression stored in the data structure after parsing.

```
X + Y * 100
```

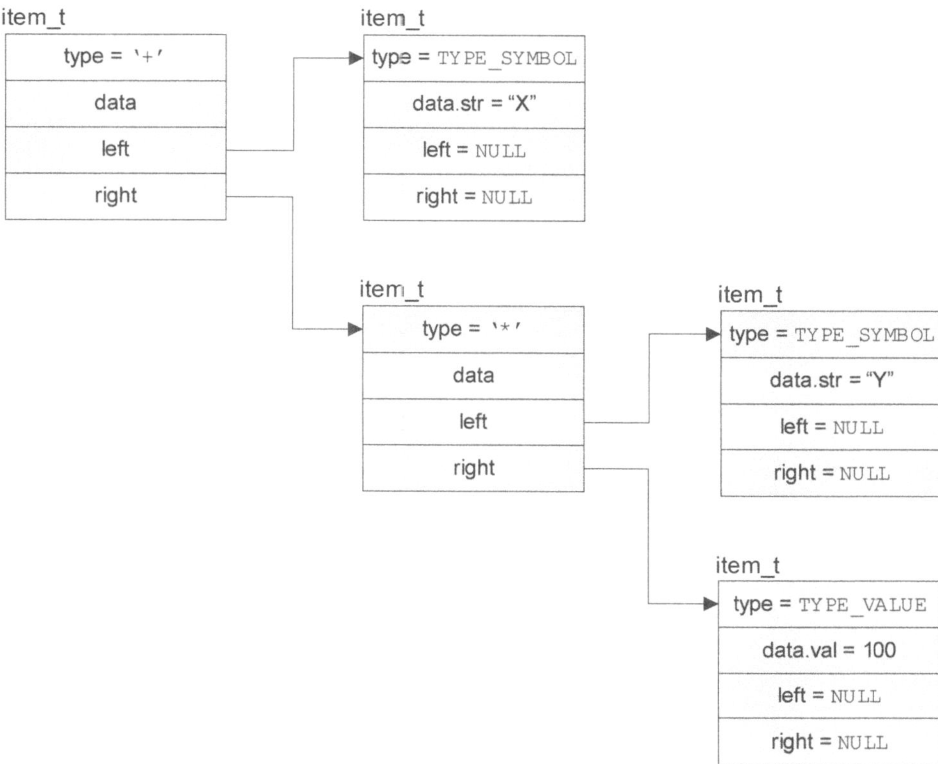

Figure-10.1

The following, Code-10.1b, is the data structure to keep an assembly code line(s).

```
25  typedef struct {
26      char        *name;
27      item_t      *addr;
28      int         isABS: 1;
29      int         isREL: 1;
30      int         isBEG: 1;
31      int         isEND: 1;
32  } attr_t;
33
34  /* ///////////////////////////////////////////////////////////////////////////////////// */
35  typedef struct line_ {
36      char        *label;     // label field
37      int         segType;    // segment type
38      int         globalLbl;  // 1 = global, 0 = local
39      int         inst;       // instruction code field
40      attr_t      *attr;      // segment descriptor
41      item_t      *oprds;     // operand field(s)
42      int         desRegF;    // destination F
43      char        *src;       // source file code
44      int         lineno;     // source file line number
45      char        *fname;     // source file name
46      struct line_ *next;
47  } line_t;
```

/p16ecc/as_source_1/common.h   Code-10.1b

- #25~33: attribute data structure. It applies to ".segment" lines.
- #36~48: assembly code line structure.

10.2 Lexer for PIC16F Assembler

The lexer script, asm.l, for the PIC16F assembler is organized the same way as cc.l.

10.2.1 Macro part of the lexer

Code-10.2 is the macro part of the lexer.

```
40   %}
41
42   sp                        [ \t\015]
43   symbol                    [_a-zA-Z][_a-zA-Z0-9$]*
44   hex                       (0[xX][0-9a-fA-F]+)
45   dec                       ([0-9]|([1-9][0-9]+))
46
47   s_char                    ([^"\r\n])
48   s_char_sequence           ({s_char}+)
49   string                    \"{s_char_sequence}?\"
50
51   %x OPCODE
52   %x OPERAND
53
54   %%
```
/p16ecc/as_source_1/asm.l   Code-10.2

- #44: regular expression for Hex values.
- #45: regular expression for Decimal values.
- #51~52: two new states are added, OPCODE and OPERAND, beyond the default (INITIAL) state.

10.2.2 Rule part of the lexer

Code-10.3a shows the rules for the INITIAL state.

```
56   ^{sp}*";".*                 {   appendStr();
57                                   if ( memcmp(yytext, "; :: ", 5) == 0 )
58                                   {
59                                       yylval.name = dupStr(yytext);
60                                       return S_COMMENT;
61                                   }
62                               }
63   ^{sp}+{symbol}:(:)?         |
64   ^{symbol}((:?):)?           {   int i = 0;
65                                   appendStr();
66                                   while ( yytext[i] <= ' ' ) i++;
67                                   yylval.name = dupStr(&yytext[i]);
68                                   BEGIN(OPCODE);
69                                   return LABEL;
70                               }
71   ^{sp}+                      {   appendStr();
72                                   BEGIN(OPCODE);
73                               }
74   \n                          {   appendStr();
75                                   return yytext[0];
76                               }
```
/p16ecc/as_source_1/asm.l   Code-10.3a

- #56~62: rule to match a comment line.
- #63~70: rules to match a label in a line, then switch to OPCODE state.
- #71~73: no label filed, switch to OPCODE state.

Code-10.3b shows the rules for the OPCODE state.

```
77    <OPCODE>(".")?{symbol}              {    inst_t *ip;
78                                             appendStr();
79                                             BEGIN(OPERAND);
80
81                                             ip = searchInst(yytext);
82                                             if ( ip != NULL )
83                                             {
84                                                 yylval.value = ip->token;
85                                                 return (ip->token == SEGMENT)? SEGMENT: PIC_INST;
86                                             }
87
88                                             printf("Line #%d: ", yylineno-1);
89                                             printf("illegal opcode - %s\n", yytext);
90                                        }
91    <OPCODE>{sp}*(";".*)?               {    appendStr(); }
92    <OPCODE>\n                          {    appendStr();
93                                             BEGIN(INITIAL);
94                                             return yytext[0];
95                                        }
/p16ecc/as_source_1/asm.l                                                        Code-10.3b
```

- #77~90: match the opcode (a PIC16F instruction or a pseudo instruction), and the value is assigned (#84). Then, switch to OPERAND state.

Code-10.3c shows the rules for the OPERAND state.

```
96    <OPERAND>{string}                  {    appendStr();
97                                             yytext[yyleng-1] = 0;
98                                             yylval.name = dupStr(yytext+1);
99                                             return STRING;
100                                       }
101   <OPERAND>{symbol}                  {    appendStr();
102                                            yylval.name = dupStr(yytext);
103                                            return SYMBOL;
104                                       }
105   <OPERAND>{hex}                     |
106   <OPERAND>{dec}                     {    appendStr();
107                                            yylval.value = convertNum((char *)yytext);
108                                            return NUMBER;
109                                       }
110   <OPERAND>">>"                      {    appendStr(); return RSHIFT; }
111   <OPERAND>"<<"                      {    appendStr(); return LSHIFT; }
112
113   <OPERAND>[-+%^:&|=,()*/~]          {    appendStr();
114                                            return yytext[0];
115                                       }
116   <OPERAND>{sp}*";".*                {    appendStr(); }
117   <OPERAND>{sp}+                     {    appendStr(); }
118   <OPERAND>"++"                      {    appendStr();
119                                            return PLUS_PLUS;
120                                       }
121   <OPERAND>"--"                      {    appendStr();
122                                            return MINUS_MINUS;
123                                       }
124   <OPERAND>"["                       {    appendStr();    return '['; }
125   <OPERAND>"]"                       {    appendStr();    return ']'; }
126   <OPERAND>"."                       {    appendStr();    return '.'; }
127   <OPERAND>\n                        {    appendStr();
128                                            BEGIN(INITIAL);
129                                            return yytext[0];
130                                       }
/p16ecc/as_source_1/asm.l                                                        Code-10.3c
```

- #96~109: rule for matching values (string, symbol, and constants).
- #110~115: operators for expressions.
- #118~125: special character(s) for PIC16F instruction operands.
- #126: the character '.' means the current address n assembly language.
- #127~129: end of the line, go back INITIAL state.

Code-10.3d is the C code part in cc.l, and it only lists the function _main() that activates the parser, as well as the lexer.

```
263   line_t *_main(char *filename)
264   {
265       int yyparse_ret;
266
267       // if the file has been parsed, skip it
268       if ( searchStrName(fileList, filename) != NULL )
269           return NULL;
270
271       fileList = addStrName(fileList, filename);
272       yyin = fopen(filename, "r");
273
274       if ( yyin == NULL )
275       {
276           printf ("can't open file - %s!", filename);
277           exit (99);
278       }
279
280       currentFile = filename;
281
282       yylineno = 0;
283       linePtr = NULL;
284
285       yyparse_ret = yyparse();   // parse the input text (file)
286
287       if ( yyparse_ret != 0 )     // if fail to parse, stop!
288       {
289           printf("yyparse stopped at #Line %d\n", yylineno);
290           exit(99);
291       }
292
293       return linePtr;
294   }
```
`/p16ecc/as_source_1/asm.l`   Code-10.3d

- #272: open the input file (.asm).
- #285: start parsing the input file.

10.3 Parser for PIC16F Assembler

The assembler parser is in file asm.y.

<u>10.3.1 Definition part of the parser</u>

Code-10.4 is the definition part of the parser. It lists all the tokens that are used in the lexer, asm.l.

```
11   %}
12
13   %union {
14       int      value;  // integer value
15       char     *name;  // any symbol/string
16       line_t   *line;  // an assembly line
17       item_t   *item;  // an operand/expr
18       attr_t   *attr;  // segment attribution
19   }
20
21   %token <name>    LABEL    SYMBOL   STRING   S_COMMENT
22   %token <value>   NUMBER   PIC_INST
23
24   %token ADDWF ANDWF CLRF COMF DECF DECFSZ INCF INCFSZ
25   %token IORWF MOVF MOVWF RLF RRF SUBWF NOP SWAPF XORWF
26   %token BCF BSF BTFSC BTFSS CLRWDT GOTO SLEEP
27   %token CALL RETFIE RETLW RETURN
28   %token ADDLW ANDLW IORLW MOVLW SUBLW XORLW CLRW
29   %token RESET CALLW BRW MOVIW MOVWI MOVLP MOVLB
30   %token ADDFSR BRA LSLF LSRF ASRF SUBWFB ADDWFC
31   %token MOVIW_OP MOVWI_OP MOVIW_OFF MOVWI_OFF
32   %token FCALL     RS       EQU      DW       DBLANK   CBLANK
33   %token END       SEGMENT INVOKE   DEVICE   PSEL      BSEL
34   %token RSHIFT LSHIFT PLUS_PLUS   MINUS_MINUS
35   %token FSR_PRE_INC FSR_POST_INC FSR_PRE_DEC FSR_POST_DEC FSR_OFFSET
36
37
38   %left '-' '+'
39   %left '*' '/' '%' RSHIFT LSHIFT
40   %left '&' '|' '^' '~'
41   %nonassoc UMINUS
42   %nonassoc INVERSE
43
44   %type <line>   prog lines line source_line
45   %type <item>   opernd items exp primary_exp addition_exp
46   %type <item>   multiply_exp shift_exp and_exp xor_exp or_exp param
47   %type <attr>   attrs attr
48   %type <item>   e_opernd
49
50   %%
```
/p16ecc/as_source_1/asm.y   Code-10.4

- #13~19: union for the actual value of tokens.

<u>10.3.2 Rule part of the parser</u>

Code-10.5a shows the rules and actions for parsing the assembly lines.

```
51   prog
52       :   lines                     {   linePtr = $1; }
53       ;
54   lines
55       :   line                      {   $$ = $1; }
56       |   lines line                {   $$ = $1;
57                                         appendLine(&$1, $2);
58                                     }
59       ;
```

```
 60  line
 61      :     source_line '\n'          {   $$ = $1;
 62                                          $$->src = dupStr(__yyline);
 63                                          $$->lineno = yylineno;
 64                                      }
 65      |     '\n'                      {   $$ = newLine(NULL, 0, NULL);
 66                                          $$->src = dupStr(__yyline);
 67                                          $$->lineno = yylineno;
 68                                      }
 69      ;
 70  source_line
 71      :     LABEL                     {   $$ = newLine($1, 0,  NULL); free($1); }
 72      |     LABEL PIC_INST            {   $$ = newLine($1, $2, NULL); free($1); }
 73      |     LABEL PIC_INST opernd     {   $$ = newLine($1, $2, $3  ); free($1); }
 74      |           PIC_INST            {   $$ = newLine(NULL, $1, NULL); }
 75      |           PIC_INST opernd     {   $$ = newLine(NULL, $1, $2); }
 76      |     S_COMMENT                 {   $$ = newLine(NULL, S_COMMENT, strItem($1));  free($1); }
 77      |     SEGMENT SYMBOL            {   $$ = newSegLine($2, NULL, NULL); }
 78      |     SEGMENT SYMBOL
 79            '(' attrs ')'             {   $$ = newSegLine($2, $4, NULL); }
 80      |     SEGMENT SYMBOL
 81            '(' attrs ')' param       {   $$ = newSegLine($2, $4, $6); }
 82      ;
/p16ecc/as_source_1/asm.y                                                       Code-10.5a
```

- #70~82: rules for all combinations/formats of assembly source lines. Among them, the segment lines have a different format from that of regular PIC16F instruction lines.

Code-10.5b shows the rules and actions for attribute items that appear in segment lines ("`.segment`").

```
 83  attrs
 84      :     attr                      {   $$ = $1; }
 85      |     attrs ',' attr            {   $$ = mergeAttr($1, $3); }
 86      ;
 87  attr
 88      :     exp                       {   $$ = newAttr($1, 0);
 89                                          if ( $$ == NULL )
 90                                              yyerror("illegal attr. specifier!");
 91                                      }
 92      |     '=' NUMBER                {   $$ = newAttr(valItem($2), '='); }
 93      ;
 94  param
 95      :     SYMBOL                    {   $$ = newItem(0);
 96                                          $$->left  = symItem($1); free($1);
 97                                      }
 98      |     SYMBOL ':' NUMBER         {   $$ = newItem(':');
 99                                          $$->left  = symItem($1); free($1);
100                                          $$->right = valItem($3);
101                                      }
102      ;
/p16ecc/as_source_1/asm.y                                                       Code-10.5b
```

Code-10.5c shows the rules and actions for parsing operand fields, which can be a combination of math/logic expressions.

```
103  opernd
104      :     items                     {   $$ = $1; }
105      |     e_opernd                  {   $$ = $1; }
106      ;
107  items
108      :     exp                       {   $$ = $1; }
109      |     items ',' exp             {   $$ = appendItem($1, $3); }
110      ;
111  primary_exp
112      :     '(' exp ')'               {   $$ = $2; }
113      |     '+' primary_exp           {   $$ = $2; }
114      |     '-' primary_exp           {   $$ = newItem(UMINUS); $$->left = $2; }
115      |     '~' primary_exp           {   $$ = newItem(INVERSE);$$->left = $2; }
116      |     SYMBOL                    {   $$ = symItem($1); free($1); }
117      |     STRING                    {   $$ = strItem($1); free($1); }
118      |     NUMBER                    {   $$ = valItem($1);   }
119      |     '.'                       {   $$ = newItem('.'); }
120      ;
```

```
121    multiply_exp
122        :    primary_exp
123        |    multiply_exp '*'
124             primary_exp                  {    $$ = newItem('*');
125                                                $$->left  = $1;
126                                                $$->right = $3;
127                                           }
128        |    multiply_exp '/'
129             primary_exp                  {    $$ = newItem('/');
130                                                $$->left  = $1;
131                                                $$->right = $3;
132                                           }
133        |    multiply_exp '%'
134             primary_exp                  {    $$ = newItem('%');
135                                                $$->left  = $1;
136                                                $$->right = $3;
137                                           }
138        ;
139    addition_exp
140        :    multiply_exp
141        |    addition_exp '+'
142             multiply_exp                 {    $$ = newItem('+');
143                                                $$->left  = $1;
144                                                $$->right = $3;
145                                           }
146        |    addition_exp '-'
147             multiply_exp                 {    $$ = newItem('-');
148                                                $$->left  = $1;
149                                                $$->right = $3;
150                                           }
151        ;
152    shift_exp
153        :    addition_exp
154        |    shift_exp LSHIFT
155             addition_exp                 {    $$ = newItem(LSHIFT);
156                                                $$->left  = $1;
157                                                $$->right = $3;
158                                           }
159        |    shift_exp RSHIFT
160             addition_exp                 {    $$ = newItem(RSHIFT);
161                                                $$->left  = $1;
162                                                $$->right = $3;
163                                           }
164        ;
165    and_exp
166        :    shift_exp
167
168        |    and_exp '&' shift_exp        {    $$ = newItem('&');
169                                                $$->left  = $1;
170                                                $$->right = $3;
171                                           }
172        ;
173
174    xor_exp
175        :    and_exp
176        |    xor_exp '^' and_exp          {    $$ = newItem('^');
177                                                $$->left  = $1;
178                                                $$->right = $3;
179                                           }
180        ;
181    or_exp
182        :    xor_exp
183        |    or_exp '|' xor_exp           {    $$ = newItem('|');
184                                                $$->left  = $1;
185                                                $$->right = $3;
186                                           }
187        ;
188    exp
189        :    or_exp                       {    $$ = $1; }
190        ;
191    e_opernd
192        :    PLUS_PLUS SYMBOL             {    $$ = newItem(FSR_PRE_INC);  $$->left = symItem($2); free($2); }
193        |    MINUS_MINUS SYMBOL           {    $$ = newItem(FSR_PRE_DEC);  $$->left = symItem($2); free($2); }
194        |    SYMBOL PLUS_PLUS             {    $$ = newItem(FSR_POST_INC); $$->left = symItem($1); free($1); }
195        |    SYMBOL MINUS_MINUS           {    $$ = newItem(FSR_POST_DEC); $$->left = symItem($1); free($1); }
196        |    exp '[' SYMBOL ']'           {    $$ = newItem(FSR_OFFSET);
197                                                $$->left = $1;
198                                                $$->right= symItem($3); free($3);
199                                           }
```

                                        Code-10.5c

- #191~199: the operand, e_opernd, have the special format for PIC16F instructions "MOVWI" and "MOVIW".

10.4 Start Running PIC16F Assembler

The main.cpp file is the starting point for running the assembler, as illustrated in Code-10.6.

```cpp
1   #include <stdio.h>
2   #include <string.h>
3   extern "C" {
4   #include "common.h"
5   #include "asm.h"
6   }
7
8   int main(int argc, char *argv[])
9   {
10      for (int i = 1; i < argc; i++)
11      {
12          curFile = argv[i];
13          printf ("assembling '%s' ...\n", curFile);
14
15          // using Lex and Bison to parse the input file ...
16          errorCnt  = 0;
17          line_t *lp = _main(curFile);
18
19          if ( !errorCnt )
20          {
21              //...
22          }
23      }
24      return 0;
25  }
```
/p16ecc/as_source_1/main.cpp   Code-10.6

- #12: get an input file (.asm), from the command line.
- #17: start parsing the file.

Regarding the project completion, Code-10.7 provides the makefile.

```makefile
1   CC  = gcc
2   CXX = g++
3   RM  = rm
4   MV  = mv
5   CP  = cp
6   EXE = as16e.exe
7   OBJ = main.o lex.yy.o asm.o common.o
8   OPTIONS= -c -Os
9
10  $(EXE): $(OBJ) asm.h makefile
11      $(CXX) -static $(OBJ) -o $(EXE)
12      $(CP) $(EXE) ../bin
13
14  %.o: %.c asm.h makefile
15      $(CC) $(OPTIONS) $<
16
17  %.o: %.cpp asm.h makefile
18      $(CXX) $(OPTIONS) $<
19
20  lex.yy.c: asm.l asm.h makefile
21      flex asm.l
22
23  asm.h: asm.y makefile
24      bison -d asm.y -o asm.c
25
26  asm.c: asm.y makefile
27      bison -d asm.y -o asm.c
28
29  clean:
30      $(RM) *.o
31      $(RM) asm.h
32      $(RM) asm.c
33      $(RM) lex.yy.c
34      $(RM) $(EXE)
```
/p16ecc/as_source_1/makefile   Code-10.7

Chapter-11

PIC16F Assembler - Generate the Outputs

<u>Project directory</u>: /p16ecc/as_source_2
<u>New Source files</u>: p16asm.h, p16asm.cpp, p16asm0.h, p16asm1.cpp, p16inst.h,
p16inst.cpp, …
<u>Updated files</u>: main.cpp

After parsing, as described in the chapter above, all the source lines in an assembly file are read and stored in the line buffer, indicated by `linePtr`, which is returned from `_main()` as shown in Code-10.3b.

11.1 Scan input lines

An assembler class, `P16E_asm`, is created. It uses a so-called 'double scan' strategy to scan the input lines. During the first scan, it searches for `.equ` instruction lines that define symbol values. In the second scan, it replaces the symbols that appear in the instruction operand fields with their actual values.

The class is described as follows: Code-11.1.

```
6     enum {ASM_PASS1, ASM_PASS2};
7
8     #define I_TYPE(ip, t)    (ip && ip->type == t)
9
10    typedef enum {VAL_EXP, STR_EXP, BAD_EXP} EXP_TYPE;
11
12    class P16E_asm {
13
14        private:
15            char    *objFile;
16            char    *lstFile;
17            Symbol  *symbolList;
18            ObjWriter *objWriter;
19            LstWriter *lstWriter;
20
21        public:
22            int   errorCount;
23
24        public:
25            P16E_asm(char *fname, const char *ext);
26            ~P16E_asm();
27            void run(line_t *lp, int pass);
28            void output(line_t *lp);
29
30        private:
31            EXP_TYPE parseItem(item_t *ip, char *buf, int *val, int instr);
32            void objOutput(char t, item_t *ip);
33            void objOutput(item_t *ip);
34
35            item_t *symbolReplace(Symbol *slist, item_t *ip);
36            item_t *mergeItems(int op, item_t *ip1, item_t *ip2);
37            void    regulateInst(line_t *lp);
38    };
```
/p16ecc/as_source_2/p16asm.h   Code-11.1

- #27: scan input lines.
- #28: generate outputs.

315

And the class is instantiated in `main()`, shown in Code-11.2.

```
 6    #include "p16asm.h"
 7
 8    static void asm0(char *filename);
 9
10    int main(int argc, char *argv[])
11    {
12        for (int i = 1; i < argc; i++)
13        {
14            asm0(argv[i]);
15        }
16        return 0;
17    }
18
19    static void asm0(char *filename)
20    {
21        curFile = filename;
22        printf ("assembling '%s' ...\n", filename);
23
24        // using Lex and Bison to parse the input file ...
25        errorCnt  = 0;
26        line_t *lp = _main(filename);
27        if ( errorCnt != 0 ) return;
28
29        P16E_asm p16asm(filename, ".obj");
30        if ( p16asm.errorCount != 0 ) return;
31
32        p16asm.run(lp, ASM_PASS1);
33        if ( p16asm.errorCount != 0 ) return;
34
35        p16asm.run(lp, ASM_PASS2);
36        if ( p16asm.errorCount > 0 ) return;
37
38        p16asm.output(lp);
39    }
```

/p16ecc/as_source_2/main.cpp   Code-11.2

- #26: parse the input file.
- #29: instantiate the class `P16E_asm`.
- #32: first scan on input lines.
- #35: second scan on input lines.
- #38: generate outputs.

The following (Code-11.3) is the constructor for class `P16E_asm`. It creates the output files, *file*.obj and *file*.lst. Also, it initializes the symbol table with the values of the PIC16F core registers.

```
12    typedef struct {
13        const char  *regName;
14        int          regAddr;
15    } CoreReg_t;
16
17    const CoreReg_t pic16_reg[] = {
18        {"INDF0",   0x00}, {"INDF1",   0x01},
19        {"PCL",     0x02}, {"STATUS",  0x03},
20        {"FSR0L",   0x04}, {"FSR0H",   0x05},
21        {"FSR1L",   0x06}, {"FSR1H",   0x07},
22        {"BSR",     0x08}, {"WREG",    0x09},
23        {"PCLATH",  0x0a}, {"INTCON",  0x0b},
24        {"FSR0",    0x00}, {"FSR1",    0x01},
25        {NULL,      0}
26    };
27
28    P16E_asm :: P16E_asm(char *fname, const char *ext)
29    {
30        memset(this, 0, sizeof(P16E_asm));
31
32        int len = strlen(fname);
33        if ( len > 4 && strcasecmp(&fname[len-4], ".asm") == 0 )
34        {
35            len -= 4;
36            objFile = new char[len+5];  memcpy(objFile, fname, len); strcpy(&objFile[len],    ext );
37            lstFile = new char[len+5];  memcpy(lstFile, fname, len); strcpy(&lstFile[len], ".lst");
38            FILE *objFout = fopen(objFile, "w");
39            FILE *lstFout = fopen(lstFile, "w");
```

```cpp
40
41        if ( objFout == NULL || lstFout == NULL )
42        {
43            errorCount++;
44            printf("open file error!\n");
45            return;
46        }
47
48        objWriter = new ObjWriter(objFout);
49        lstWriter = new LstWriter(lstFout);
50
51        // add core registers's definitions...
52        for(const CoreReg_t *rp = pic16_reg; rp->regName; rp++)
53        {
54            Symbol *sp = addSymbol(&symbolList, new Symbol((char*)rp->regName));
55            sp->item    = valItem(rp->regAddr);
56            sp->type    = EQU;
57        }
58    }
59    else
60    {
61        errorCount++;
62        printf("file name error!\n");
63    }
64 }
```
/p16ecc/as_source_2/p16asm.cpp   Code-11.3

- #36~37: create output files (.obj and .lst files).
- #48~49: instantiate file writers for both .obj and .lst files.
- #52~57: initialize the symbol table with PIC16F core registers' values.

The following code, Code-11.4, is the `P16E_asm::run()` function, which scans the input lines.

```cpp
79  void P16E_asm :: run(line_t *lp, int pass)
80  {
81      for (; lp; lp = lp->next)
82      {
83          if ( pass == ASM_PASS1 && lp->inst == EQU )
84          {
85              if ( lp->label && lp->oprds )
86              {
87                  lp->oprds   = symbolReplace(symbolList, lp->oprds);
88                  Symbol *sym = addSymbol(&symbolList, new Symbol(lp->label));
89                  sym->type   = EQU;
90                  sym->item   = cloneItem(lp->oprds);
91                  sym->global = lp->globalLbl? true: false;
92              }
93          }
94
95          if ( pass == ASM_PASS2 && lp->inst != EQU )
96          {
97              regulateInst(lp);    // validate instruction format
98              if ( lp->oprds )
99              {
100                 lp->oprds = symbolReplace(symbolList, lp->oprds);
101                 for (item_t *ip = lp->oprds; ip->next; ip = ip->next)
102                     ip->next = symbolReplace(symbolList, ip->next);
103             }
104         }
105     }
106 }
```
/p16ecc/as_source_2/p16asm.cpp   Code-11.4

- #83~93: first scan, add the symbols, defined by .equ instructions, to the symbol table with their value.
- #95~104: second scan, replace the symbol with its value, for the operand(s).

317

11.2 Output File Format

This section only describes the format for **.obj** file, which contains PIC16F instruction binary codes. The .obj file format presented here is plain text. That means it can be read and modified using a regular text editor.

Each line in a **.obj** file can be referred to as a 'record', and the first letter of a line determines the type of the record. Table-11.1 below gives all types of records.

Type letter	Description
'S'	`.segment` line
'T'	`.invoke` line
'P'	`.device` line
'U'	Function start line
'G'	Label line (global)
'L'	Label line (local)
'F'	Function call reference line
'K'	`.bsel` line
'J'	`.psel` line
'N'	`.dblank` line
'M'	`.cblank` line
'R'	`.rs` line
'E'	`.equ` line (for global labels)
'W'	Instruction codes or ROMed data (can hold multiple items in the same line)
';'	Comments line

Table-11.1

11.3 Instruction Format and Translation Table

Every PIC16F instruction takes 14 bits (or a word). The assembler needs to translate the instruction mnemonic into the PIC16F instruction binary format. Typically, an instruction combines the opcode and operand into a 14-bit binary format, and the instruction format varies depending on the instruction form.

The following (Code-11.5) classifies the formats of the enhanced PIC16F CPU instruction set by operand type.

```
4     // PIC16E instruction formats...
5     enum {
6         P16_FD_I = 1,      // ADDWF x, W
7         P16_F_I,           // CLRF x
8         P16_BIT_I,         // BTFSS x, n
9         P16_LIT5_I,        // MOVLB N
10        P16_LIT8_I,        // MOVLW N
11        P16_LIT9_I,        // BRA lbl
12        P16_LIT11_I,       // GOTO, CALL
13        P16_FSR_OP_I,      // ADDFSR
14        P16_MOV_OP_I,      // MOVIW ++FSR0
15        P16_MOV_OFF_I,     // MOVIW n[FSR0]
16     };
```
`/p16ecc/as_source_2/p16inst.h`
Code-11.5

Additionally, in Code-11.6, a lookup table, `P16_inst_code_tbl[]`, containing the instruction binary codes, and the routine P16inst() are provided to facilitate translation.

```cpp
const P16inst_t P16_inst_code_tbl[] = {
    {ADDWF,      0x0700, P16_FD_I},
    {ANDWF,      0x0500, P16_FD_I},
    {CLRF,       0x0180, P16_F_I},
    {CLRW,       0x0100, 0},
    {COMF,       0x0900, P16_FD_I},
    {DECF,       0x0300, P16_FD_I},
    {DECFSZ,     0x0b00, P16_FD_I},
    {INCF,       0x0a00, P16_FD_I},
    {INCFSZ,     0x0F00, P16_FD_I},
    {IORWF,      0x0400, P16_FD_I},
    {MOVF,       0x0800, P16_FD_I},
    {MOVWF,      0x0080, P16_F_I},
    {NOP,        0x0000, 0},
    {RLF,        0x0d00, P16_FD_I},
    {RRF,        0x0c00, P16_FD_I},
    {SUBWF,      0x0200, P16_FD_I},
    {SWAPF,      0x0e00, P16_FD_I},
    {XORWF,      0x0600, P16_FD_I},

    {BCF,        0x1000, P16_BIT_I},
    {BSF,        0x1400, P16_BIT_I},
    {BTFSC,      0x1800, P16_BIT_I},
    {BTFSS,      0x1c00, P16_BIT_I},

    {ADDLW,      0x3e00, P16_LIT8_I},
    {ANDLW,      0x3900, P16_LIT8_I},
    {CALL,       0x2000, P16_LIT11_I},
    {GOTO,       0x2800, P16_LIT11_I},
    {IORLW,      0x3800, P16_LIT8_I},
    {MOVLW,      0x3000, P16_LIT8_I},

        . . .

    {RESET,      0x0001, 0},
    {CALLW,      0x000a, 0},
    {BRW,        0x000b, 0},
    {ADDWFC,     0x3d00, P16_FD_I},
    {SUBWFB,     0x3b00, P16_FD_I},
    {MOVLB,      0x0020, P16_LIT5_I},
    {ADDFSR,     0x3100, P16_FSR_OP_I},
    {MOVLP,      0x3180, P16_F_I},
    {LSLF,       0x3500, P16_FD_I},
    {LSRF,       0x3600, P16_FD_I},
    {ASRF,       0x3700, P16_FD_I},
    {BRA,        0x3200, P16_LIT9_I},
    {MOVIW_OP,   0x0010, P16_MOV_OP_I},
    {MOVWI_OP,   0x0018, P16_MOV_OP_I},
    {MOVIW_OFF,  0x3f00, P16_MOV_OFF_I},
    {MOVWI_OFF,  0x3f80, P16_MOV_OFF_I},
    {0, 0, 0}
};

//////////////////////////////////////////////////////////////////////////////
P16inst_t *P16inst(int inst)
{
    const P16inst_t *p = P16_inst_code_tbl;
    while ( p->inst != 0 )
    {
        if ( p->inst == inst )
            return (P16inst_t *)p;

        p++;
    }
    return NULL;
}
```


319

11.4 Generate the Outputs

In function `P16E_asm::output()`, as shown in Code-11.7a~Code-11.7d, it translates the pseudo (or directive) instructions and generates the output accordingly.

```
17   void P16E_asm :: output(line_t *lp)
18   {
19       unsigned int addr = 0;
20       for (bool done = false; !done && lp; lp = lp->next)
21       {
22           if ( lp->label && lp->inst != EQU )
23           {
24               if ( lp->globalLbl )
25                   objWriter->output('G', lp->label);
26               else
27                   objWriter->output('L', lp->label);
28
29               objWriter->flush();
30           }
```
/p16ecc/as_source_2/p16asm1.cpp   Code-11.7a

- #20: scan all the lines in the buffer.
- #22~30: for the label field, if it exists, output the label record.

```
32           item_t *ip0 = lp->oprds;              // fisrt operand
33           item_t *ip1 = ip0? ip0->next: NULL;   // second operand
34           P16inst_t *p16inst;
35           char buf[4096];
36           int  v, mask;
37           char type = 0;
38           int  incAddr = 0;
39
40           switch ( lp->inst )
41           {
42               case S_COMMENT: // comments (C source)
43               case DEVICE:    // device name, RAM & ROM size
44               case INVOKE:    // invoke library file
45               case CBLANK:    // code space blank
46               case DBLANK:    // data space blank
47               case FCALL:     // function call index
48               case BSEL:      // .bsel
49               case PSEL:      // .psel
50               case RS:        // .rs
51                   if ( ip0 )
52                   {
53                       char t = (lp->inst == S_COMMENT)?   ';':
54                                (lp->inst == DEVICE)?      'P':
55                                (lp->inst == INVOKE)?      'I':
56                                (lp->inst == CBLANK)?      'N':
57                                (lp->inst == DBLANK)?      'M':
58                                (lp->inst == FCALL)?       'F':
59                                (lp->inst == BSEL)?        'K':
60                                (lp->inst == PSEL)?          'J': 'R';
61                       objOutput(t, ip0);
62
63                       for (item_t *ip = ip1; ip; ip = ip->next)
64                           objOutput(ip);
65                       objWriter->flush();
66                   }
67                   if ( (lp->inst == BSEL || lp->inst == PSEL) && !ip1 )
68                   {
69                       int code = (lp->inst == BSEL)? 0x0020: 0x3180;
70                       lstWriter->output(addr, code, '?');
71                       incAddr = 1, type = '?';
72                   }
73                   lstWriter->output(lp);
74                   break;
```
/p16ecc/as_source_2/p16asm1.cpp   Code-11.7b

- #42~66: generate .obj output for ';'/'P'/'I'/.../'R' type records, with the operand fields (#63~64).
- #67~73: generate .lst output.

```
76                          case SEGMENT:
77                              addr = 0;
78                              objWriter->output('S', lp->attr->name);
79                              if ( isREL(lp) )
80                                  objWriter->output("REL");
81                              else if ( isABS(lp) )
82                              {
83                                  objWriter->output("ABS");
84                                  if ( lp->attr && lp->attr->addr )
85                                  {
86                                      item_t *ip = lp->attr->addr;
87                                      objOutput(ip);
88                                      addr = ip->data.val;
89                                  }
90                              }
91
92                              if ( isBEG(lp) )
93                                  objWriter->output("BEG");
94                              else if ( isEND(lp) )
95                                  objWriter->output("END");
96                              if ( I_TYPE(ip0, ':') )
97                              {
98                                  objOutput('U', ip0->left);   // function_name
99                                  objOutput(ip0->right);       // function RAM cost
100                             }
101                             objWriter->flush();
102                             lstWriter->output(lp);
103                             break;
```
/p16ecc/as_source_2/p16asm1.cpp   Code-11.7c

- #76~103: generate .obj output for .segment lines ('S' type records).

```
105                         case EQU:
106                             if ( lp->label && lp->globalLbl && ip0 )
107                             {
108                                 objWriter->output('E', lp->label);
109                                 objOutput(ip0);
110                                 objWriter->flush();
111                             }
112                             lstWriter->output(lp);
113                             break;
114
115                         case DW:
116                             for(item_t *ip = lp->oprds; ip; ip = ip->next)
117                             {
118                                 if ( parseItem(ip, buf, &v, DW) == VAL_EXP )
119                                     objWriter->outputW(v & 0x3fff);
120                                 else
121                                     objWriter->outputW(buf, 0x3fff);
122
123                                 lstWriter->output(addr++, v, type);
124                             }
125                             lstWriter->output(lp);
126                             break;
127
128                         case END:
129                             lstWriter->output(lp);
130                             done = true;
131                             break;
```
/p16ecc/as_source_2/p16asm1.cpp   Code-11.7d

- #105~113: generate the .obj output for .equ instruction lines ('E' record).
- #115~126: generate the obj output for .dw instruction lines (which usually are the ROMed data), in 'W' type records. Note, the function P16E_asm::parseItem() evaluates the expression in the operand field. The result type (VAL_EXP vs. STR_EXP) tells that the result is either a constant or a string (unsolved value).

The following (Code-11.8a~Code-11.8f) are for regular PIC16F instruction translation and output. The translation and production are done based on the operand forms (listed in Code-10.12).

1) Generate the output for instruction with operand type `P16_FD_I` and `P16_F_I`, which contains a 7-bit RAM bank address:

```
133              default:    // all regular instructions ...
134                  p16inst = P16inst(lp->inst);
135                  if ( p16inst )
136                  {
137                      int code= p16inst->code;
138                      incAddr = 1;
139                      switch ( p16inst->form )
140                      {
141                          case P16_FD_I:      // ADDWF x, W  or  ADDWF x, F
142                              if ( lp->desRegF ) code |= 0x0080;
143                          case P16_F_I:       // MOVWF x
144                              switch ( parseItem(ip0, buf, &v, lp->inst) )
145                              {
146                                  case VAL_EXP:
147                                      code |= (v & 0x7f);
148                                      objWriter->outputW(code);
149                                      break;
150                                  case STR_EXP:
151                                      objWriter->outputW(code, buf, 0x7f);
152                                      type = '?';
153                                      break;
154                                  default:
155                                      break;
156                              }
157                              break;
```
/p16ecc/as_source_2/p16asm1.cpp   Code-11.8a

- #134: get the instruction translation from the look-up table (Code-10.13).
- #137: get the instruction opcode.
- #141~156: generate the code for the instruction.

2) Generate the output for instruction with operand type `P16_LIT5_I`, `P16_LIT8_I`, `P16_LIT9_I`, `P16_LIT11_I`:

```
159                          case P16_LIT5_I:    // MOVLB
160                          case P16_LIT8_I:    // MOVLW
161                          case P16_LIT9_I:    // BRA
162                          case P16_LIT11_I:   // GOTO, CALL
163                              mask = (p16inst->form == P16_LIT5_I)? 0x01f:
164                                     (p16inst->form == P16_LIT8_I)? 0x0ff:
165                                     (p16inst->form == P16_LIT9_I)? 0x1ff: 0x7ff;
166
167                              switch ( parseItem(ip0, buf, &v, lp->inst) )
168                              {
169                                  case VAL_EXP:
170                                      code |= (v & mask);
171                                      objWriter->outputW(code);
172                                      break;
173                                  case STR_EXP:
174                                      type = '?';
175                                      objWriter->outputW(code, buf, mask);
176                                      break;
177                                  default:
178                                      break;
179                              }
180                              break;
```
/p16ecc/as_source_2/p16asm1.cpp   Code-11.8b

3) Generate the output for instruction with operand type P16_BIT_I,

```
182                                     case P16_BIT_I:       // BCF x, 1
183                                         if ( parseItem(ip1, buf, &v, lp->inst) == VAL_EXP )
184                                             code |= (v & 7) << 7;
185
186                                         switch ( parseItem(ip0, buf, &v, lp->inst) )
187                                         {
188                                             case VAL_EXP:
189                                                 code |= (v & 0x7f);
190                                                 objWriter->outputW(code);
191                                                 break;
192                                             case STR_EXP:
193                                                 objWriter->outputW(code, buf, 0x7f);
194                                                 type = '?';
195                                                 break;
196                                             default:
197                                                 break;
198                                         }
199                                         break;
/p16ecc/as_source_2/p16asm1.cpp                                      Code-11.8c
```

4) Generate the output for instruction with operand type P16_FSR_OP_I,

```
201                                     case P16_FSR_OP_I:   // ADDFSR INDF0, 1
202                                         if ( parseItem(ip0, buf, &v, lp->inst) == VAL_EXP )
203                                             code |= (v&1) << 6;
204                                         if ( parseItem(ip1, buf, &v, lp->inst) == VAL_EXP )
205                                             code |= v & 0x3f;
206                                         objWriter->outputW(code);
207                                         break;
/p16ecc/as_source_2/p16asm1.cpp                                      Code-11.8d
```

Note, it doesn't accept a symbol value.

5) Generate the output for instructions with operand type P16_MOV_OP_I and P16_MOV_OFF_I,

```
209                                     case P16_MOV_OP_I:   // MOVWI ++FSR0, MOVWI --FSR0, ...
210                                         switch ( ip0->type )
211                                         {
212                                             case FSR_PRE_INC:    // MOVIW/MOVWI   ++INDF0
213                                             case FSR_PRE_DEC:    // MOVIW/MOVWI   --INDF0
214                                             case FSR_POST_INC:   // MOVIW/MOVWI   INDF0++
215                                             case FSR_POST_DEC:   // MOVIW/MOVWI   INDF0--
216                                                 if ( parseItem(ip0->left, buf, &v, lp->inst) == VAL_EXP )
217                                                 {
218                                                     code |= (v&1) << 2;
219                                                     code |= (ip0->type == FSR_PRE_INC )? 0:
220                                                             (ip0->type == FSR_PRE_DEC )? 1:
221                                                             (ip0->type == FSR_POST_INC)? 2: 3;
222                                                     objWriter->outputW(code);
223                                                 }
224                                         }
225                                         break;
226
227                                     case P16_MOV_OFF_I: // MOVWI n[FSR0]
228                                         if ( ip0->type == FSR_OFFSET )
229                                         {
230                                             if ( parseItem(ip0->left, buf, &v, lp->inst) == VAL_EXP )
231                                                 code |= v & 0x3f;    // offset
232
233                                             if ( parseItem(ip0->right, buf, &v, lp->inst) == VAL_EXP )
234                                                 code |= (v&1) << 6; // FSR index
235
236                                             objWriter->outputW(code);
237                                         }
238                                         break;
/p16ecc/as_source_2/p16asm1.cpp                                      Code-11.8e
```

6) Generate the output for the instructions that have no operand,

```
240                                        default:     // NOP, RETURN, CLRW, ...
241                                            objWriter->outputW(code);
242                                            break;
243                                    }
244                                    lstWriter->output(addr, code, type);
245                                    lstWriter->output(lp);
246                                }
247                                else if ( lp->inst )
248                                    printf("unknown instruction - %d\n", lp->inst);
249                                else
250                                    lstWriter->output(lp);
251                                break;
252                        }
253                    addr += incAddr;
254                }
255
256        objWriter->flush();
257        objWriter->close();
258        lstWriter->close();
259    }
```

/p16ecc/as_source_2/p16asm1.cpp	Code-11.8f

Table-11.2a and Table-11.2b are the test results for this chapter.

Assembly source file (ch8_t10.asm)	Object file (ch8_t10.obj)
<pre>;***
; Microchip Enhanced PIC16F1xxx C Compiler (CC16E), v0.09.02
; Fri Sep 12 17:37:28 2025
;***

    .device "pic16e"

    .segment   BANKi (REL)
p::   .rs   2
    .segment   BANKi (REL)
n::   .rs   2
    .segment   BANKi (REL)
m::   .rs   2

    .segment   CODE2 (REL) foo:0
foo::
; :: ch8_t10.c #6: *p += 0xff000000;
    .bsel p
    movf   p, W
    movwf FSR0L
    .bsel p, p+1
    movf   p+1, W
    movwf FSR0H
    addfsr 0, 3
    decf   INDF0, F
; :: ch8_t10.c #7: n += 1;
    .bsel p+1, n
    incf   n, F
    .bsel n, n+1
    btfsc 3, 2
    incf   n+1, F
; :: ch8_t10.c #8: m -= 0xff00;
    .bsel n+1, m+1
    incf   m+1, F
    return

    .end</pre> | <pre>P "pic16e"
S BANKi REL
G p
R 2
S BANKi REL
G n
R 2
S BANKi REL
G m
R 2
S CODE2 REL
U foo 0
G foo
; "; :: ch8_t10.c #6: *p += 0xff000000;"
K p
W 0x0800:p 0x0084
K p (p+1)
W 0x0800:(p+1) 0x0085 0x3103 0x0380
; "; :: ch8_t10.c #7: n += 1;"
K (p+1) n
W 0x0A80:n
K n (n+1)
W 0x1903 0x0A80:(n+1)
; "; :: ch8_t10.c #8: m -= 0xff00;"
K (n+1) (m+1)
W 0x0A80:(m+1) 0x0008</pre> |

	Table-11.2a

Note, the operator ' : ' is used in .obj files for 'W' record lines, and it's equivalent to the logic operator ' | '. It indicates that the value followed is a RAM bank address.

List file (ch8_t10.lst)

```
    1 ;***********************************************************
    2 ; Microchip Enhanced PIC16F1xxx C Compiler (CC16E), v0.09.02
    3 ; Fri Sep 12 17:37:28 2025
    4 ;***********************************************************
    5
    6  .device "pic16e"
    7
    8  .segment    BANKi (REL)
    9 p::    .rs   2
   10  .segment    BANKi (REL)
   11 n::    .rs   2
   12  .segment    BANKi (REL)
   13 m::    .rs   2
   14
   15
   16  .segment    CODE2 (REL) foo:0
   17 foo::
   18 ; :: ch8_t10.c #6: *p += 0xff000000;
00000: 0x0020?     19  .bsel  p
00001: 0x0800?     20  movf  p, W
00002: 0x0084      21  movwf  FSR0L
                   22  .bsel  p, p+1
00003: 0x0800?     23  movf  p+1, W
00004: 0x0085      24  movwf  FSR0H
00005: 0x3103      25  addfsr 0, 3
00006: 0x0380      26  decf  INDF0, F
                   27 ; :: ch8_t10.c #7: n += 1;
                   28  .bsel  p+1, n
00007: 0x0A80?     29  incf  n, F
                   30  .bsel  n, n+1
00008: 0x1903      31  btfsc  3, 2
00009: 0x0A80?     32  incf  n+1, F
                   33 ; :: ch8_t10.c #8: m -= 0xff00;
                   34  .bsel  n+1, m+1
0000A: 0x0A80?     35  incf  m+1, F
0000B: 0x0008      36  return
                   37
                   38  .
```

Table-11.2b

PIC16F Assembler – Support Extended PIC16F1xxxx

As mentioned earlier in this book, the latest products in the PIC16F family from Microchip have extended the architecture to support up to 4Kbytes of RAM. In fact, some newly released products (such as PIC16F15213, PIC16F15214, …) that have only a few hundred bytes of RAM on chip also adopt the new structure.

The PIC16F chips with the new structure can be called 'extended' PIC1F16 CPU. It has only one instruction (MOVLB), changed from the enhanced PIC16F family. It extends the operand field from 5 to 6 bits to load register BSR, which indicates the current RAM bank. In theory, this extended architecture can support up to $64{\times}80 + 16 = 5136$ (bytes) of RAM. The mnemonics of instruction MOVLB remain the same as its predecessor, but its machine code format has changed (see the details below). That means it's not compatible with the enhanced PIC16F.

MOVLB format of enhanced PIC16F	MOVLB format of extended PIC16F
00 0000 001k $kkkk$	00 0001 01kk $kkkk$

Where $k...k$ are the operand bits that are loaded to register BSR.

Thus, the RAM address spaces for the extended PIC16F devices are:

    Regular address space: 0x0000 ~ 0x1FFF
    Linear address space: 0x2000 ~ 0x33FF

12.1 Option for Running Assembler

<u>Project directory</u>: /p16ecc/as_source_3
<u>Updated files</u>: main.cpp

The assembler, as16e.exe, needs to support both enhanced PIC16F and extended PIC16F devices. That means it must be aware of which family of PIC16F code needs to be generated.

A running option '-X' is used in the command line when it needs to generate the code for an extended PIC16F device, like:

    as16e **-X** *file*.asm

Code-12.1, shown below, has the corresponding modification to support the option setting in the function main().

```
 8    int useBsr6 = 0;
 9
10    static void asm0(char *filename);
11
12    int main(int argc, char *argv[])
13    {
14        for (int i = 1; i < argc; i++)
15        {
16            if ( argv[i][0] == '-' )      // option
17            {
18                if ( strcmp(argv[i], "-X") == 0 )
19                    useBsr6 = 1;
20                else
21                    printf("unknown option '%s'!\n", argv[i]);
22                continue;
23            }
24
25            int length = strlen(argv[i]);
26            if ( !(length > 4 && strcasecmp(&argv[i][length-4], ".asm") == 0) )
27            {
28                printf("improper file type/name '%s'!\n", argv[i]);
29                continue;
30            }
31
32            asm0(argv[i]);
33        }
34        return 0;
35    }
```
`/p16ecc/as_source_3/main.cpp`   Code-12.1

- #18~19: get the running option '-X', and set the flag variable useBsr6.

12.2 Expand Operand Types and Translation Table

<u>Updated files</u>: p16inst.h, p16inst.cpp, p16asm1.cpp
The following, Code-12.2, has the operand type expanded to involve the new instruction.

```
 1    #ifndef _P16E_INST_H
 2    #define _P16E_INST_H
 3
 4    // PIC16E instruction formats...
 5    enum {
 6        P16_FD_I = 1,     // ADDWF x, W
 7        P16_F_I,          // CLRF x
 8        P16_BIT_I,        // BTFSS x, n
 9        P16_LIT5_I,       // MOVLB N       - enhanced PIC16F
10        P16_LIT6_I,       // MOVLB N       - extended PIC16F
11        P16_LIT8_I,       // MOVLW N
12        P16_LIT9_I,       // BRA lbl
13        P16_LIT11_I,      // GOTO, CALL
14        P16_FSR_OP_I,     // ADDFSR
15        P16_MOV_OP_I,     // MOVIW ++FSR0
16        P16_MOV_OFF_I,    // MOVIW n[FSR0]
17    };
```
`/p16ecc/as_source_3/p16inst.h`   Code-12.2

- #10: new type operand added.

The translation table and the translation routine, shown in Code-12.3a and Code-12.3b, have the following changes, as well.

```cpp
11   const P16inst_t P16_inst_code_tbl[] = {
12       {ADDWF,      0x0700, P16_FD_I},
13       {ANDWF,      0x0500, P16_FD_I},
14       {CLRF,       0x0180, P16_F_I},
15       {CLRW,       0x0100, 0},
16       {COMF,       0x0900, P16_FD_I},
17       {DECF,       0x0300, P16_FD_I},
18       {DECFSZ,     0x0b00, P16_FD_I},
19       {INCF,       0x0a00, P16_FD_I},
20       {INCFSZ,     0x0f00, P16_FD_I},
21       {IORWF,      0x0400, P16_FD_I},
22       {MOVF,       0x0800, P16_FD_I},
23       {MOVWF,      0x0080, P16_F_I},
24       {NOP,        0x0000, 0},
25       {RLF,        0x0d00, P16_FD_I},
26       {RRF,        0x0c00, P16_FD_I},
27       {SUBWF,      0x0200, P16_FD_I},
28       {SWAPF,      0x0e00, P16_FD_I},
29       {XORWF,      0x0600, P16_FD_I},
30
31       {BCF,        0x1000, P16_BIT_I},
32       {BSF,        0x1400, P16_BIT_I},
33       {BTFSC,      0x1800, P16_BIT_I},
34       {BTFSS,      0x1c00, P16_BIT_I},
35
36       {ADDLW,      0x3e00, P16_LIT8_I},
37       {ANDLW,      0x3900, P16_LIT8_I},
38       {CALL,       0x2000, P16_LIT11_I},
39       {GOTO,       0x2800, P16_LIT11_I},
40       {IORLW,      0x3800, P16_LIT8_I},
41       {MOVLW,      0x3000, P16_LIT8_I},
 .    .  .
52       {RESET,      0x0001, 0},
53       {CALLW,      0x000a, 0},
54       {BRW,        0x000b, 0},
55       {ADDWFC,     0x3d00, P16_FD_I},
56       {SUBWFB,     0x3b00, P16_FD_I},
57       {MOVLB,      0x0020, P16_LIT5_I},     // enhanced FIC16F
58       {MOVLB,      0x0140, P16_LIT6_I},     // extebded FIC16F
59       {ADDFSR,     0x3100, P16_FSR_OP_I},
60       {MOVLP,      0x3180, P16_F_I},
61       {LSLF,       0x3500, P16_FD_I},
62       {LSRF,       0x3600, P16_FD_I},
63       {ASRF,       0x3700, P16_FD_I},
64       {BRA,        0x3200, P16_LIT9_I},
65       {MOVIW_OP,   0x0010, P16_MOV_OP_I},
66       {MOVWI_OP,   0x0018, P16_MOV_OP_I},
67       {MOVIW_OFF,  0x3f00, P16_MOV_OFF_I},
68       {MOVWI_OFF,  0x3f80, P16_MOV_OFF_I},
69       {0, 0, 0}
70   };
```

/p16ecc/as_source_3/p16inst.cpp   Code-12.3a

- #58: translation of instruction MOVLB for extended PIC16F.

```cpp
73   P16inst_t *P16inst(int inst)
74   {
75       const P16inst_t *p = P16_inst_code_tbl;
76       while ( p->inst != 0 )
77       {
78           if ( p->inst == inst )
79           {
80               if ( !(p->inst == MOVLB && useBsr6 && p->form != P16_LIT6_I) )
81                   return (P16inst_t *)p;
82           }
83
84           p++;
85       }
86       return NULL;
87   }
```

/p16ecc/as_source_3/p16inst.cpp   Code-12.3b

- #80~81: skip the MOVLB item that is for enhanced PIC16F.

Finally, in `P16E_asm::output()` routine, as illustrated in Code-12.4a and Code-12.4b, the changes are made to support extended PIC16F devices.

```
40              switch ( lp->inst )
41              {
42                  case S_COMMENT: // comments (C source)
43                  case DEVICE:    // device name, RAM & ROM size
44                  case INVOKE:    // invoke library file
45                  case CBLANK:    // code space blank
46                  case DBLANK:    // data space blank
47                  case FCALL:     // function call index
48                  case BSEL:      // .bsel
49                  case PSEL:      // .psel
50                  case RS:        // .rs
51                      if ( ip0 )
52                      {
53                          char t = (lp->inst == S_COMMENT)?   ';':
54                                   (lp->inst == DEVICE)?      'P':
55                                   (lp->inst == INVOKE)?      'I':
56                                   (lp->inst == CBLANK)?      'N':
57                                   (lp->inst == DBLANK)?      'M':
58                                   (lp->inst == FCALL)?       'F':
59                                   (lp->inst == BSEL)?        'K':
60                                   (lp->inst == PSEL)?        'J': 'R';
61                          objOutput(t, ip0);
62
63                          for (item_t *ip = ip1; ip; ip = ip->next)
64                              objOutput(ip);
65                          objWriter->flush();
66                      }
67                      if ( (lp->inst == BSEL || lp->inst == PSEL) && !ip1 )
68                      {
69                          int code = (lp->inst == PSEL)? 0x3180:
70                                     (useBsr6  == 0   )? 0x0020: 0x0140;
71                          lstWriter->output(addr, code, '?');
72                          incAddr = 1, type = '?';
73                      }
74                      lstWriter->output(lp);
75                      break;
```
/p16ecc/as_source_3/p16asm1.cpp   Code-12.4a

- #70: choose the opcode for MOVLB, 0x0020 (enhanced PIC16F) vs. 0x0140 (extended PIC16F).

```
159                         case P16_LIT5_I:    // MOVLB    - enhanced PIC16F
160                         case P16_LIT6_I:    // MOVLB    - extended PIC16F
161                         case P16_LIT8_I:    // MOVLW
162                         case P16_LIT9_I:    // BRA
163                         case P16_LIT11_I:   // GOTO, CALL
164                             mask = (p16inst->form == P16_LIT5_I)? 0x01f:
165                                    (p16inst->form == P16_LIT6_I)? 0x03f:
166                                    (p16inst->form == P16_LIT8_I)? 0x0ff:
167                                    (p16inst->form == P16_LIT9_I)? 0x1ff: 0x7ff;
```
/p16ecc/as_source_3/p16asm1.cpp   Code-12.4b

- #160, #165: new operand type added to support extended PIC16F.

Note that MOVLB instructions are inherent within the `.bsel` instructions, and the code generation for MOVLB instructions actually happens in the linking procedure. That means the linker (lk16e.exe) needs to support extended PIC16F devices (with the command option "`-X`").

PART-IV:
PIC16F Linker (lk16e.exe)

As the last tool in the compiler toolchain, the linker (lk16e) combines all the .obj files of a user application into the final output, a HEX file, which is usually 'burned' to the PIC16F chip for running.

As described earlier, the contents of .obj files are either relocatable or floating-point. One of the tasks that a linker performs is to assign addresses to all the pieces of records or segments in .obj files. Further, it has to link all the 'default' library files required for the application to run.

As in the design of cc16e and as16e, the command tools flex and bison are also used during the design of the linker lk16e.

Chapter-13

PIC16F Linker - Lexer and Parser

Project directory: /p16ecc/lk_source_1
Source files: main. cpp, lnk. l, lnk.y, common.h, common.c, ...

13.1 Basic Data Structures

Almost identical to that in as16e, common.h, as shown in Code-13.1a~Code-13.1b, defines the data structure used to store the output from .obj lines (or records) after parsing.

```
4    extern int useBSR6;
5
6    // item types ...
7    enum {
8        TYPE_VALUE=1,
9        TYPE_SYMBOL,
10       TYPE_STRING
11   };
12
13   /* ///////////////////////////////////////////////////////////////////////// */
14   typedef union data_ {
15       int    val;
16       char *str;
17   } data_t;
18
19   /* ///////////////////////////////////////////////////////////////////////// */
20   typedef struct item_ {
21       int            type;
22       data_t         data;
23       struct item_ *left;
24       struct item_ *right;
25       struct item_ *next;
26   } item_t;
```
/p16ecc/lk_source_1/common.h   Code-13.1a

- #7~11: record item value type.
- #14~17: record item value.
- #20~26: record item holder (it supports math/logic expressions).

```
28    /* ///////////////////////////////////////////////////////////////////// */
29    typedef struct line_ {
30        int            type;        // line type
31        item_t         *items;
32        char           *src;        // source file code
33        int            lineno;      // source file line number
34        char           *fname;      // source file name
35        char           insert;
36        char           retry;       // new in v2.1
37        char           pclath;
38        struct line_   *next;
39    } line_t;
```
/p16ecc/lk_source_1/common.h   Code-13.1b

- #29~39: data structure used to keep a line (or record) from the .obj file.

13.2 Lexer for PIC16F Linker

The script or text of the lexer, lnk.l, is similar to that in asm.l.

13.2.1 Macro part of the lexer

Code-13.2 is the macro part for the lexer.

```
24    %}
25
26    sp                      [ \t\015]
27    symbol                  [_a-zA-Z][_a-zA-Z0-9$]*
28    hex                     (0[xX][a-fA-F0-9]+)
29    dec                     ([0-9]|[1-9][0-9]+)
30
31    s_char                  ([^"\r\n])
32    s_char_sequence         ({s_char}+)
33    string                  \"{s_char_sequence}?\"
34
35    %x OPERAND
36    %x ESCAPE
37
38    %%
```
/p16ecc/lk_source_1/lnk.l   Code-13.2

- #28: regular expression for Hex values.
- #29: regular expression for Decimal values.
- #35~36: two new states are added, OPCODE and ESCAPE, beyond the default (INITIAL) state.

13.2.2 Rule part of the lexer

Code-13.3a shows the rules for the INITIAL state, which identifies the type of a line (record).

```
37   ^";"{sp}+.*                       {   char *p = yytext;
38                                          appendStr();
39                                          for(p++; *p == ' '; p++);
40                                          p[strlen(p)-1] = 0;
41                                          yylval.syml = dupStr(p+1);
42                                          return COMMENT;
43                                      }
44   ^.{sp}+                           {   appendStr();
45                                          if ( isAnOpcode(yytext[0]) )
46                                          {
47                                              BEGIN(OPERAND);
48                                              yylval.value = yytext[0];
49                                              return TYPE;
50                                          }
51                                          BEGIN(ESCAPE);
52                                      }
53   {sp}+                             {   appendStr(); }
54   \n                                {   appendStr();
55                                          return yytext[0];
56                                      }
57
58   <ESCAPE>.*                        {   /*ignore the line */ }
59   <ESCAPE>\n                        {   appendStr();
60                                          BEGIN(INITIAL);
61                                      }
```
/p16ecc/lk_source_1/lnk.l   Code-13.3a

- #37~43: rule to match a comment line.
- #44~52: rules to identify the type of a line, and then enter OPERAND state.
- #58~61: ignore the unknown types of lines.

Code-13.3b shows the rules for the OPERAND state.

```
63   <OPERAND>{symbol}                 {   appendStr();
64                                          yylval.syml = dupStr(yytext);
65                                          return SYMBOL;
66                                      }
67   <OPERAND>{string}                 {   int len = strlen(yytext);
68                                          appendStr();
69                                          yytext[len-1] = 0;
70                                          yylval.syml = dupStr(yytext+1);
71                                          return STRING;
72                                      }
73   <OPERAND>{hex}                    |
74   <OPERAND>{dec}                    {   appendStr();
75                                          yylval.value = convertNum((char *)yytext);
76                                          return NUMBER;
77                                      }
78   <OPERAND>">>"                     {   appendStr(); return RSHIFT; }
79   <OPERAND>"<<"                     {   appendStr(); return LSHIFT; }
80   <OPERAND>[-+%^:&|()*/~.]          {   appendStr();
81                                          return yytext[0];
82                                      }
83   <OPERAND>{sp}+                    {   appendStr(); }
84   <OPERAND>\n                       {   appendStr();
85                                          BEGIN(INITIAL);
86                                          return yytext[0];
87                                      }
```
/p16ecc/lk_source_1/lnk.l   Code-13.3b

- #63~77: rules and actions for a single value
- #78~82: rules for matching operators.
- #84~87: end of a line, go back to INITIAL state.

13.3 Parser for PIC16F Linker

The parser for the assembler is in file asm.y.

<u>10.3.1 Definition part of the parser</u>

Code-13.4 is the definition part of the parser. It lists all the tokens that are used in lnk.l.

```
 7    %union {
 8        int       value;
 9        char      *syml;
10        line_t    *line;
11        item_t    *item;
12    }
13
14    %token <value> TYPE RSHIFT LSHIFT UMINUS NUMBER
15    %token <syml> SYMBOL COMMENT STRING
16
17    %left '-' '+'
18    %left '*' '/' '%' RSHIFT LSHIFT
19    %left '&' '|' '^' '~'
20    %nonassoc UMINUS
21    %nonassoc INVERSE
22
23
24    %type <line>  prog lines line obj_line
25    %type <item>  items item
26    %type <item>  multiplicative_expr additive_expr shift_expr and_expr
27    %type <item>  exclusive_or_expr inclusive_or_expr primary_expr
28
29    %%
```
`/p16ecc/lk_source_1/lnk.y`   Code-13.4

- #7~12: union for the actual value of tokens.

<u>13.3.2 Rule part of the parser</u>

Code-13.5a shows the rules and actions for parsing the lines from an .obj file.

```
29    %%
30    prog
31        :    lines                          {    linePtr = $1; }
32        ;
33    lines
34        :    line                           {    $$ = $1; }
35        |    lines line                      {    $$ = $1;
36                                                  appendLine(&$1, $2);
37                                             }
38        ;
39    line
40        :    '\n'                            {    $$ = newLine(0, NULL);
41                                                  $$->src = dupStr(__yyline);
42                                                  $$->lineno = yylineno;
43                                             }
44        |    obj_line '\n'                   {    $$ = $1;
45                                                  $$->src = dupStr(__yyline);
46                                                  $$->lineno = yylineno;
47                                             }
48        |    COMMENT '\n'                    {    $$ = newLine(';', strItem($1));
49                                                  free($1);
50                                                  $$->src = dupStr(__yyline);
51                                                  $$->lineno = yylineno;
52                                             }
53        ;
54    obj_line
55        :    TYPE items                      {    $$ = newLine($1, $2); }
56        |    TYPE                            {    $$ = newLine($1, NULL); }
57        ;
```
`/p16ecc/lk_source_1/lnk.y`   Code-13.5a

Code-13.5b illustrates the rules and actions for items in a (record) line of the .obj file. Mostly, it's identical to that in designing the assembler.

```
58   items
59     :     item                              {    $$ = $1; }
60     |     items item                        {    $$ = appendItem($1, $2); }
61     ;
62   item
63     :     inclusive_or_expr                 {    $$ = $1; }
64     |     NUMBER ':' inclusive_or_expr      {    $$ = newItem(':');
65                                                  $$->left = valItem($1);
66                                                  $$->right = $3;
67                                             }
68     ;
69   primary_expr
70     :    '(' inclusive_or_expr ')'          {    $$ = $2; }
71     |    '~'  primary_expr                  {    $$ = newItem('~');
72                                                  $$->left = $2;
73                                             }
74     |    SYMBOL                             {    $$ = symItem($1); free($1); }
75     |    STRING                             {    $$ = strItem($1); free($1); }
76     |    NUMBER                             {    $$ = valItem($1);   }
77     |    '.'                                {    $$ = newItem('.'); }
78     ;


138  inclusive_or_expr
139    : exclusive_or_expr
140    | inclusive_or_expr '|'
141      exclusive_or_expr                     {    $$ = newItem('|');
142                                                 $$->left  = $1;
143                                                 $$->right = $3;
144                                             }
145    ;
146
147  %%
```
`/p16ecc/lk_source_1/lnk.y`   Code-13.5b

- #64: the operator ':' works as '|'. It indicates that the lower 7-bit value is a bank address.

13.4 Segmentize Record Lines

When an .obj file is being parsed, all the record lines are kept in the line buffer, indicated by `linePtr`, as shown in Code-13.5a.

All lines need to be segmentized or reorganized into segments. Every segment starts with a `segment` record line, and each segment will be packed to one of the following segment groups, shown in Table-13.1, at the right sequential place.

codeSegGroup	CONST and CODEn segments
dataSegGroup	DATAn segments
fuseSegGroup	FUSEn segments
miscSegGroup	unknown segments

Table-13.1

Code-13.6 defines class `Segment`.

```
4    enum {
5         CODE_SEGMENT,
6         DATA_SEGMENT,
7         CONST_SEGMENT,
8         FUSE_SEGMENT,
9         MISC_SEGMENT
10   };
11
```

```cpp
class Segment {
    public:
        line_t  *lines;           // line pointer
        char    *fileName;        // file name
        bool    isLIB;            // lib code
        int     isBEG: 1;         // begin of segment link
        int     isEND: 1;         // end of segment link
        int     isREL: 1;         // relocatable segment
        int     isABS: 1;         // absolute address seg.
        int     isUsed: 1;        // code will be used
        int     memAddr;          // assigned memory address
        int     dataSize;         // function data size (local)

        Segment *next;

    public:
        Segment(char *file_name);
        ~Segment();
        void init(void);
        void addLine(line_t *lp);
        int  lineCount(void);
        int  type(void);          // segment type
        int  size(void);          // segment size
        char *name(void);         // segment name
        void print(void);
        bool isName(const char *_name) { return strcmp(name(), _name) == 0; }
};

extern Segment  *codeSegGroup;
extern Segment  *dataSegGroup;
extern Segment  *fuseSegGroup;
extern Segment  *miscSegGroup;

void addSegment(Segment *seg);
void deleteSegments(Segment *list);
void printfSegments(Segment *list);
```

/p16ecc/lk_source_1/segment.h   Code-13.6

- #4~10: segment types.
- #40~43: segment groups.

Firgure-13-1 below illustrates the structure of the segment group `codeSegGroup`.

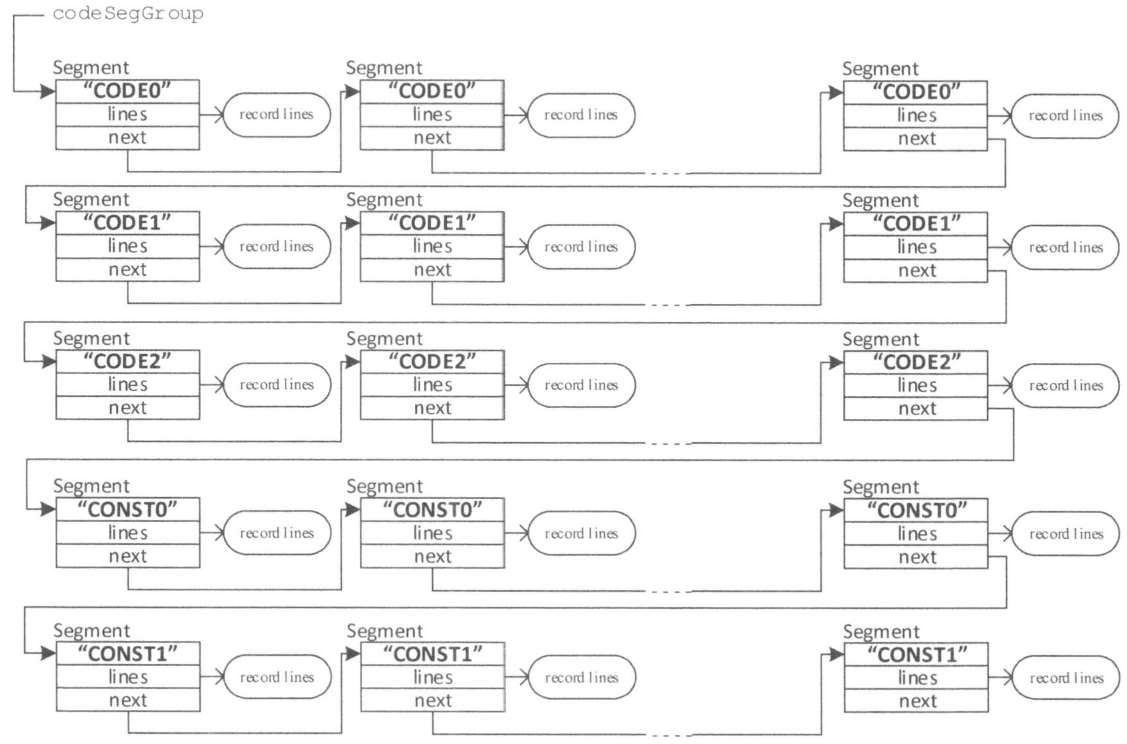

Figure-13-1

Firgure-13-2 below illustrates how the segments are reorganized, based on their attributes `isBEG` and `isEND`.

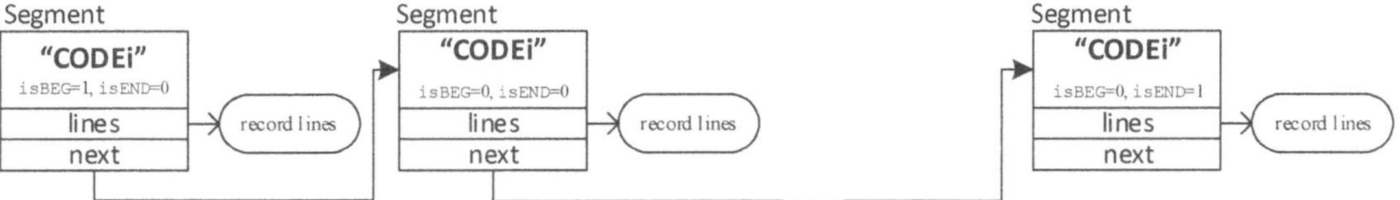

Figure-13-2

13.5 Read in Inputs and Reorganizing Segments

In main.cpp, it reads and processes all the .obj files of an application, as shown in Code-13.7.

```
21   int main(int argc, char *argv[], char *env[])
22   {
23       int start_parsing = 0;
24       Path *path = new Path(env);
25       libPath = path->get();
26
27       printf ("PIC16E linker, %s\n", VERSION);
28
29       for (int i = 1; i < argc; i++)
30       {

   .  .  .

46           {
47               link0(argv[i]);    // parse source file
48               start_parsing++;
49           }
50       }
```
/p16ecc/lk_source_1/main. cpp   Code-13.7

- #47: start parsing an input (.obj) file.

Code-13.8a and Code-13.8b illustrate the operation of parsing the input file and adding lines to the segment.

```
68   static void link0(char *filename, bool lib_f)
69   {
70       curFile = filename;
71
72       printf("linking '%s' ...\n", filename);
73
74       // using Lex and Bison to parse the source ...
75       errorCnt = 0;
76       line_t *lp = _main(filename);
77       if ( errorCnt > 0 ) exit(1);
78
79       Segment *seg = NULL;
80       while ( lp )
81       {

   .  .  .

94               case 'S':   // SEGMENT line...
95                   seg = NULL;
96               default:
97                   if ( seg == NULL )
98                   {   // create a segment...
99                       seg = new Segment(filename);
100                      seg->isLIB = lib_f;
101                      add_seg = true;
102                  }
103                  save_line = true;
104                  break;
```
/p16ecc/lk_source_1/main. cpp   Code-13.8a

- #76: parse the input file and read all the lines (records) in.
- #80: start scanning all the lines in the buffer.
- #94~95: reach to a .segment line, ready to generate a new segment.
- #97~102: create a new segment.

336

```
106
107                 line_t *next = lp->next;
108                 lp->next = NULL;
109
110                 if ( save_line )
111                 {
112                     seg->addLine(lp);          // add the line into segment...
113                     if ( add_seg )             // new segment, add the segment
114                     {
115                         seg->init();
116                         addSegment(seg);
117                     }
118                 }
119                 else
120                     freeLine(lp);
121
122                 lp = next;
123             }
124         }
```
/p16ecc/lk_source_1/main.cpp   Code-13.8b

- #112: add the current line to the segment.
- #113~117: add the new segment to a proper group, at the right place, by calling `addSegment()` after initialization.

Thus, all the input .obj files are read and stored in the segment groups, shown in Table-13.1, in the proper sequential order.

Regarding project completion, Code 13.9 provides the makefile.

```
1    CC  = gcc
2    CXX = g++
3    RM  = rm
4    MV  = mv
5    CP  = cp
6    EXE = lk16e.exe
7    OBJ = main.o lex.yy.o lnk.o common.o path.o segment.o
8
9    OPTIONS= -c -Os
10
11   $(EXE): $(OBJ) lnk.h makefile
12       $(CXX) -static $(OBJ) -o $(EXE)
13       $(CP) $(EXE) ../bin
14
15   %.o: %.c lnk.h makefile
16       $(CC) $(OPTIONS) $<
17
18   %.o: %.cpp lnk.h makefile
19       $(CXX) $(OPTIONS) $<
20
21   lex.yy.c: lnk.l lnk.h makefile
22       flex lnk.l
23
24   lnk.h: lnk.y makefile
25       bison -d lnk.y -o lnk.c
26
27   lnk.c: lnk.y makefile
28       bison -d lnk.y -o lnk.c
29
30   clean:
31       $(RM) *.o
32       $(RM) lnk.h
33       $(RM) lnk.c
34       $(RM) lex.yy.c
35       $(RM) $(EXE)
```
/p16ecc/lk_source_1/makefile   Code-13.9

Project directory: /p16ecc/lk_source_2
New Source files: memory.h, memory.cpp, symbol.cpp, symbol.cpp
Updated files: main.cpp

14.1 Memory Class

One of the tasks of the linking operations is to allocate all codes (data and instruction codes) to the target CPU's memories. The memory management class, Memory, is created for the purpose.

Code-14.1 shows the definition for the class.

```
4    typedef enum {DATA_MEMORY, CODE_MEMORY} MEMORY_TYPE;
5
6    #define BANKED_ADDRESS(a)    ((a&0x7f) >= 32 && (a&0x7f) < 0x70)
7    #define TO_LINEAR_ADDR(a)    (0x2000 + (a>>7)*80 + (a&0x7f) - 32)
8    #define TO_BANKED_ADDR(a)    ((((a&0x1fff)/80)<<7) + (a&0x1fff)%80 + 32)
9
10   #define FIXED_ADDR_FLAG      (1 << 0)
11   #define LINEAR_DATA_FLAG     (1 << 1)
12   #define PAGE_CHECK_FLAG      (1 << 2)
13
14   class Memory {
15       public:
16           Memory(MEMORY_TYPE type);
17           ~Memory();
18           void resize(int size);
19           void init(void);
20           void reset(void);
21           void blank(int start, int length);
22           bool getSpace(int init_addr, int *act_addr, int req_size, int flags=0);
23           int  memUsed(int *total);
24           void printSpace(FILE *);
25           void display(int);
26
27           MEMORY_TYPE type;
28       private:
29           char *mem;
30           int  maxSize;
31           int  memSize;
32
33           int  maxAddr(void);
34           void fillSpace(int start, int size, char val);
35   };
```

/p16ecc/lk_source_2/memory.h   Code-14.1

- #4: memory types, RAM or ROM.
- #18: the member function that sizes the memory space.
- #29: the array that indicates the status of memory locations. Byte values in it: -1 = forbidden, 0 = unused, 1 = used.

Code-14.2 shows the class's constructor and initialization.

```cpp
 9    Memory :: Memory(MEMORY_TYPE _type)
10    {
11        switch ( type = _type )
12        {
13            case DATA_MEMORY: maxSize = (useBSR6? 64:32)*80+16; break;  // max RAM size
14            case CODE_MEMORY: maxSize = 1024*32;                break;  // max ROM size
15        }
16
17        memSize = maxSize;
18        mem = new char[maxAddr()];
19        init();
20    }

22    Memory :: ~Memory()
23    {
24        delete [] mem;
25    }
26
27    void Memory :: resize(int _size)
28    {
29        if ( _size <= maxSize )
30        {
31            memSize = _size;
32            init();
33        }
34    }
35
36    void Memory :: init(void)
37    {
38        if ( type == DATA_MEMORY )
39            for (int a = 0; a < maxAddr(); a++)
40                mem[a] = BANKED_ADDRESS(a)? 0: -1;
41        else
42            memset(mem, 0, memSize);
43    }
```

/p16ecc/lk_source_2/memory.cpp   Code-14.2

- #11~18: initialize the memory array size (maximum).
- #36~43: initialize the memory array. The memory locations cannot be allocated for use when marked with the value -1.

The Memory classes for both data and code need to be instantiated before parsing any input file. See Code-14.3.

```cpp
16    static char *outputFile  = (char*)"_OUTPUT_";   // default output
17    static Symbol *objList = NULL;                   // list of .obj files
18    static Symbol *libList = NULL;                   // list of .lib files
19    static Memory *dataMem = NULL;
20    static Memory *codeMem = NULL;
21    static Symbol *fileList= NULL;
22    int     useBSR6 = 0;
23    char    *libPath;
24
25    static void link0(char *filename, bool lib_f=false);
26    static void addSegment(Segment *seg);
27
28    ///////////////////////////////////////////////////////////////////////////
29    int main(int argc, char *argv[], char *env[])
30    {
31        int start_parsing = 0;
32        Path *path = new Path(env);
33        libPath = path->get();
34
35        printf ("PIC16E linker, %s\n", VERSION);
36
37        for (int i = 1; i < argc; i++)
38        {
          . . . .
```

339

```
 53              if ( !start_parsing )
 54              {
 55                  dataMem = new Memory(DATA_MEMORY);
 56                  codeMem = new Memory(CODE_MEMORY);
 57                  libList = new Symbol((char*)"crt0");
 58                  start_parsing++;
 59              }
 60
 61              link0(argv[i]);    // parse source file
 62          }
```
/p16ecc/lk_source_2/main.cpp   Code-14.3

- #55~56: instantiate Memory class for both data and code memories.

#57: log in the system library file name – "crt0" (see the description in the next section).

After parsing an input file, the lines of type 'P', 'N', and 'M' will initialize the memory classes, as shown in Code-14.4, in routine link0().

```
 85  static void link0(char *filename, bool lib_f)
 86  {
        .  .  .
 96      errorCnt = 0;
 97      line_t *lp = _main(filename);
 98      if ( errorCnt > 0 ) exit(1);
 99
100      Segment *seg = NULL;
101      while ( lp )
102      {
103          bool save_line = false;
104          bool add_seg   = false;
105          item_t *ip0, *ip1;
106
107          switch ( lp->type ) // input line type?
108          {
109              case 'P':   // processor
110                  ip0 = itemPtr(lp->items, 1);
111                  ip1 = itemPtr(lp->items, 2);
112                  if ( ip0 ) dataMem->resize(ip0->data.val);
113                  if ( ip1 ) codeMem->resize(ip1->data.val);
114                  break;

133              case 'N':   // RAM blank
134              case 'M':   // ROM blank
135                  ip0 = itemPtr(lp->items, 0);    // 1st item in line
136                  ip1 = itemPtr(lp->items, 1);    // 2nd item in line
137                  if ( ip0 && ip0->type == TYPE_VALUE &&
138                       ip1 && ip1->type == TYPE_VALUE )
139                  {
140                      if ( lp->type == 'N' ) codeMem->blank(ip0->data.val, ip1->data.val);
141                      if ( lp->type == 'M' ) dataMem->blank(ip0->data.val, ip1->data.val);
142                  }
143                  break;
```
/p16ecc/lk_source_2/main.cpp   Code-14.4

- #109~114: for 'P' line, in the input file, resizes the memory size.
- #133~143: for 'N' and 'M' lines, in the input file, set the forbidden locations.

14.2 Symbol Class

The class `Symbol` is defined as in Code-14.5.

```
  3     #include "segment.h"
  4
  5     class Symbol {
  6         public:
  7             int             type;
  8             char            *name;
  9             Segment         *segment;
 10             int             value;  // value of the symbol
 11             int             size;   // for data
 12             Symbol          *next;
 13
 14         public:
 15             Symbol(char *sname);
 16             Symbol(char *sname, int stype, Segment *seg);
 17             ~Symbol();
 18     };
 19
 20     void    addSymbol(Symbol **slist, Symbol *sp);
 21     void    addSymbol(Symbol **slist, char *file);
 22     void    addSymbol(Symbol **slist, Symbol *sp, int value);
 23     void    deleteSymbols(Symbol **slist);
 24     Symbol  *searchSymbol(Symbol *slist, char *str, Segment *seg=NULL);
 25     Symbol  *logSymbol(Symbol **slist, int type, char *name, Segment *, int val);
 26     void    printSymbol(Symbol *slist);
```
`/p16ecc/lk_source_2/symbol.h`   Code-14.5

The class has multiple uses; one of them is to keep track of file names. There are two file streams, `libList` and `fileList`, which are used to keep track of library names and input file names. The operations are shown in Code-14.6.

```
 85     static void link0(char *filename, bool lib_f)
 86     {
 87         if ( searchSymbol(fileList, filename) )
 88             return;
 89
 90         addSymbol(&fileList, filename);
 91         curFile = filename;
 92
 93         printf("linking '%s' ...\n", filename);
 94
 95         // using Lex and Bison to parse the source ...
 96         errorCnt = 0;
 97         line_t *lp = _main(filename);
 98         if ( errorCnt > 0 ) exit(1);
 99
100         Segment *seg = NULL;
101         while ( lp )
102         {
103             bool save_line = false;
104             bool add_seg   = false;
105             item_t *ip0, *ip1;

            .  .  .
```

```cpp
116         case 'I':   // include lib file
117             ip0 = itemPtr(lp->items, 0);
118             if ( ip0 && ip0->type == TYPE_STRING )
119             {
120                 std::string str = ip0->data.str;
121                 int pos = str.find_last_of("/");
122                 if ( pos == std::string::npos ) pos = str.find_last_of("\\");
123                 if ( pos != std::string::npos ) str = str.substr(pos+1);
124
125                 pos = str.find(".");
126                 if ( pos != std::string::npos ) str = str.substr(0, pos);
127
128                 if ( !searchSymbol(libList, (char*)str.c_str()) )
129                     addSymbol(&libList, new Symbol((char*)str.c_str()));
130             }
131             break;
```

/p16ecc/lk_source_2/main.cpp   Code-14.6

- #87~90: before parsing an input file, check if it has been involved (logged in the list).
- #116~131: for 'I' line, log in the library file name (stripped from the string).

Chapter-15

PIC16F Linker – Including Lib Files and Evaluator

Project directory: /p16ecc/lk_source_3
New Source files: exp.h, exp.cpp
Updated files: main.cpp

15.1 Including Lib Files

As mentioned earlier, all the library files (names) are kept in the list, `libList`. Basically, library files are in C source format. That means they need to be compiled into the .obj file format for the linking operation as well.

After processing all the user application .obj files discussed in the last chapter, the list needs to be scanned and processed. Code-15.1 and Code-15.2 show the operation.

```
14    #define VERSION      "v0.1.3"
15    #define TEMP_FOLDER "_tmp_"
16
17    static char *outputFile  = (char*)"_OUTPUT_";    // default output
18    static Symbol *objList = NULL;                    // list of .obj files
19    static Symbol *libList = NULL;                    // list of .lib files
20    static Memory *dataMem = NULL;
21    static Memory *codeMem = NULL;
22    static Symbol *fileList= NULL;
23    int     useBSR6 = 0;
24    char   *libPath;
25
26    static void link0(char *filename, bool lib_f=false);
27    static char *getLibFile(char *libfile);
28    static void cleanup(void);
29
30    ////////////////////////////////////////////////////////////////////////////
31    int main(int argc, char *argv[], char *env[])
32    {
33        int start_parsing = 0;
34        Path *path = new Path(env);
35        libPath = path->get();
36
37        printf ("PIC16E linker, %s\n", VERSION);
38
39        for (int i = 1; i < argc; i++)
40        {
63            link0(argv[i]);    // parse source file
64        }
65
66        if ( start_parsing )
67        {
68            std::string tmp_folder = TEMP_FOLDER;
69            system(("mkdir " + tmp_folder).c_str());     // create a temp folder
70            for (Symbol *sp = libList; sp; sp = sp->next)
71            {
72                char *ssp = getLibFile(sp->name);
73                if ( ssp ) link0(ssp, 1);
74            }
75            system(("rm -r " + tmp_folder).c_str());    // remove temp folder & files
76
77            cleanup();
78        }
```

                                                          Code-15.1

- #68~69: create a temporary folder for holding the results of compiling the library files.
- #72: calling the routine, `getLibFIle()`, to get and compile a library file.
- #73: read in the .obj file.

```cpp
186  static char *getLibFile(char *libfile)
187  {
188      static char buf[4096];
189      std::string file_name;
190      std::string cc_cmd = "cc16e ";
191      std::string as_cmd = useBSR6? "as16e -X ": "as16e ";
192      std::string cp_cmd = "cp ";
193      std::string tmp_path = TEMP_FOLDER;
194      char *p0 = libfile, *p1;
195      FILE *fd;
196
197      while ( (p1 = strchr(p0, '/'))  ) p0 = p1 + 1;
198      while ( (p1 = strchr(p0, '\\')) ) p0 = p1 + 1;
199
200      if ( libPath != NULL )
201          sprintf(buf, "%slib/%s", libPath, p0);
202      else
203          sprintf(buf, "%s", p0);
204
205      file_name = buf; file_name += ".c";
206      fd = fopen(file_name.c_str(), "r"); // check if the source exists
207      if ( fd == NULL ) return NULL;
208      fclose(fd);
209
210      // copy the file to current folder...
211      system((cp_cmd + file_name + " " + tmp_path).c_str());
212      file_name = tmp_path + "/" + libfile + ".c";
213
214      // compile the source
215      system((cc_cmd + file_name).c_str());
216
217      // and assemble it
218      file_name = tmp_path + "/" + libfile + ".asm";
219      system((as_cmd + file_name).c_str());
220
221      // return obj file name...
222      file_name = tmp_path + "/" + libfile + ".obj";
223      fd = fopen(file_name.c_str(), "r");
224      if ( fd == NULL ) return NULL;
225
226      fclose (fd);
227      sprintf(buf, "%s", file_name.c_str());
228      return buf;
229  }
```

/p16ecc/lk_source_3/main.cpp   Code-15.2

- #190~192: compiling, assembling, and copying commands.
- #197~203: generate the complete file path for the library file.
- #205~208: check if the file exists.
- #211: copy the library file to the temporary folder.
- #215: compile the file, using the command "cc16e *file*.c".
- #219: assemble the file, using the command "as16e *file*.asm".
- #224: confirm the operation.

15.2 Item Evaluator

The items in a line or record of the .obj file are stored in the structure `item_t`, witch described in `common.h`, as shown in Figure-10.1. An item can involve mathematical or logical expressions that may contain symbols representing data or function addresses. During the linking procedure, each item must converge to a real value to generate the linking result.

The routine shown in Code-15.3 serves as the evaluator to determine the value of an item.

```cpp
17  char *expValue(item_t *ip, Segment *seg, Symbol *slist, int *val, int addr, int obj_line)
18  {
19      char *lbl0, *lbl1;
20      int  val0, val1;
21
22      if ( ip == NULL ) return NULL;
23
24      switch ( ip->type )
25      {
26          case ':':
27              lbl0 = expVal(ip->left,  seg, slist, addr, &val0, obj_line);
28              lbl1 = expVal(ip->right, seg, slist, addr, &val1, obj_line);
29              if ( val1 >= 0x2000 ) val1 = TO_BANKED_ADDR(val1);
30              *val = (val0 & ~0x7f) | (val1 & 0x7f);
31              return lbl0? lbl0: lbl1;
32
33          default:
34              return expVal(ip, seg, slist, addr, val, obj_line);
35      }
36  }
```

`/p16ecc/lk_source_3/exp.cpp`   Code-15.3

- #26~31: operator ':' expression, it means it's an instruction with a 7-bit bank address.

Note:

    `ip` – the item to be evaluated.
    `slist` – symbol list that keeps the values for symbols.
    `val` – the pointer used for holding the return value.
    `addr` – the current RAM or ROM address

The function, `expVal()`, is called to achieve the result (see Code-15.4). A return value of NULL indicates that the evaluation succeeded.

```cpp
39  static char *expVal(item_t *ip, Segment *seg, Symbol *slist, int addr, int *val, int obj_line)
40  {
41      Symbol *symb;
42      char *lbl0, *lbl1;
43      int  val0, val1;
44
45      *val = 0;
46      if ( ip == NULL ) return NULL;
47
48      switch ( ip->type )
49      {
50          case TYPE_VALUE:
51              *val = ip->data.val;
52              return NULL;
53
54          case TYPE_SYMBOL:
55              symb = searchSymbol(slist, ip->data.str, seg);
56              if ( symb )
57                  *val = symb->value;
58              else if ( obj_line > 0 )
59                  printf("can't find '%s' definition in \"%s\" Line #%d!\n",
60                          ip->data.str, seg->fileName, obj_line);
61
62              // if 'symbol' not found in the 'slist', report it.
63              return symb? NULL: ip->data.str;
64
65          case '.':
66              *val = addr;
67              return NULL;
68
69          case '+':   case '-':   case '*':   case '/':   case '%':
70          case '|':   case '^':   case '&':   case RSHIFT:    case LSHIFT:
71              lbl0 = expVal(ip->left,  seg, slist, addr, &val0, obj_line);
72              lbl1 = expVal(ip->right, seg, slist, addr, &val1, obj_line);
73              *val = expVal(val0, val1, ip->type);
74              return lbl0? lbl0: lbl1;
75
76          case '~':
77              lbl0 = expVal(ip->left,  seg, slist, addr, val, obj_line);
78              *val = ~(*val);
79              return lbl0;
80      }
81      return NULL;
82  }
```

345

```cpp
83
84    static int expVal(int v0, int v1, int type)
85    {
86        switch ( type )
87        {
88            case '+':        return v0 + v1;
89            case '-':        return v0 - v1;
90            case '*':        return v0 * v1;
91            case '/':        return v0 / v1;
92            case '%':        return v0 % v1;
93            case '&':        return v0 & v1;
94            case '|':        return v0 | v1;
95            case '^':        return v0 ^ v1;
96            case RSHIFT:     return v0 >> v1;
97            case LSHIFT:     return v0 << v1;
98            default:         return 0;
99        }
100   }
```

Chapter-16

PIC16F Linker Class – Memory Allocation

<u>Project directory</u>: /p16ecc/lk_source_4
<u>New Source files</u>: p16link.h, p16link.cpp, p16link1.cpp, p16link2.cpp
<u>Updated files</u>: main.cpp

16.1 Instantiate the Class & Start Running

The linker class, P16link, is defined in **p16link.h**, as shown in Code-16.1.

```
6   enum {
7       RAM_LOC_ABS,
8       RAM_LOC_LINEAR,
9       RAM_LOC_FLOAT,
10      ROM_LOC_ABS,
11      ROM_LOC_FLOAT,
12  };
13
14  class P16link {
15
16      private:
17          Memory   *codeMem;
18          Memory   *dataMem;
19          Symbol   *symbList;
20          FILE     *fout;
21
22      public:
23          int      errorCount;
24
25      public:
26          P16link(Memory *ram, Memory *rom);
27          ~P16link();
28
29          bool scanInclusion(void);
30          void assignSegmentsAddress(void);
31
32      private:
33          // p16link.cpp
34          void searchUsedSegment(Symbol *slist);
35          bool searchUsedSegment(Segment *segp, item_t *ip, Symbol *slist);
36          void removeUnusedSegment(Segment **segp);
37
38          // p16link1.cpp
39          void scanFuncLocalData(void);
40          void logSegmentSymbols(Segment *seglist, Symbol **symlist, bool incl_func_name=false);
41          void assignFuncLocalData(Symbol **symlist);
42
43          // p16link2.cpp
44          void assignSegmentMem(Segment *sp, int mode);
45          void assignSegmentAddr(Memory *mem, Segment *seg, int addr, int flag);
46          bool confirmSegmentMem(void);
47
48          // p16link3.cpp
49          void scanFcall(Segment *segp);
50  };
```
/p16ecc/lk_source_4/p16link.h
Code-16.1

It is instantiated in main(), as illustrated in Code-16.2.

```
27   static void link0(char *filename, bool lib_f=false);
28   static char *getLibFile(char *libfile);
29   static void cleanup(void);
30
31   ///////////////////////////////////////////////////////////////////////
32   int main(int argc, char *argv[], char *env[])
33   {

       . . .

67       if ( start_parsing )
68       {
69           std::string tmp_folder = TEMP_FOLDER;
70           system(("mkdir " + tmp_folder).c_str());    // create a temp folder
71           for (Symbol *sp = libList; sp; sp = sp->next)
72           {
73               char *ssp = getLibFile(sp->name);
74               if ( ssp ) link0(ssp, 1);
75           }
76           system(("rm -r " + tmp_folder).c_str());    // remove temp folder & files
77
78           P16link p16link(dataMem, codeMem);
79           p16link.scanInclusion();
80           p16link.assignSegmentsAddress();
81
82           cleanup();
83       }
```

/p16ecc/lk_source_4/main.cpp   Code-16.2

- #78: instantiate the class.
- #79: scan all the segments to remove unused segments.
- #80: assign memory address for segments.

16.2 Remove Unused Segments

The code or data (variables) in segments might not be used. For example, the functions in crt0.c, system library, may or may not be used. Those code or data segments should be removed from the segment groups before further processing. The following, Code-16.3, outlines the overall operation for this.

```
26   bool P16link :: scanInclusion(void)
27   {
28       Symbol *sym_list = NULL;
29
30       logSegmentSymbols(codeSegGroup, &sym_list, true);
31       logSegmentSymbols(dataSegGroup, &sym_list);
32       logSegmentSymbols(miscSegGroup, &sym_list);
33
34       // remove unused segments (code & data)
35       searchUsedSegment(sym_list);
36       removeUnusedSegment(&codeSegGroup);
37       removeUnusedSegment(&dataSegGroup);
38
39       deleteSymbols(&sym_list);
40       return (errorCount == 0);
41   }
```

/p16ecc/lk_source_4/p16link.cpp   Code-16.3

- #28: create a temp symbol list.
- #30~32: log in the symbols, including the function names, in the list.
- #35: determine those segments that are being used.
- #36: remove unused segments in group codeSegGroup.
- #37: remove unused segments in group dataSegGroup.

The algorithm used in the following, and as shown in Code-16.4, describes how to determine if a segment is used.

(1) First of all, the segments in `codeSegGrcup` in type `CODE0`, `CODE1`, or `CODEi` are considered as the 'seed' segments and marked as a 'used' segment unconditionally.

(2) If a (label) symbol in **segment**$_j$ appears in a 'W' type line of **segment**$_i$, and **segment**$_i$ has been marked as 'used', then **segment**$_j$ will be marked as 'used', as well; and the whole search process will restart again.

```
103   void P16link :: searchUsedSegment(Symbol *slist)
104   {
105       for (bool done = false; !done;)
106       {
107           done = true;
108           for (Segment *segp = codeSegGroup; segp; segp = segp->next)
109           {
110               // those segments must be included...
111               if ( (segp->isName("CODE0") ||
112                     segp->isName("CODE1") ||
113                     segp->isName("CODEi")  ) && !segp->isUsed )
114               {
115                   segp->isUsed = 1;
116                   done = false;
117               }
118
119               if ( segp->isUsed )
120               {
121                   for (line_t *lp = segp->lines; lp; lp = lp->next)
122                   {
123                       if ( lp->type == 'W' )
124                           for (item_t *ip = lp->items; ip; ip = ip->next)
125                           {
126                               if ( searchUsedSegment(segp, ip, slist) )
127                                   done = false;
128                           }
129                   }
130               }
131           }
132       }
133   }
134
135   bool P16link :: searchUsedSegment(Segment *segp, item_t *ip, Symbol *slist)
136   {
137       Symbol *symb;
138       switch ( ip->type )
139       {
140           case TYPE_SYMBOL:
141               symb = searchSymbol(slist, ip->data.str, segp);
142               if ( symb && symb->segment && !symb->segment->isUsed )
143               {
144                   symb->segment->isUsed = 1;
145                   return true;
146               }
147               return false;
148
149           case ':':   case RSHIFT:   case LSHIFT:
150           case '+':   case '-':   case '*':   case '/':   case '%':
151           case '&':   case '|':   case '^':
152               return ( searchUsedSegment(segp, ip->left,  slist) ||
153                        searchUsedSegment(segp, ip->right, slist) );
154
155           case '~':
156               return searchUsedSegment(segp, ip->left,  slist);
157       }
158       return false;
159   }
```
/p16ecc/lk_source_4/p16link.cpp   Code-16.4

- #111~117: detect the 'seed' segments.
- #119~130: if a segment, marked as 'used', has a symbol (in a 'W' type line) that appears in an unmarked segment.

16.3 Allocate Memory Addresses for Segments

Every segment in `codeSegGroup` can be considered a complete function. And all the codes in a segment are arranged in 'W' lines, such as

```
W  N₁ N₂ … Nₙ
```

Each item on a line occupies one location (word), and each segment in `codeSegGroup` spans consecutive locations in ROM space.

The 'R' line of `dataSegGroup`, which is in the form

```
R  N
```

Which means it takes consecutive locations (bytes) in RAM space.

Additionally, functions with local variables should also be assigned RAM space, as indicated in the 'U' lines of `codeSegGroup`.

Three more considerations should be paid:

1) 'J' and 'K' lines in `codeSegGroup` dynamically generate code for either "MOVLP" or "MOVLB" instruction, depending on the count and values of the items in a line.

   ```
   J  N            - generate the code of instruction "MOVLP N".
   K  N            - generate the code of instruction "MOVLB N".
   J  N₁ N₂        - generate the code of instruction "MOVLP N₂" only when N₁ and N₂ indicate
                     different pages in ROM space.
   K  N₁ N₂        - generate the code of instruction "MOVLB N₂" only when N₁ and N₂ indicate
                     different banks in RAM space.
   ```

   That means, (1) RAM address allocation for `dataSegGroup` should be done first; (2) Memory address allocation may require multiple adjustments until every 'J' and 'K' line has been determined whether the instruction code needs to be generated.

2) As shown in Code-13.6, the segments with `isABS=1` have the highest priority. They should be allocated first. After that, it's the turn for those segments with the "BANKn" name, and then followed by the rest of the segments.

3) The segments with the name "CODE1" or "CODEi", mentioned in Figure-13.1 and Figure-13.2, shall be allocated in a back-to-back sequence for the segments with the same name, with no gap between.

Code-16.5 illustrates the top-level operation for memory allocation in both RAM and ROM.

```cpp
14    void P16link :: assignSegmentsAddress(void)
15    {
16        scanFuncLocalData();
17
18        while ( !errorCount )
19        {
20            codeMem->reset();
21            dataMem->reset();
22            deleteSymbols(&symbList);
23
24            // assign data addresses...
25            assignSegmentMem(dataSegGroup, RAM_LOC_ABS);     // assign address for ABS data
26            assignSegmentMem(dataSegGroup, RAM_LOC_LINEAR);  // assign address for linear data
27            assignSegmentMem(dataSegGroup, RAM_LOC_FLOAT);   // assign address for float data
28            logSegmentSymbols(dataSegGroup, &symbList);      // log data symbols
29
30            assignFuncLocalData(&symbList);                  // assign address for func local data
31
32            // assign code addresses...
33            assignSegmentMem(codeSegGroup, ROM_LOC_ABS);     // assign address for ABS code
34            assignSegmentMem(codeSegGroup, ROM_LOC_FLOAT);   // assign address for float code
35            logSegmentSymbols(codeSegGroup, &symbList);      // log code symbols
36
37            logSegmentSymbols(miscSegGroup, &symbList);
38
39            if ( confirmSegmentMem() ) break;                // confirm code assignments
40        }
41    }
```

  Code-16.5

- #16: scan all function segments, to validate and get the amount of local data.
- #25~27: allocate RAM locations for segments in `dataSegGroup`.
- #28: add the symbols in segments of `dataSegGroup` to the symbol list.
- #30: allocate RAM locations for all function local data.
- #33~34: allocate ROM memory locations for segments in `codeSegGroup`.
- #35: add the symbols in segments of `codeSegGroup` to the symbol list.
- #39: confirm ROM memory allocation for 'J' and 'K' lines (see the details later).

As illustrated in Code-16.6, `P16link::assignSegmentMem()` performs the middle-level operation for memory allocation. It resets the searching address and filters the segment to be allocated.

```cpp
19    void P16link :: assignSegmentMem(Segment *seg_p, int mode)
20    {
21        for (; seg_p; seg_p = seg_p->next)
22        {
23            int addr = 0;    // default start adddress
24            int flag = 0;    // condition flags.
25
26            if ( seg_p->isABS )
27            {
28                flag |= FIXED_ADDR_FLAG;
29                addr  = seg_p->memAddr;
30            }
31
32            switch ( seg_p->type() )
33            {
34                case CODE_SEGMENT:
35                    flag |= PAGE_CHECK_FLAG;
36                case CONST_SEGMENT:
37                    if ( !(flag & FIXED_ADDR_FLAG) && mode == ROM_LOC_ABS )
38                        continue;
39                    if (  (flag & FIXED_ADDR_FLAG) && mode != ROM_LOC_ABS )
40                        continue;
41
42                    assignSegmentAddr(codeMem, seg_p, addr, flag);
43                    break;
44
45                case DATA_SEGMENT:
46                    if ( seg_p->isName("BANKn") )
47                        flag |= LINEAR_DATA_FLAG;
48
```

```
49                        switch ( mode )
50        =             {
51                          case RAM_LOC_ABS:
52                              if ( !(flag & FIXED_ADDR_FLAG) )
53                                  continue;
54                              break;
55                          case RAM_LOC_LINEAR:
56                              if ( (flag & FIXED_ADDR_FLAG) || !(flag & LINEAR_DATA_FLAG) )
57                                  continue;
58                              break;
59                          case RAM_LOC_FLOAT:
60                              if ( (flag & FIXED_ADDR_FLAG) || (flag & LINEAR_DATA_FLAG) )
61                                  continue;
62                          }
63                          assignSegmentAddr(dataMem, seg_p, addr, flag);
64                          break;
65                  }
66          }
67      }
```
/p16ecc/lk_source_4/p16link2.cpp   Code-16.6

- #23~24: reset the initial values. Memory allocation starts at address '0'.
- #26~30: update the starting address, `addr`, for the segment that starts at a fixed address.
- #34~43: allocate memory for a `CODE` or `CONST` segment in ROM space.
- #45~64: allocate memory for a `DATA` segment in RAM space.

At the lower level, as illustrated in Code-16.7, the routine `P16link::assignSegmentAddr()` requests the memory space from the `Memory` class and sets the start address in the segment.

```
69      void P16link :: assignSegmentAddr(Memory *mem, Segment *seg, int addr, int flag)
70      ={
71          char *seg_name = seg->name();    // segment name
72          bool multi_seg = INIT_CODE(seg) || ISR_CODE(seg);
73
74          if ( multi_seg && !seg->isBEG ) // init/code   code
75              return;
76
77          int req_size = seg->size();
78          for (Segment *ssp = seg->next; multi_seg && ssp && ssp->isName(seg_name); ssp = ssp->next)
79              req_size += ssp->size();
80
81          if ( !mem->getSpace(addr, &addr, req_size, flag) )  // don't have memory space !!!
82        =   {
83              line_t *lp = seg->lines;
84              printf("%s #%d, out of memory! --- %s\n", lp->fname, lp->lineno, seg_name);
85              errorCount++;
86              return;
87          }
88
89          seg->memAddr = addr;
90          addr += seg->size();
91          for (Segment *ssp = seg->next; multi_seg && ssp && ssp->isName(seg_name); ssp = ssp->next)
92        =   {
93              ssp->memAddr = addr;
94              addr += ssp->size();
95          }
96      }
```
/p16ecc/lk_source_4/p16link2.cpp   Code-16.7

- #72: check if the segment has the name "CODEi" or "CODE1".
- #77~79: sum up for the total space required for all the segments that have the name "CODEi" or "CODE1".
- #81~87: if the memory request failed, report the error.
- #89~90: obtain the start address for the segment.
- #91~95: if the segment has the name "CODEi" or "CODE1", assign the addresses for the following segments that have the same name.

The routine `P16link::confirmSegmentMem()`, as illustrated in Code-16.8, is to confirm whether 'J' and 'K' lines in segments have altered their behavior, whether they generate the instruction code or not. The flag, `'insert'`, in each 'J' and 'K' line indicates whether the line should create the corresponding

352

instruction code ($\mathtt{insert} = 1$ means that the instruction is required). In contrast, changing the flag will affect the address assigned to other segments. The flag, 'insert', has the default value '0', and changing any of these flags will cause the memory allocation to restart. The flag 'retry', when set, prevents the 'insert' value from changing '1' to '0', thereby avoiding the memory allocation from being stuck in an endless operation.

```cpp
98    bool P16link :: confirmSegmentMem(void)
99    {
100       bool done = true;
101       for (Segment *seg = codeSegGroup; seg; seg = seg->next)
102       {
103           int addr = seg->memAddr;    // segment start address
104           int obj_line = seg->lines->lineno;
105           for (line_t *lp = seg->lines; lp; lp = lp->next, obj_line++)
106           {
107               int ltype   = lp->type;
108               item_t *ip = lp->items;
109               int item_count = itemCount(ip);
110               int v, v0, v1;
111               switch ( ltype )
112               {
 . . .
126                   case 'J':
127                   case 'K':
128                       if ( ip )
129                       {
130                           item_t *ip1 = ip->next;
131                           char   *lbl0 = expValue(ip,  seg, symbList, &v0, addr, obj_line);
132                           char   *lbl1 = expValue(ip1, seg, symbList, &v1, addr, obj_line);
133                           if ( lbl0 ) errorCount++;
134                           if ( lbl1 ) errorCount++;
135                           if ( ip1 && !lbl0 && !lbl1 )
136                           {
137                               if ( ltype == 'J' ) // 'J' - rom page switch
138                               {
139                                   int switch_page = (v0 ^ v1) & (0xf << 11);
140                                   if ( switch_page && !lp->insert )
141                                   {
142                                       lp->insert = 1;
143                                       done = false;
144                                   }
145                                   if ( !switch_page && lp->insert && !lp->retry )
146                                   {
147                                       lp->insert = 0;
148                                       done = false;
149                                       lp->retry++;
150                                   }
151                               }
152                               else                // 'K' - ram bank switch
153                               {
154                                   v0 = (v0 >= 0x2000)? (v0 % 0x2000)/80: v0 >> 7;
155                                   v1 = (v1 >= 0x2000)? (v1 % 0x2000)/80: v1 >> 7;
156                                   if ( v0 != v1 && !lp->insert )
157                                   {
158                                       lp->insert = 1;
159                                       done = false;
160                                   }
161                               }
162                           }
163
164                           if ( !ip1 || lp->insert )
165                               addr++;
166                       }
167                       break;
168               }
169           }
170       }
171       return done;
172    }
```

/p16ecc/lk_source_4/p16link2.cpp   Code-16.8

- #137~151: confirm if 'J' line needs to generate instruction code of "MOVLP".
- #153~161: confirm if 'K' line needs to generate instruction code of "MOVLB".
- #164~165: increase the address if the instruction code needs to be generated.

PIC16F Linker Class – Generate Outputs

Two output files will be generated during the linking operation:

(1) .hex file   - the file that contains all the target CPU's executable code for download.
(2) .map file   - the file that lists all the code in assembly format for reference.

17.1 Generate HEX File

Project directory: /p16ecc/lk_source_5
New Source files: p16link3.cpp, fwriter_hex.h, fwriter_hex.cpp
Updated files: main.cpp

The .hex file is a pure text file that presents the instruction code for the target CPU to run, in hexadecimal values. Each line in a .hex file is called a 'record', and has the following format.

```
:ccaaaattdddddd…ddddss↵
```

Where each line in the file starts with the letter ':' and ends with a carriage return.

cc   – the count of 'data' part in bytes.
aaaa   – lower 16-bit address of a segment (64KB/segment).
tt   – the type of a record:
                00 = data record,
                01 = ending record of a file,
                04 = set the higher 16-bit address for a (new) segment.
dddd…dd – the data part of a record.
ss   – the checksum of a record.

The following, Code-17.1, is the definition of the hex file writer.

```
4     #define HEX_BUF_SIZE      128
5
6     class HexWriter {
7
8         private:
9             int startAddr;
10            int currentAddr;
11            int length;
12            unsigned char buf[HEX_BUF_SIZE];
13            FILE *fout;
14
15        public:
16            HexWriter(char *filename);
17            ~HexWriter();
18
19            void outputWord(int addr, int word);      // output an instruction word
20
21        private:
22            void outputLine(void);                     // output a data line
23            void outputAddr(int addr);                 // set Hi 16-bit address of a segment
24     };
```
/p16ecc/lk_source_5/fwriter_hex.h   Code-17.1

As illustrated in Code-17.2, the modifications in `main()` routine are made to involve the hex file output.

```
32   int main(int argc, char *argv[], char *env[])
33   {
34       int start_parsing = 0;
35       Path *path = new Path(env);
36       libPath = path->get();
37
38       printf ("PIC16E linker, %s\n", VERSION);
39
       . . .
80           P16link p16link(dataMem, codeMem);
81           p16link.scanInclusion();
82           p16link.assignSegmentsAddress();
83
84           if ( p16link.errorCount == 0 )
85           {
86               std::string hex_output = outputFile;    hex_output += ".hex";
87               p16link.outputHex((char*)hex_output.c_str());
88
89               int m_used, m_total;
90               m_used = dataMem->memUsed(&m_total);
91               printf("RAM used: %d bytes (%.2f%c)\n", m_used, (float)m_used*100/(float)m_total, '%');
92               m_used = codeMem->memUsed(&m_total);
93               printf("ROM used: %d words (%.2f%c)\n", m_used, (float)m_used*100/(float)m_total, '%');
94           }
95
96           cleanup();
97       }
98       delete path;
99       return errorCnt;
100  }
```
/p16ecc/lk_source_5/main.cpp   Code-17.2

- #86: generate the output file name.
- #87: start generating .hex file.
- #89~93: report the total memory used for the application.

Only two segment groups, `codeSegGroup` and `fuseSegGroup`, are used to generate the .hex file. The following, Code-17.3a and Code-17.3b, illustrate the operation of `P16link::outputHex()`.

```
16   void P16link :: outputHex(char *filename)
17   {
18       HexWriter hexWriter(filename);
19
20       for (Segment *seg = codeSegGroup; seg; seg = seg->next)
21       {
22           int addr = seg->memAddr & 0x7fff;
23           for (line_t *lp = seg->lines; lp; lp = lp->next)
24           {
25               item_t *ip = lp->items;
26               int w, v0, v1;
27               switch ( lp->type )
28               {
29                   case 'W':
30                       for (; ip; ip = ip->next)
31                       {
32                           expValue(ip, seg, symbList, &w, addr);
33                           hexWriter.outputWord(addr*2, w);
34                           addr++;
35                       }
36                       break;
37
```

```cpp
38                        case 'J':
39                        case 'K':
40                            if ( ip )
41                            {
42                                item_t *ip1 = ip->next;
43                                expValue(ip,  seg, symbList, &v0, addr);
44                                expValue(ip1, seg, symbList, &v1, addr);
45                                if ( !ip1 || lp->insert )
46                                {
47                                    int v = (ip1 == NULL)? v0: v1;
48
49                                    if ( lp->type == 'J' )  // rom page
50                                        w = 0x3180 | ((v >> 8) & 0x7f);
51                                    else                    // ram bank
52                                    {
53                                        int movlb_code = useBSR6? 0x0140: 0x0020;
54                                        int movlb_mask = useBSR6? 0x003f: 0x001f;
55                                        int bank_addr  = (v >= 0x2000)? (v - 0x2000)/80: v >> 7;
56                                        w = movlb_code | (bank_addr & movlb_mask);
57                                    }
58                                    hexWriter.outputWord(addr*2, w);
59                                    addr++;
60                                }
61                            }
62                            break;
63                        }
64                    }
65                }
```

/p16ecc/lk_source_5/p16link3.cpp   Code-17.3a

- #18: instantiate the class `HexWriter`.
- #20: go through all the lines in `codeSegGroup`.
- #29~36: output the items in 'W' lines.
- #38~62: output the items in 'J' and 'K' lines.

```cpp
67        for (Segment *seg = fuseSegGroup; seg; seg = seg->next)
68        {
69            int addr = seg->memAddr;
70            for (line_t *lp = seg->lines; lp; lp = lp->next)
71            {
72                item_t *ip = lp->items;
73                int w;
74                switch ( lp->type )
75                {
76                    case 'W':
77                        for (; ip; ip = ip->next)
78                        {
79                            expValue(ip, seg, symbList, &w, addr);
80                            hexWriter.outputWord(addr*2, w);
81                            addr++;
82                        }
83                        break;
84                }
85            }
86        }
87    }
```

/p16ecc/lk_source_5/p16link3.cpp   Code-17.3b

- #67: go through all the lines in `fuseSegGroup`.
- #76~82: output the items in 'W' lines.

17.2 Generate MAP File

<u>Project directory</u>: /p16ecc/lk_source_6
<u>New Source files</u>: fwriter_map.h, fwriter_map.cpp, disasm.h, disasm.cpp, fcall.h, fcall.cpp
<u>Updated files</u>: main.cpp, p16link1.cpp, p16link3.cpp

The .map file contains the following outputs:
1) RAM usage that reveals both external data variables and internal data variables for each function.
2) All the functions' code in the assembly code format.
3) Function calling map (caller and callee), and calling chains, which indicate the calling depth.

Similar to that in the last section, the .map file output starts in main(), as illustrated in Code-17.4. It needs to collect the information in all 'F' lines in codeSegGroup, to create a calling table (or matrix)

```
32    int main(int argc, char *argv[], char *env[])
33    {
34        int start_parsing = 0;
35        Path *path = new Path(env);
36        libPath = path->get();
37
38        printf ("PIC16E linker, %s\n", VERSION);


68        if ( start_parsing )
69        {
70            std::string tmp_folder = TEMP_FOLDER;
71            system(("mkdir " + tmp_folder).c_str());      // create a temp folder
72            for (Symbol *sp = libList; sp; sp = sp->next)
73            {
74                char *ssp = getLibFile(sp->name);
75                if ( ssp ) link0(ssp, true);
76            }
77            system(("rm -r " + tmp_folder).c_str());      // remove temp folder & files
78
79            P16link p16link(dataMem, codeMem);
80            p16link.scanInclusion();             // remove unused segments
81            p16link.assignSegmentsAddress();     // allocate memories
82
83            if ( p16link.errorCount == 0 )
84            {
85                std::string hex_output = outputFile;    hex_output += ".hex";
86                std::string map_output = outputFile;    map_output += ".map";
87                p16link.outputHex((char*)hex_output.c_str());
88                p16link.outputMap((char*)map_output.c_str());
89
90                int m_used, m_total;
91                m_used = dataMem->memUsed(&m_total);
92                printf("RAM used: %d bytes (%.2F%c)\n", m_used, (float)m_used*100/(float)m_total, '%');
93                m_used = codeMem->memUsed(&m_total);
94                printf("ROM used: %d words (%.2F%c)\n", m_used, (float)m_used*100/(float)m_total, '%');
95            }
96
97            cleanup();
98        }
```
/p16ecc/lk_source_6/main.cpp   Code-17.4

* #88: start to generate the .map file.

In P16link::assignSegmentAddress(), before starting .map file output operation, it needs to generate a calling table or matrix, as illustrated in Code-17.5.

```
15    void P16link :: assignSegmentsAddress(void)
16    {
17        scanFuncLocalData();
18        scanFuncCalls();      // generate calling table
19
20        while ( !errorCount )
21        {
```
/p16ecc/lk_source_6/p16link1.cpp   Code-17.5

- #18: scan all the 'F' lines in `codeSegGroup` (see Code-17.6).

```
65   void P16link :: scanFuncCalls(void)
66   {
67       for (Segment *sp = codeSegGroup; sp; sp = sp->next)
68       {
69           bool add_list = false;
70           for (line_t *lp = sp->lines; lp; lp = lp->next)
71               switch ( lp->type )
72               {
73                   case 'U':   // segment data specifier
74                       fcallMgr.addFunc(lp, sp);
75                       add_list = true;
76                       break;
77                   case 'F':   // function call
78                       if ( add_list )
79                           fcallMgr.addFcall(lp);
80                       break;
81                   case 'G':   // global label
82                       if ( add_list )
83                           fcallMgr.setGlobal(lp, lp->items->data.str);
84                       break;
85               }
86       }
87
88       fcallMgr.createBasicMatrix();
89  //   fcallMgr.printFunc();
90   }
```
`/p16ecc/lk_source_6/p16link1.cpp`   Code-17.6

- #67~86: collect the information (caller and callee), logged in .
- #88: create the calling table (matrix).

The class `FcallMgr` is responsible for managing function calls. Assume there are N functions, F_1 ... F_N, in the application code logged in, and the calling matrix $\mathcal{M}(N, N)$ is to indicate the calling relationship,

$$\mathcal{M}(i, j) = 1 \qquad \rightarrow F_i \text{ calls } F_j$$

Code-17.7 illustrates the creation of the matrix.

```
11   void FcallMgr :: createBasicMatrix(void)
12   {
13       basicFcallMatrix = new char *[funcCount];
14       for (int i = 0; i < funcCount; i++)
15       {
16           basicFcallMatrix[i] = new char[funcCount];
17           memset(basicFcallMatrix[i], 0, funcCount);
18       }
19
20       // build up the basic calling chain ...
21       for (Fcall *fp = fcallList; fp; fp = fp->next)
22       {
23           FuncAttr *fp1 = getFunc(fp->fileName, fp->caller, IS_CALLER);
24           FuncAttr *fp2 = getFunc(fp->fileName, fp->callee, IS_CALLEE);
25
26           if ( fp1 && fp2 && fp1->funcIndex != fp2->funcIndex )
27               FNODE(fp1->funcIndex, fp2->funcIndex) = 1;
28       }
29   }
```
`/p16ecc/lk_source_6/fcall1.cpp`   Code-17.7

- #13~18: create the calling matrix M, `basicFcallMatrix,` and zero it out.
- #21~28: $M(i, j) = 1$ if F_i calls F_j.

The following, Code-17.8, is the top-level routine that generates the outputs to the .map file,

```
95    void P16link :: outputMap(char *filename)
96    {
97        int usedMemSize, totalMemSize;
98
99        MapWriter mapWriter(filename);
100
101       // output RAM map ...
102       usedMemSize = dataMem->memUsed(&totalMemSize);
103       mapWriter.outputSeg(symbList, usedMemSize, totalMemSize);
104
105       // output data memory map ...
106       dataMem->printSpace(mapWriter.fout);
107
108       // output ROM map ...
109       usedMemSize = codeMem->memUsed(&totalMemSize);
110       mapWriter.outputSeg(codeSegGroup, symbList, usedMemSize, totalMemSize);
111
112       // output calling map ...
113       fcallMgr.outputCallPath(mapWriter.fout);
114   }
```
/p16ecc/lk_source_6/p16link3.cpp   Code-17.8

- #99: instantiate the MAP file writer.
- #102~103: output the data memory (RAM) usage information.
- #109~110: output the code memory (ROM) usage information.
- #113: output the calling list.

At the low-level, `MapWriter::outputSeg()` outputs the data & code segment memory allocations, as illustrated in Code-17.9 and Code-17.10.

```
26    void MapWriter :: outputSeg(Symbol *slist, int used, int total)
27    {
28        float pct = (float)used*100/(float)total;
29        fprintf(fout, "- DATA MEMORY %d bytes(%.2f%c) used\n", used, pct, '%');
30
31        for (; slist; slist = slist->next)
32        {
33            Segment *seg = slist->segment;
34            char *fname = seg->lines->fname;       // file name
35            char *name  = slist->name;             // symbol name
36            int  laddr;
37
38            if ( seg->type() == DATA_SEGMENT && !strstr(name, "$sizeof$") && !strstr(name, "$init$") )
39            {
40                int size = seg->size();
41                int addr = slist->value;
42                if ( addr >= 0x2000 )              // in linear space..
43                {
44                    laddr = addr;
45                    addr  = ((addr&0x1fff)/80)*128 + 0x20 + (addr&0x1fff)%80;
46                }
47                else if ( (addr & 0x7f) < 0x20 )    // in register space..
48                    laddr = addr;
49                else if ( (addr & 0x7f) < 0x70 )    // in generic RAM space..
50                    laddr = (addr>>7)*80 + (addr&0x7f) - 0x20 + 0x2000;
51
52                fprintf(fout, "%s  %s (%d): 0x%X", fname, name, size, laddr);
53                fprintf(fout, " [%d:0x%02X]\n", addr>>7, addr&0x7f);
54            }
55
56            if ( seg->type() == CODE_SEGMENT && seg->dataSize && slist->type == 'U' && strstr(name, "_$data$") )
57            {
58                int size = seg->dataSize;
59                int addr = slist->value;
60                if ( addr >= 0x2000 )
61                {
62                    laddr = addr;
63                    addr  = ((addr&0x1fff)/80)*128 + 0x20 + (addr&0x1fff)%80;
64                }
65                else
66                    laddr = (addr>>7)*80 + (addr&0x7f) - 0x20 + 0x2000;
67
68                fprintf(fout, "%s  %s (%d): 0x%X", fname, name, size, laddr);
69                fprintf(fout, " [%d:0x%02X]\n", addr>>7, addr&0x7f);
70            }
71        }
72    }
```
/p16ecc/lk_source_6/fwriter_map.cpp   Code-17.9

- #38~54: external data allocation.
- #56~70: function internal data allocation.

```cpp
74   void MapWriter :: outputSeg(Segment *seg, Symbol *slist, int used, int total)
75   {
76       float pct = (float)used*100/(float)total;
77       fprintf(fout, "\n- CODE MEMORY %d words(%.2f%c) used\n", used, pct, '%');
78
79       for (; seg; seg = seg->next)
80       {
81           int addr = seg->memAddr;
82           int type = seg->type();
83
84           if ( type == CODE_SEGMENT || type == CONST_SEGMENT )
85               addr &= 0x7fff;
86
87           for (line_t *lp = seg->lines; lp; lp = lp->next)
88           {
89               item_t *ip = lp->items;
90               int v;
91               switch ( lp->type )
92               {
93                   case 'S':   // segment..
94                       outputSkip0();
95                       fprintf(fout, "(%s line#%d, %d words)\n",
96                               seg->fileName, lp->lineno, seg->size());
97                       break;
98                   case 'G':
99                       outputSkip0();
100                      fprintf(fout, "%s::\n", ip->data.str);
101                      break;
102                  case 'L':
103                      outputSkip0();
104                      fprintf(fout, "%s:\n", ip->data.str);
105                      break;
106                  case ';':
107                      if ( !seg->isLIB )
108                      {
109                          outputSkip();
110                          fprintf(fout, "%s\n", ip->data.str);
111                      }
112                      break;
113                  case 'W':
114                      for (; ip; ip = ip->next, addr++)
115                      {
116                          expValue(ip, seg, slist, &v, addr);
117                          outputInst(addr, v);
118                      }
119                      break;
120                  case 'R':
121                      addr += seg->size();
122                      break;
123                  case 'J':   // page select
124                  case 'K':   // bank select
125                      if ( lp->insert || itemCount(ip) == 1 )
126                      {
127                          if ( itemCount(ip) == 1 )
128                              expValue(ip, seg, slist, &v, addr);
129                          else
130                              expValue(ip->next, seg, slist, &v, addr);
131
132                          if ( lp->type == 'J' )  // rom page select
133                              outputInst(addr, 0x3180|((v >> 8) & 0x7f));
134                          else                    // ram bank select
135                          {
136                              int movlb_code = useBSR6? 0x0140: 0x0020;
137                              int movlb_mask = useBSR6? 0x003f: 0x001f;
138                              int bank_addr  = (v >= 0x2000)? (v - 0x2000)/80: v >> 7;
139                              outputInst(addr, movlb_code | (bank_addr & movlb_mask));
140                          }
141                          addr++;
142                      }
143                      break;
144                  }
145              }
146          }
147   }
```

The following table, Table-17.1, shows a MAP file (partial) of an example.

```
>>>>>>>>>>> MAP OUTPUT <<<<<<<<<<<<

- DATA MEMORY 103 bytes(42.92%) used
timer0.obj tmr0Count (1): 0x2000 [0:0x20]
key.obj keyValue (1): 0x2001 [0:0x21]
key.obj keyTimer (1): 0x2002 [0:0x22]
key.obj keyState (1): 0x2003 [0:0x23]
key.obj keyRead (1): 0x2004 [0:0x24]
key.obj keyLast (1): 0x2005 [0:0x25]
key.obj keyElapse (2): 0x2006 [0:0x26]
radio.obj RADIO_mode (1): 0x2008 [0:0x28]
radio.obj RADIO_volume (1): 0x2009 [0:0x29]
radio.obj RADIO_fmFreq (2): 0x200A [0:0x2A]
radio.obj RADIO_amFreq (2): 0x200C [0:0x2C]
. . .
- MEMORY MAP  'X'=forbidden, '+'=used, '.'=unused
0000: XXXXXXXXXXXXXXXXXXXXXXXXXXXXXXXX++++++++++++++++++++++++++++++++
0040: ++++++++++++++++++++++++++++++++++++++++++++++++XXXXXXXXXXXXXXXX
0080: XXXXXXXXXXXXXXXXXXXXXXXXXXXXXXXX+++++++++++++++++++++.........
00C0: ..................................................XXXXXXXXXXXXXXXX
0100: XXXXXXXXXXXXXXXXXXXXXXXXXXXXXXXX.........................
0140: ..................................................XXXXXXXXXXXXXXXX
. . .
- CODE MEMORY 3423 words(83.57%) used
        (main.obj line#10, 3 words)
0000: 0000        nop
0001: 3180        movlp  0x00
0002: 2812        goto  0x012
        (main.obj line#17, 3 words)
0004: 018A        clrf   PCLATH
0005: 0870        movf  0x70, W
0006: 001D        movwi  [--FSR1]
        (timer0.obj line#43, 8 words)
          ; :: timer0.c #21: interrupt tmr0_isr() if ( INTCONbits->TMR0IF )
0007: 1D0B        btfss  INTCON, 2
0008: 280F        goto  0x00F
          ; :: timer0.c #23: TMR0 = 6;
0009: 3006        movlw  0x06
000A: 0020        movlb  0x00
000B: 0095        movwf  0x15
. . .
*********** CALLING PATH ***********
[0]tmr0_isr
[1] KEY_scan

[0]main
[1] TMR0_init

[0]main
[1] I2C_init

[0]main
[1] OLED_init
[2] TMR0_delayMs

[0]main
[1] OLED_init
[2] OLED_clr
[3]  OLED_command
[4]  I2C_start
[5]   I2C_delay
```

Table-17.1

17.3 Optimization of Memory Usage

As described in the previous section, because of the PIC16F CPU's architecture, each function is assigned a static memory area for its internal data, including temporary data and function parameters.

Assume:

$F_1, F_2, \ldots, F_N$	- All N functions in the application code.
$M_1, M_2, \ldots, M_N$	- Internal data memory areas of N functions.
$L_1, L_2, \ldots, L_N$	- The length of the internal data memory for functions.

Thus, the total memory cost for internal memory, in this case, is $\sum L_i$.

In an 'ideal' environment that supports the C language, the internal data are allocated on the stack. Every time a function starts running (in an active state), it dynamically allocates data memory from the stack and releases it when it finishes running (in an inactive state). That means, F_i doesn't own the memory for its internal data until it becomes active.

In our case, M_i is statically assigned to function F_i. That means M_i is not in use when F_i is in inactive status, becoming a waste. It makes excellent sense if M_i can be used for other functions during periods when it's not in use, or M_i will be shared by multiple functions, to reduce the total cost for internal memory.

17.3.1 Function exclusive

The following figure (Figure 17.1a) illustrates the function-calling chain.

$$F_1 \rightarrow F_2 \rightarrow F_3$$

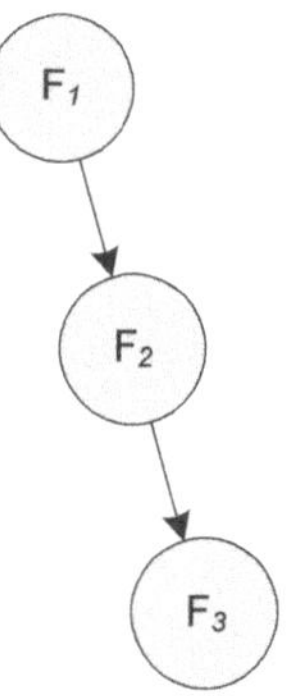

Figure-17.1a

When F_3 is in active status, both F_1 and F_2 are also active. That means that no memory sharing can happen among the functions.

In Figure-17.2b, it has two calling chains

$$F_1 \rightarrow F_2 \qquad \text{and} \qquad F_1 \rightarrow F_3$$

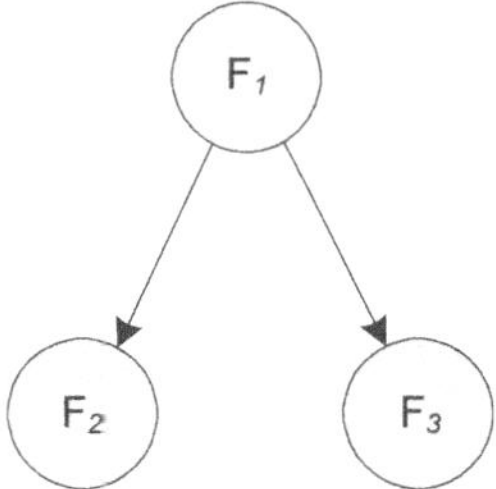

Figure-17.1b

F_3 will not be in active status if F_2 is active, vice versa. So F_2 and F_3 can share the memory, as $M_{2,3}$, and $L_{2,3} = \max(L_2, L_3)$. In this case, F_2 and F_3 are in the same 'sharing group'.

In Figure-17.2c, it has two calling chains

$$F_1 \rightarrow F_2 \qquad \text{and} \qquad F_1 \rightarrow F_3 \rightarrow F_4$$

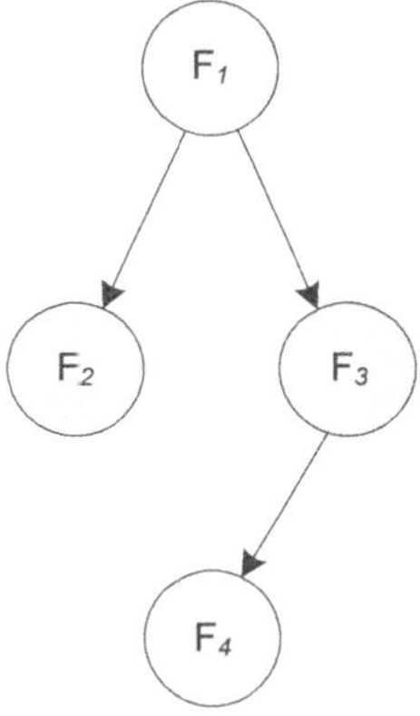

Figure-17.1c

In this case, there are two possible sharing choices available: (1) F_2 and F_3; (2) F_2 and F_4, but only one of them can be chosen. Assume $L_1 = 0$, $L_2 = 3$, $L_3 = 2$, and $L_4 = 4$.

Then the total memory cost can be one of the following choices:

$$L_1 + L_2 + L_3 + L_4 = 0 + 3 + 2 + 4 = 9$$

$$L_1 + L_{2,3} + L_4 = 0 + \max(3, 2) + 4 = 0 + 3 + 4 = 7$$

$$L_1 + L_{2,4} + L_3 = 0 + \max(3, 4) + 2 = 0 + 4 + 2 = 6$$

Furthermore, as illustrated in Figure 17.1d, more functions can share memory if they are all mutually exclusive from each other.

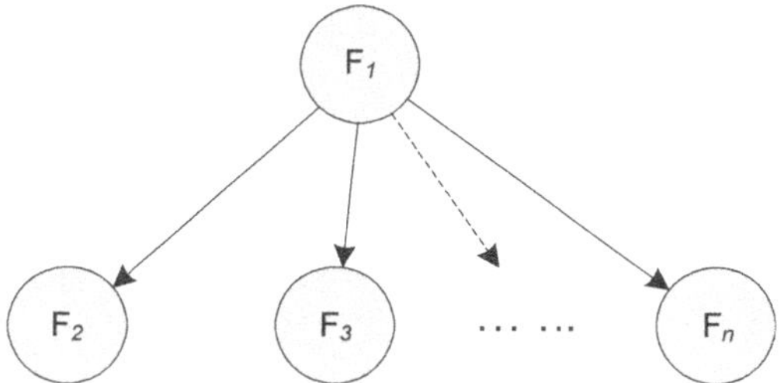

Figure-17.1d

Thus, the total memory cost $= L_1 + L_{2,3,\ldots n} = L_1 + \max(L_2, L_3, \ldots L_n)$.

Conclusions:
- F_i and F_j cannot share internal memory if they are in the same calling chain. In other words, F_i and F_j can share internal memory only if they are mutually exclusive of being active.
- Any function can only appear in one sharing group.
- Different sharing options can result in different benefits.
- F_i should not be ignored for the sharing if $L_i = 0$.

17.3.2 Thread exclusive

In general, all embedded applications belong to the 'multithreading'. There are two types of processes co-occurring, and they are independent of each other:

1) Foreground process, which originates from the `main()` function.
2) Background process or interrupt service routine (ISR), which is driven by interrupt events.

Figure-17.2 illustrates the situation.

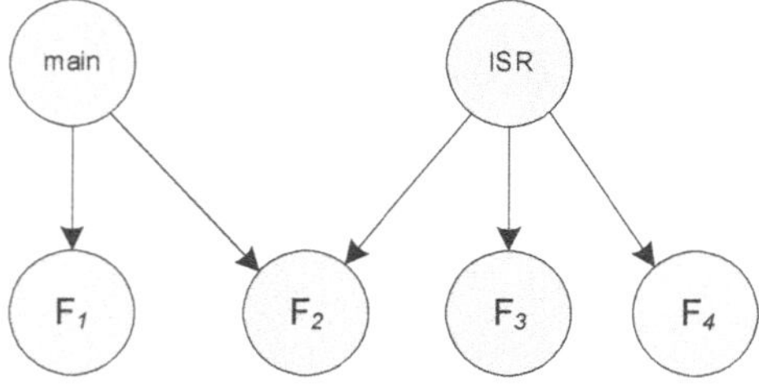

Figure-17.2

Where function F_2 is called by both `main` and ISR processes (no matter direct or indirect), in this case, only F_3 and F_4 can share the memory, forming a sharing group; however, F_2 cannot be in any sharing group since it's called from different types of processes. Here, both `main` and ISR functions can be called 'root' of the calling chain.

In fact, there could be more than one ISR process in an application, as shown in Figure-17.3.

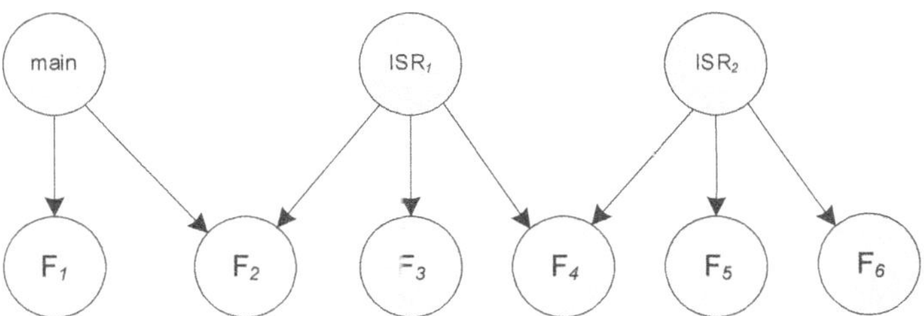

Figure-17.3

Due to the structure of the PIC16F CPU, there is only one interrupt vector, and no interrupt priority or override can happen. That means, interrupt events are handled one after another, and no overlapping will occur. So, F_3, F_4, F_5, and F_6 can share the memory as a group in this case.

Conclusions:
- F_i can not be in any sharing group if it's called from different types of processes (different types of root function).
- All interrupt service routines are in parallel, and they are naturally exclusive, under the PIC16F structure environment.

Furthermore, for safety purposes, functions that are only called using function pointers cannot join any sharing group either.

17.3.3 Implementation of memory sharing

```
Project directory: /p16ecc/lk_source_7
Updated files: p16link1.cpp,p16link2.cpp,p16link3.cpp,fcall1.cpp
```

In `P16link::scanFuncCalls()`, as shown in Code-17.11, it adds the expansion that extends the fundamental calling matrix to generate the extended calling matrix for functions.

```cpp
65   void P16link :: scanFuncCalls(void)
66   {
67       for (Segment *sp = codeSegGroup; sp; sp = sp->next)
68       {
69           bool add_list = false;
70           for (line_t *lp = sp->lines; lp; lp = lp->next)
71               switch ( lp->type )
72               {
73                   case 'U':   // segment data specifier
74                       fcallMgr.addFunc(lp, sp);
75                       add_list = true;
76                       break;
77                   case 'F':   // function call
78                       if ( add_list )
79                           fcallMgr.addFcall(lp);
80                       break;
81                   case 'G':   // global label
82                       if ( add_list )
83                           fcallMgr.setGlobal(lp, lp->items->data.str);
84                       break;
85               }
86       }
87
88       fcallMgr.createBasicMatrix();
89       fcallMgr.createExtendMatrix();
90   //  fcallMgr.printFunc();
91   }
```

/p16ecc/lk_source_7/p16link1.cpp   Code-17.11

- #88: generate the basic calling matrix, `basicFcallMatrix`.
- #89: generate the extended calling matrix, `extendFcallMatrix`.

Code-17.12 shows the operations of `FcallMgr::createExtendMatrix()`. First of all, it copies the contents from basicFcallMatrix to form `extendFcallMatrix`, and then it extends the calling chain:

$$\text{if } M(i, j) \mathrel{!=} 0 \text{ and } M(j, k) \mathrel{!=} 0, \text{ then mark } M(i, k) = 2 \text{ that indicates the indirect calling relationship.}$$

After that, it classifies all the functions into the following types (sharing can only happen within the same type of functions):

```
MAIN_FUNC       - main process (thread) root function.
ISR_FUNC        - ISR process (thread) root function.
SUB_FUNC        - function called by some other function(s).
NULL_FUNC       - function not called by any other function.
```

And finally, it finds the root for all the functions.

```
33    void FcallMgr :: createExtendMatrix(void)
34    {
35        extendFcallMatrix = new char *[funcCount];
36        for (int i = 0; i < funcCount; i++)
37        {
38            extendFcallMatrix[i] = new char[funcCount];
39            memcpy(extendFcallMatrix[i], basicFcallMatrix[i], funcCount);
40        }
41
42        // extending the calling relationship...
43        //   if N(i,j) and N(j,k), then N(i,k) = 2 which means it's extended link.
44        for (bool done = false; !done;)
45        {
46            done = true;
47            for (int i = 0; i < funcCount; i++)
48            for (int j = 0; j < funcCount; j++)
49            {
50                if ( ENODE(i, j) )
51                    for (int k = 0; k < funcCount; k++)
52                        if ( ENODE(j, k) && !ENODE(i, k) && i != k )
53                        {
54                            ENODE(i, k) = 2;
55                            done = false;
56                        }
57            }
58        }
59
60        for (int i = 0; i < funcCount; i++) // assign function type
61        {
62            FuncAttr *fp = getFunc(i);
63            fp->funcType = FuncType(i);
64        }
65
66        for (int i = 0; i < funcCount; i++) // assign function root
67            findRoot(i);
68    }
```
/p16ecc/lk_source_7/fcall1.cpp   Code-17.12

- #35~40: copy the contents from `basicFcallMatrix`.
- #44~58: extend the contents of `extendFcallMatrix`.
- #60~64: assign the types for the functions.
- #66~67: find the 'root' type for each function.

Code-17.13 illustrates how to find the root (type) of the function. It also detects if the function is in multiple calling chains.

```
136   void FcallMgr :: findRoot(int f_index)
137   {
138       FuncAttr *fp = getFunc(f_index);
139
140       fp->multiRoots = false;
141       fp->root = (funcType(f_index) == NULL_FUNC)? ((f_index+1) << 8): 0;
142
143       for (int i = 0; i < funcCount; i++)
144       {
145           if ( (i != f_index) && ENODE(i, f_index) )
146               switch ( funcType(i) )
147               {
148                   case MAIN_FUNC:
149                       if ( fp->root && fp->root != MAIN_FUNC ) fp->multiRoots = true;
150                       fp->root = MAIN_FUNC;
151                       break;
152                   case ISR_FUNC:
153                       if ( fp->root && fp->root != ISR_FUNC ) fp->multiRoots = true;
154                       fp->root = ISR_FUNC;
155                       break;
156                   case NULL_FUNC:
157                       if ( fp->root && fp->root != ((i+1) << 8) ) fp->multiRoots = true;
158                       fp->root = (i+1) << 8;
159                       break;
160               }
161       }
162   }
```
/p16ecc/lk_source_7/fcall1.cpp   Code-17.13

In `P16link::assignFuncLocalData()`, which allocates RAM addresses for function internal data, it groups functions that meet the specified criteria and assigns the same starting address to them. The procedure is illustrated as in Code-17.14,

```
174   void P16link :: assignFuncLocalData(Symbol **symlist)
175   {
176       char func_list[fcallMgr.funcCount];
177       int group_list[fcallMgr.funcCount];
178       memset(func_list, 0, fcallMgr.funcCount);
179
180       for (bool done = false; !done;)
181       {
182           int addr, length = 0;
183           int max_size = 0;
184
185           for (int i = 0; i < fcallMgr.funcCount; i++)
186           {
187               FuncAttr *fp = fcallMgr.getFunc(i);
188               if ( !func_list[i] && fp->dataSize > 0 )
189               {
190                   if ( fcallMgr.joinInGroup(group_list, length, i) )
191                   {
192                       group_list[length++] = i;
193                       func_list[i] = 1;
194                       if ( max_size < fp->dataSize ) max_size = fp->dataSize;
195                   }
196               }
197           }
198
199           if ( max_size && dataMem->getSpace(0, &addr, max_size) )
200           {
201               for (int i = 0; i < length; i++)
202               {
203                   FuncAttr *fp = fcallMgr.getFunc(group_list[i]);
204                   std::string f_name = fp->funcName;  f_name += "_$data$";
205                   logSymbol(symlist, 'U', (char*)f_name.c_str(), fp->segment, addr);
206               }
207           }
208           done = (max_size == 0);
209       }
210   }
```
/p16ecc/lk_source_7/p16link2.cpp   Code-17.14

- #176: the buffer to keep tracking if all the functions have been treated.
- #177: the buffer to hold the function index as a group.
- #190: check if the function meets the criteria to join the group.
- #191~195: add the function to the group (#192), adjust the memory size required (#194).
- #199~207: request the memory and log the function local data name into the symbol list with the address.

The function `FcallMgr::joinInGroup()`, illustrated in Code-17.15, is to confirm the criteria against the current group members.

```
115   bool FcallMgr :: joinInGroup(int *group, int group_length, int f_index0)
116   {
117       FuncAttr *fp0 = getFunc(f_index0);
118       int      type0 = fp0->funcType;
119
120       for (int i = 0; i < group_length; i++)
121       {
122           int  f_index1 = group[i];
123           FuncAttr *fp1 = getFunc(f_index1);
124           int      type1 = fp1->funcType;
125
126           if ( f_index1 == f_index0 ) continue;
127
128           if ( fp0->multiRoots ) return false;
129           if ( fp0->root != fp1->root ) return false;
130           if ( SAME_CHAIN(f_index0, f_index1) ) return false;
131           if ( type0 != type1 ) return false;
132       }
133       return true;
134   }
```
/p16ecc/lk_source_7/fcall1.cpp   Code-17.15

- #128: check if the function has multiple roots.
- #129: check if the function has the same root (type) as that in the group.
- #130: check if the function is in the calling chain with any member of the group.
- #131: check if the function has the same type as other members in the group.

The following table, Table-17.2, shows the MAP file (partial) that is generated using the optimized linker. It shows that the local data memory overlaps, saving 46.6% of memory (103 vs. 55 bytes).

```
>>>>>>>>>>>> MAP OUTPUT <<<<<<<<<<<<

- DATA MEMORY 55 bytes(22.92%) used
timer0.obj tmr0Count (1): 0x2000 [0:0x20]
key.obj keyValue (1): 0x2001 [0:0x21]
key.obj keyTimer (1): 0x2002 [0:0x22]
key.obj keyState (1): 0x2003 [0:0x23]
key.obj keyRead (1): 0x2004 [0:0x24]
key.obj keyLast (1): 0x2005 [0:0x25]
key.obj keyElapse (2): 0x2006 [0:0x26]
radio.obj RADIO_mode (1): 0x2008 [0:0x28]
radio.obj RADIO_volume (1): 0x2009 [0:0x29]
radio.obj RADIO_fmFreq (2): 0x200A [0:0x2A]
radio.obj RADIO_amFreq (2): 0x200C [0:0x2C]
i2c.obj I2C_delay_$data$ (1): 0x200E [0:0x2E]
i2c.obj I2C_start_$data$ (1): 0x2011 [0:0x31]
i2c.obj I2C_readByte_$data$ (3): 0x2011 [0:0x31]
i2c.obj I2C_writeByte_$data$ (2): 0x2014 [0:0x34]
oled.obj OLED_setPos_$data$ (3): 0x2016 [0:0x36]
oled.obj OLED_clr_$data$ (4): 0x2016 [0:0x36]
si47xx.obj SI47xx_write_$data$ (3): 0x2016 [0:0x36]
si47xx.obj SI47xx_read_$data$ (3): 0x2016 [0:0x36]
oled.obj OLED_init_$data$ (3): 0x201A [0:0x3A]
oled.obj OLED_displayChar_8x6_str_$data$ (6): 0x201A [0:0x3A]
oled.obj OLED_displayChar_32x19_$data$ (5): 0x201A [0:0x3A]
oled.obj OLED_displayChar_16_$data$ (7): 0x201A [0:0x3A]
si47xx.obj SI47xx_waitCTS_$data$ (2): 0x201A [0:0x3A]
oled.obj OLED_command_$data$ (1): 0x2021 [0:0x41]
oled.obj OLED_data_$data$ (5): 0x2021 [0:0x41]
si47xx.obj SI47xx_command_$data$ (3): 0x2021 [0:0x41]
key.obj KEY_scan_$data$ (2): 0x2026 [0:0x46]
radio.obj RADIO_poll_$data$ (4): 0x2028 [0:0x48]
radio.obj RADIO_dispFreq_$data$ (8): 0x202C [0:0x4C]
si47xx.obj SI47xx_property_$data$ (10): 0x202C [0:0x4C]
si47xx.obj SI47xx_setFreq_$data$ (8): 0x202C [0:0x4C]
si47xx.obj SI47xx_setVolume_$data$ (1): 0x2036 [0:0x56]

- MEMORY MAP  'X'=forbidden, '+'=used, '.'=unused
0000: XXXXXXXXXXXXXXXXXXXXXXXXXXXXXXXX++++++++++++++++++++++++++++++++
0040: ++++++++++++++++++++++........................XXXXXXXXXXXXXXXX
0080: XXXXXXXXXXXXXXXXXXXXXXXXXXXXXXXX................................
00C0: ......................................XXXXXXXXXXXXXXXX
0100: XXXXXXXXXXXXXXXXXXXXXXXXXXXXXXXX................................
0140: ......................................XXXXXXXXXXXXXXXX
...
```

Table-17.2

PART-V: Application Examples

Chapter-18

Application Examples Using cc16e Tools

The chapter presents two application examples that use the compiler toolchain discussed in previous chapters.

18.1 Example 1 – Digital AM/FM Radio

Silicon Labs developed a series of radio chips for AM/FM radio receivers. This application uses a digital radio module that integrates a si4730 chip from Silicon Labs, which supports AM and FM reception, stereo auto output, and can be purchased online. A 0.91-inch 128×32 LED display is used as well (both are available on eBay and AliExpress).

<u>18.1.1 Hardware design and implementation</u>

si4730 radio module

The module has integrated all the components onto a PCB (about 14 x 14 mm^2), in the shape of a stamp. There are 12 pins, breaking out for connections, shown in Table-18.1:

Pin	Function	Pin	Function
1	FMI – FM antenna	12	OutL – audio output (left)
2	GND	11	OutR – audio output (right)
3	AMI – AM antenna	10	GND
4	RST – reset (effective low)	9	SDA – I^{2}C bus data line
5	SEN – sensitivity mode	8	VDD – 3.3V power supply
6	SCL – I^{2}C bus clock line	7	GPIO

Table-18.1

Note:
- To switch from AM to FM mode, or vice versa, for radio operation, a reset signal should be applied to the RST pin first.
- High voltage on SEN pin sets the module in high sensitivity mode.
- The model is controlled through I^{2}C interface (SCL and SDA pins), in 7-bit address mode (device address = 0xC6/0xC7).

LED display module

The display model has only four pins, shown in Table-18.2:

Pin	Function	Pin	Function
1	GND	3	SCL – I^{2}C bus clock line
2	VDD – 3.3V power supply	4	SDA – I^{2}C bus data line

Table-18.2

The model is controlled through I^{2}C interface (SCL and SDA pins), in 7-bit address mode (device address = 0x78/0x79).

PIC16F CPU chip

A PIC12F1840 CPU from Microchip is used in this application. The chip has only eight pins. Only five pins are used for I/O connections. The connection and function are shown in Table-18.3. The basic specifications of the chip are:

- Enhanced PIC16Fxxxx structure.
- 4096 words of flash ROM.
- 256 bytes of RAM.
- Maximum running speed = 32MHz.
- Power supply range = 2.8~5.5V.

Pin	Function	Pin	Function
1	Vss – ground	8	Vdd – power supply
2	PA5 – button for AM/FM selection	7	PA0 – button for decreasing freq.
3	PA4 – button for increasing freq.	6	PA1 – I²C bus clock line (SCL)
4	PA3/MCLR	5	PA2 – I²C bus data line (SDA)

Table-18.3

Schematic of the design

Figure-18.1 gives the schematic of the application.

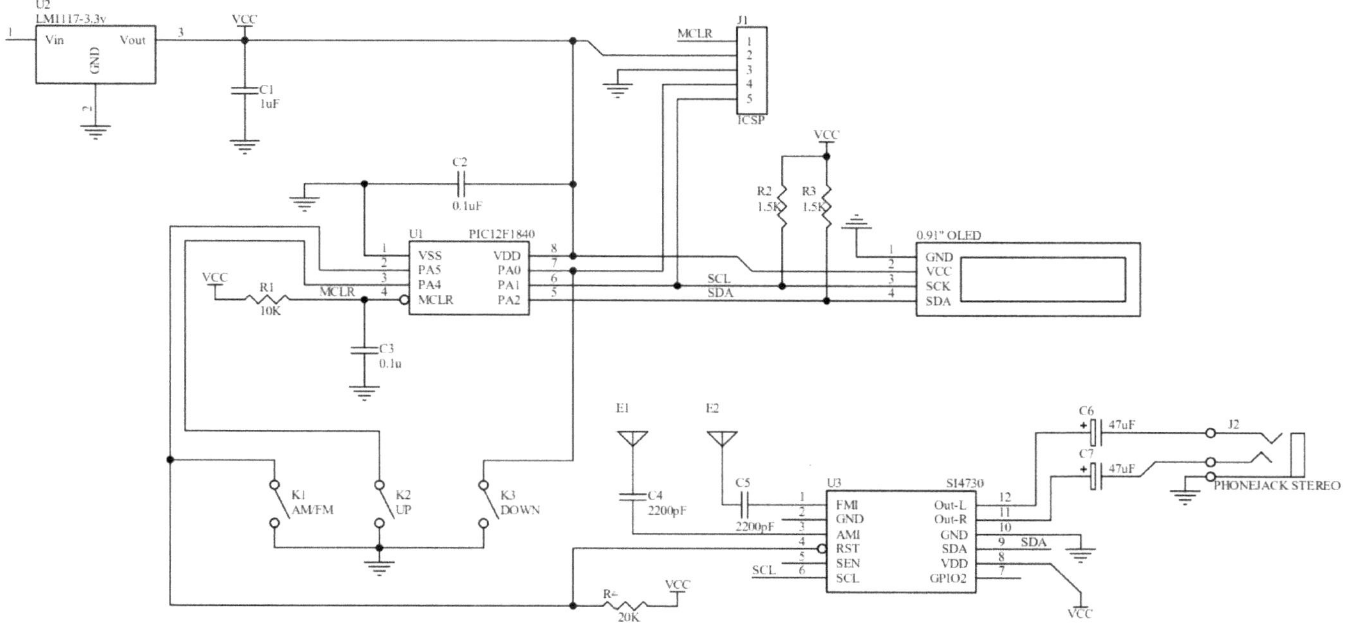

Figure-18.1

In addition to the components mentioned above, three buttons have been added to support radio running.

<u>18.1.2 Software design and implementation</u>

<u>Project directory</u>: /p16ecc/applications/si4730radio_pic12f1840

Under the project folder, it contains the following source files:

```
F:\p16ecc\applications\si4730radio_pic12f1840>ls -lsa
total 80
 8 drwxr-xr-x 3 mengj Administrators 8192 Dec 2 12:37 .
 4 drwxr-xr-x 16 mengj Administrators 4096 Nov 30 12:16 ..
25 -rw-r--r-- 1 mengj Administrators 24936 Feb 17 2022 font.c
 1 -rw-r--r-- 1 mengj Administrators  715 Feb 17 2022 font.h
 2 -rw-r--r-- 1 mengj Administrators 1139 Feb 13 2022 i2c.c
 1 -rw-r--r-- 1 mengj Administrators  486 Feb 13 2022 i2c.h
 1 -rw-r--r-- 1 mengj Administrators  215 Sep 27 10:02 includes.h
 2 -rw-r--r-- 1 mengj Administrators 1701 Sep 27 10:08 key.c
 1 -rw-r--r-- 1 mengj Administrators  562 Feb 1 2023 key.h
 1 -rw-r--r-- 1 mengj Administrators  507 Sep 27 10:05 main.c
 1 -rw-r--r-- 1 mengj Administrators  311 Oct 13 11:14 makefile
 4 -rw-r--r-- 1 mengj Administrators 3473 Feb 16 2022 oled.c
 1 -rw-r--r-- 1 mengj Administrators  602 Feb 16 2022 oled.h
 5 -rw-r--r-- 1 mengj Administrators 4389 Feb 1 2023 radio.c
 1 -rw-r--r-- 1 mengj Administrators  229 Jan 30 2022 radio.h
 3 -rw-r--r-- 1 mengj Administrators 2672 Feb 20 2022 si47xx.c
13 -rw-r--r-- 1 mengj Administrators 12868 Feb 6 2022 si47xx.h
 1 -rw-r--r-- 1 mengj Administrators  532 Sep 27 10:03 timer0.c
 1  -rw-r--r-- 1 mengj Administrators  142 Nov 18 2021 timer0.h
```

makefile for the project

Code-18.1 shows the contents of the project's makefile.

```
 1    CC=cc16e
 2    AS=as16e
 3    LK=lk16e
 4    RM=rm
 5
 6    HEX=radio
 7
 8    OBJ=main.obj timer0.obj i2c.obj font.obj oled.obj key.obj radio.obj si47xx.obj
 9
10    $(HEX).hex: $(OBJ) makefile
11        $(LK) $(OBJ) -o $(HEX)
12
13    %.obj: %.asm
14        $(AS) $<
15
16    %.asm: %.c
17        $(CC) $<
18
19    clean:
20        $(RM) $(OBJ)
21        $(RM) *.lst
22        $(RM) *.hex
23        $(RM) *.map
```
/p16ecc/applications/si4730radio_pic12f1840/makefile   Code-18.1

main.c file of the project

Code-18.2 shows **main.c** source code that contains the entry function `main()` of the project.

```
 1    #include <string.h>
 2    #include "includes.h"
 3
 4    #pragma isr_no_stack
 5    #pragma acc_save         1
 6
 7    #pragma FUSE0    _FOSC_INTOSC & _WDT_DIS
 8    #pragma FUSE1    0xffff                          // _LVP_HV_ONLY | 0x100
 9
10    void main()
11    {
12        OSCCON = 0x70;
13        OSCTUNE = 0x00;
14        BORCON = 0x00;
15        while ( !OSCSTATbits->PLLRDY );
16
17        INTCONbits->GIE = 1;     // enable interrupt
18        ANSELA = 0;              // all inputs are digital
19
20        TMR0_init();
21        I2C_init();
22        OLED_init();
23        KEY_init();
24        RADIO_init();
25
26        for (;;)
27            RADIO_poll();
28    }
```
/p16ecc/applications/si4730radio_pic12f1840/main.c   Code-18.2

- #4~5: pragma statements that define application configurations (only save ACC0 during interrupt).
- #12~13: set the internal RC oscillator's frequency (32MHz).
- #15: wait for stabilization of the internal RC oscillator.
- #17: enable the global interrupt.
- #18: set the PORTA to digital mode.
- #20~24: initializations of all components:
  TIMER0 – generate interrupt every 1ms (#20);
  I²C – run at the rate of 200Kbits (#21);
  LED – blank the display (#22);
  SI4730 radio module – set to FM mode/99.9MHz (#24).
- #26~27: keep polling the key inputs (switch mode/set the receiving frequency).

radio.c file of the project

Code-18.3 shows the `RADIO_init()` function in **radio.c** that initializes the radio.

```
10    void RADIO_init(void)
11    {
12        TRISAbits->b5 = 0;        // PA5 is input
13
14        LATAbits->b5  = 0;        // RST = 0
15        TMR0_delayMs(100);
16        TRISAbits->b5 = 1;        // RST = 1
17        TMR0_delayMs(100);
18
19        KEY_init();               // put PA5 to input
20        startFM();
21        startFM();
22    }
```
/p16ecc/applications/si4730radio_pic12f1840/radio.c   Code-18.3

- #14~17: generate a reset signal to si4730 module, before setting the radio mode.
- #20~21: set to FM mode (twice).

Code 18.4 shows the `startFM()` function in radio.c that sets the radio to FM mode.

```
24    void startFM(void)
25    {
26        while ( MODE_KEY_HOLD ) TMR0_delayMs(5);
27        KEY_read();              // clear up key buffer
28
29        RADIO_mode = FM_MODE;
30        SI47xx_init();
31        SI47xx_setVolume(RADIO_volume = 63);
32        SI47xx_setFreq(RADIO_fmFreq);
33
34        RADIO_dispFreq();
35
36        OLED_displayChar_16( 7-2, 0, verdana_16x11ptBitmaps_F, 11);
37        OLED_displayChar_16(20-2, 0, verdana_16x16ptBitmaps_M, 16);
38
39        OLED_displayChar_8x6_str(12-2, 3, "MHz");
40    }
```
/p16ecc/applications/si4730radio_pic12f1840/radio.c   Code-18.4

- #27~28: wait for button K3 unpressed (exit the reset mode).
- #32: set si4730 radio mode's volume to maximum.
- #33: set the frequency for FM mode.
- #34~39: set the display.

There are five types of events, listed in Table-18.4, generated when any key/button is pressed:

Event	Action
MODE_KEY	Toggle/Alter the radio operation mode (AM vs. FM)
INC_KEY	Increase the frequency (1KHz for AM, 100KHz for FM)
DEC_KEY	Decrease the frequency (1KHz for AM, 100KHz for FM)
FAST_INC	Increase the frequency (10KHz for AM, 500KHz for FM)
FAST_DEC	Decrease the frequency (10KHz for AM, 500KHz for FM)

Table-18.4

The function RADIO_poll() in **radio.c**, as shown in Code-18.5a~Code-18.5b, monitors and handles key/button events.

```
60    void RADIO_poll(void)
61    {
62        unsigned char key = KEY_read();
63        unsigned char delta;
64
65        switch ( key )
66        {
67            case MODE_KEY:  // FM/AM switch
68                OLED_clr();
69                if ( RADIO_mode == FM_MODE )
70                    startAM();
71                else
72                    startFM();
73                break;
74
```
/p16ecc/applications/si4730radio_pic12f1840/radio.c   Code-18.5a

- #62: read a key-in event.
- #67~72: handle the event MODE_KEY, toggling the radio mode (AM vs. FM).

```
75              case INC_KEY:
76              case FAST_INC:
77                  if ( RADIO_mode == FM_MODE )
78                  {
79                      delta = RADIO_fmFreq % 50;
80                      if ( key == FAST_INC )
81                          RADIO_fmFreq += 50 - delta; // inc 0.5 MHz
82                      else
83                          RADIO_fmFreq += 10;          // inc .1 MHz
84
85                      if ( RADIO_fmFreq > 10850 )
86                          RADIO_fmFreq = 10850;
87
88                      SI47xx_setFreq(RADIO_fmFreq);
89                  }
90                  else
91                  {
92                      delta = RADIO_amFreq % 10;
93                      if ( key == FAST_INC )
94                          RADIO_amFreq += 10 - delta; // inc 10KHz
95                      else
96                          RADIO_amFreq++;              // inc 1 KHz
97
98                      if ( RADIO_amFreq > 1650 )
99                          RADIO_amFreq = 1650;
100
101                     SI47xx_setFreq(RADIO_amFreq);
102                 }
103                 RADIO_dispFreq();
104                 break;
```

/p16ecc/applications/si4730radio_pic12f1840/radio.c   Code-18.5b

- #75~104: increase and update frequency for radio for event `INC_KEY` and `FAST_INC`.

timer0.c of the project

TIMER0 in PIC12F1840 is configured to generate an interrupt every 1ms. And it will scan the buttons during the interrupt. Below, Code-18.6, shows the functions in the file.

```
1   #include "includes.h"
2
3   unsigned char tmr0Count = 0;
4
5   void TMR0_init(void)
6   {
7       OPTION_REG = (OPTION_REG & 0xC0) | (0xD4 & 0x3F);   // 32MHz
8
9       INTCONbits->TMR0IF = 0;          // clear timer0 interrupt
10      INTCONbits->TMR0IE = 1;          // enable timer0 interrupt
11  }
12
13  void TMR0_delayMs(unsigned char ms)
14  {
15      unsigned char t = tmr0Count;
16      while ( (unsigned char)(tmr0Count - t) < ms );
17  }
18
19  interrupt tmr0_isr()
20  {
21      if ( INTCONbits->TMR0IF )
22      {
23          TMR0 = 6;
24          INTCONbits->TMR0IF = 0;
25          tmr0Count++;
26
27          if ( TRISAbits->b5 )
28              KEY_scan();
29      }
30  }
```

/p16ecc/applications/si4730radio_pic12f1840/timer0.   Code-18.6

- #5~11: configure TIMER0 (enable interrupt).
- #19~30: interrupt service routine for TIMER0. It scans the keys/buttons by calling `KEY_scan()`.

18.1.3 Compile the project

The following is the compiling command from the terminal, as well as the result:

```
F:\p16ecc\applications\si4730radio_pic12f1840>make
cc16e main.c
parse ... main.c
as16e main.asm
assembling 'main.asm' ...
cc16e timer0.c
parse ... timer0.c
as16e timer0.asm
assembling 'timer0.asm' ...
cc16e i2c.c
parse ... i2c.c
as16e i2c.asm
assembling 'i2c.asm' ...
cc16e font.c
parse ... font.c
as16e font.asm
assembling 'font.asm' ...
cc16e oled.c
parse ... oled.c
as16e oled.asm
assembling 'oled.asm' ...
cc16e key.c
parse ... key.c
as16e key.asm
assembling 'key.asm' ...
cc16e radio.c
parse ... radio.c
as16e radio.asm
assembling 'radio.asm' ...
cc16e si47xx.c
parse ... si47xx.c
as16e si47xx.asm
assembling 'si47xx.asm' ...
lk16e main.obj timer0.obj i2c.obj font.obj oled.obj key.obj radio.obj si47xx.obj -o radio
PIC16E linker, v0.1.7
linking 'main.obj' ...
linking 'timer0.obj' ...
linking 'i2c.obj' ...
linking 'font.obj' ...
linking 'oled.obj' ...
linking 'key.obj' ...
linking 'radio.obj' ...
linking 'si47xx.obj' ...
parse ... _tmp_/crt0.c
assembling '_tmp_/crt0.asm' ...
linking '_tmp_/crt0.obj' ...
parse ... _tmp_/crt1.c
assembling '_tmp_/crt1.asm' ...
linking '_tmp_/crt1.obj' ...
parse ... _tmp_/string.c
assembling '_tmp_/string.asm' ...
linking '_tmp_/string.obj' ...
RAM used: 55 bytes (22.92%)
ROM used: 3427 words (83.67%)
rm si47xx.asm main.asm font.asm radio.asm i2c.asm key.asm timer0.asm oled.asm
```

The output shows the result (RAM = 55 bytes, ROM = 3427 words).

18.1.4 Conclusions and considerations:

- The FM band works very well (signal sensitivity, L/R channel separation, audio fidelity, …).
- The AM band doesn't work well since the tuning circuit coil is missing.
- There is no volume control since PIC12F1840 doesn't supply enough I/O pins.
- Buttons can be replaced with a mechanical rotary encoder.
- On-chip EEPROM can be used to keep the radio settings.
- Another implementation is made that uses PIC16F13115 ('extended' PIC16Fxxxx structure), under the folder "/p16ecc/applications/si4730radio_pic16f13115".

18.2 Example 2 – USB to UART Adapter

USB-to-UART adapters are widely used in the IT industry. And there are several solutions for that. This example was adopted from a design developed under Microchip's MPLAB X IDE, with a few modifications. In this application, the chip's USB interface implements the CDC class, which mimics a serial interface (virtual serial port).

This application is implemented based on a PIC16F1455 (or PIC16F1454) microprocessor from Microchip. PIC16F1455/PIC16F1454 has the following features/specifications.

- Enhanced structure of PIC16Fxxxx CPU family.
- Max running speed = 48MHz.
- Integrated a USB interface, and can run at the full speed (12Mbits/sec).
- 1K bytes RAM (the lower 512 bytes are in dual-port structure).
- 8K words flash ROM.

The adapter has the following features/functions (which are limited by PIC16F1455/PIC16F1454):
- USB interface runs at full speed (12Mbits/sec).
- UART baudrate supported: 110~921600.
- UART serial data formats supported: 8/N/1, 8/N/2, 8/E/1, 8/O/1.
- Maximum data package size: 64 bytes.

18.2.1 Hardware design and implementation

The following, Figure-18.2, is the schematic of the design.

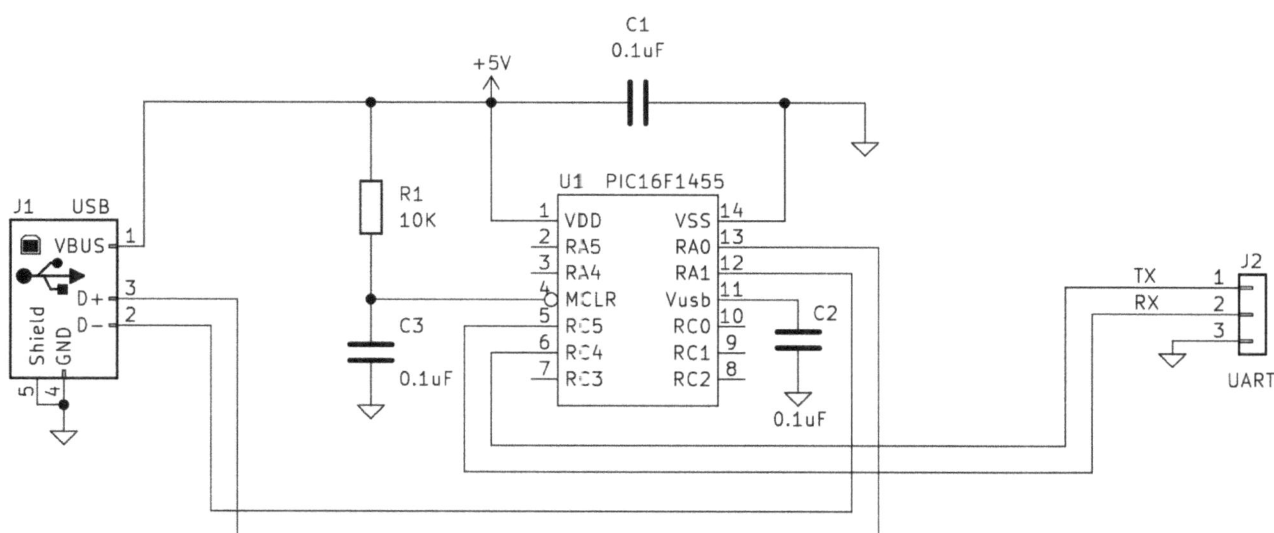

Figure-18.2

Only a few components, aside from the PIC16F1455 CPU, are used. This design/implementation works only with a 5V UART interface. The voltage shifter should be included in the design if it is needed to support a wider voltage range at the UART interface.

Project directory: /p16ecc/applications/USB2uart

The project involves the following files under the folder:

```
F:\p16ecc\applications\USB2uart>ls -sal
 2 -rw-r--r-- 1 mengj Administrators  1457 May 19 2020 GenericTypeDefs.h
10 -rw-r--r-- 1 mengj Administrators  9412 Oct 10 20:43 main.c
 1 -rw-r--r-- 1 mengj Administrators   326 Nov 6 2021 makefile
 4 -rw-r--r-- 1 mengj Administrators  3929 Oct 10 20:39 tool.c
 1 -rw-r--r-- 1 mengj Administrators   363 May 27 2018 tool.h
 4 -rw-r--r-- 1 mengj Administrators  3301 Oct 13 10:12 uart.c
 1 -rw-r--r-- 1 mengj Administrators   140 Mar 3 2020 uart.h
 2 -rw-r--r-- 1 mengj Administrators  1413 Feb 28 2018 usb.h
24 -rw-r--r-- 1 mengj Administrators 23795 Feb 28 2018 usb_ch9.h
22 -rw-r--r-- 1 mengj Administrators 21767 Feb 28 2018 usb_common.h
 3 -rw-r--r-- 1 mengj Administrators  3006 Apr 2 2018 usb_config.h
 6 -rw-r--r-- 1 mengj Administrators  5225 Mar 23 2018 usb_descriptors.c
66 -rw-r--r-- 1 mengj Administrators 66703 Oct 13 10:36 usb_device.c
 5 -rw-r--r-- 1 mengj Administrators  4521 Mar 1 2018 usb_device.h
 3 -rw-r--r-- 1 mengj Administrators  2378 Feb 28 2018 usb_device_local.h
14 -rw-r--r-- 1 mengj Administrators 13787 Oct 10 20:44 usb_function_cdc.c
18 -rw-r--r-- 1 mengj Administrators 17827 Sep 29 2020 usb_function_cdc.h
17 -rw-r--r-- 1 mengj Administrators 16841 Feb 28 2018 usb_hal.h
 8 -rw-r--r-- 1 mengj Administrators  7590 Oct 13 10:40 usb_hal_pic16f1.h
```

makefile for the project

Code-18.7 shows the contents of the project's makefile.

```
1    CC=cc16e
2    AS=as16e
3    LK=lk16e
4    RM=rm
5
6    HEX=usb2uart
7
8    OBJ=main.obj usb_descriptors.obj usb_function_cdc.obj usb_device.obj uart.obj tool.obj
9
10
11   $(HEX).hex: $(OBJ) makefile
12       $(LK) $(OBJ) -o $(HEX)
13
14   %.obj: %.asm
15       $(AS) $<
16
17   %.asm: %.c
18       $(CC) $<
19
20   clean:
21       $(RM) $(OBJ)
22       $(RM) *.lst
23       $(RM) *.hex
24       $(RM) *.map
```

/p16ecc/applications/USB2uart/makefile   Code-18.7

main.c file of the project

Code-18.8 shows the entry function, `main()`, of the project.

```c
1    /** INCLUDES *************************************************/
2    #include <pic16f1455.h>
3    #include "GenericTypeDefs.h"
4    #include "usb_hal_pic16f1.h"
5    #include "usb_device.h"
6    #include "usb_function_cdc.h"
7    #include "uart.h"
8
9    #pragma acc_save 0
10   #pragma isr_no_stack
11
12   // configure fuses ...
13   #pragma FUSE0    0x09e4
14   #pragma FUSE1    0x1fcf
   . . .
29   void main(void)
30   {
31       WPUA  = 0xff;
32       TRISA = 0xff;
33       OPTION_REGbits->WPUEN = 0;   // enable weak pull-up
34
35       InitializeSystem();
36
37       for (UART_init();;)
38       {
39           USBDeviceTasks();
40           UART_task();
41       }
42   }
```
/p16ecc/applications/USB2uart/main.c   Code-18.8

- #9~10: pragma statements that define application configurations (only save ACC0 during interrupt).
- #13~14: fuse settings.
- #35: USB interface initialization.
- #37~41: initialization of the UART interface on-chip, and it keeps polling both the USB and the UART interfaces:
  `USBDeviceTasks()` – handle events on USB interface.
  `UART_task()` – handle data communications on USB/UART interface.

Code-18.9a and Code-18.9b show the function, `mySetLineCodingHandler()`, that sets the baudrate and the data format for the UART running.

```c
125  void mySetLineCodingHandler(void)
126  {
127      unsigned short baud;
128
129      baud = cdc_notice.GetLineCoding.info.dwDTERate;
130
131      //If the request is not in a valid range
132      if ( baud >= 110 )
133      {
134          //Update the baudrate info in the CDC driver
135          CDCSetBaudRate(baud);
136
137          if ( baud >= 200 )
138          {
139              //Update the baudrate of the UART
140              baud = (48000000/4)/baud - 1;
141              TXSTAbits->BRGH = 1;
142          }
143          else
144          {
145              //Update the baudrate of the UART
146              baud = (48000000/16)/baud - 1;
147              TXSTAbits->BRGH = 0;
148          }
149          SPBRGL = baud;
150          SPBRGH = baud >> 8;
151
```
/p16ecc/applications/USB2uart/main.c   Code-18.9a

- #137~150: calculate and set the baudrate configuration in register pair SPBRGL/SPBRGH.
-

```
152            TXSTAbits->TX9 = 0;
153            uartFormat = 0;
154            if ( cdc_notice.GetLineCoding.info.bCharFormat )     // stop > 1-bit
155            {
156                uartFormat = 3;
157                TXSTAbits->TX9 = 1;
158                TXSTAbits->TX9D= 1; // extra STOP bit = 1
159            }
160
161            if ( cdc_notice.GetLineCoding.info.bParityType &&
162                 cdc_notice.GetLineCoding.info.bParityType < 3 )
163            {
164                uartFormat = cdc_notice.GetLineCoding.info.bParityType;
165                TXSTAbits->TX9 = 1;
166            }
167        }
168    }
```
/p16ecc/applications/USB2uart/main.c   Code-18.9b

- #152~166: configure serial data format (8/N/1, 8/N/2, 8/E/1, 8/O/1).

Assign the addresses for USB data memory

The following source codes, shown in Code-18.10~Code-18.12, define the data location since only the lower 512 bytes of RAM on PIC16F1455 have the dual-port structure to support USB interface.

```
111    //----- Defintions for BDT address ---------------------------------------
112    #define BDT_ENTRY_SIZE           4
113    #define BDT_BASE_ADDR            0x2000
114    #define CTRL_TRF_SETUP_ADDR      (BDT_BASE_ADDR + BDT_ENTRY_SIZE*BDT_NUM_ENTRIES)
115    #define CTRL_TRF_DATA_ADDR       (CTRL_TRF_SETUP_ADDR + USB_EP0_BUFF_SIZE)
```
/p16ecc/applications/USB2uart/usb_hal_pic16f1.h   Code-18.10

```
42    volatile BDT_ENTRY BDT[BDT_NUM_ENTRIES]      @ BDT_BASE_ADDR;
43    volatile CTRL_TRF_SETUP SetupPkt             @ CTRL_TRF_SETUP_ADDR;
44    volatile BYTE CtrlTrfData[USB_EP0_BUFF_SIZE] @ CTRL_TRF_DATA_ADDR;
```
/p16ecc/applications/USB2uart/usb_device.c   Code-18.11

```
10    /** V A R I A B L E S *****************************************************/
11    #define IN_DATA_BUFFER_ADDRESS        0x2140
12    #define OUT_DATA_BUFFER_ADDRESS       0x2190
13    #define LINE_CODING_ADDRESS           0x20A0
14    #define NOTICE_ADDRESS                (LINE_CODING_ADDRESS + LINE_CODING_LENGTH)
15
16    unsigned char cdc_data_tx[CDC_DATA_IN_EP_SIZE]  @IN_DATA_BUFFER_ADDRESS;
17    unsigned char cdc_data_rx[CDC_DATA_OUT_EP_SIZE] @OUT_DATA_BUFFER_ADDRESS;
18
19    LINE_CODING line_coding @LINE_CODING_ADDRESS;      // Buffer to store line coding information
20    CDC_NOTICE cdc_notice   @NOTICE_ADDRESS;
```
/p16ecc/applications/USB2uart/usb_function_cdc.c   Code-18.12

uart.c file that controls the communication flow over UART

A lookup table, `parityTable[]`, that stores the parity property for each byte value, is created and stored in ROM, as shown in Code-18.13.

```
7
8    const BYTE parityTable[256] = {
9        0,1,1,0,1,0,0,1,1,0,0,1,0,1,1,0,1,0,0,1,0,1,1,0,0,1,1,0,1,0,0,1,
10       1,0,0,1,0,1,1,0,0,1,1,0,1,0,0,1,0,1,1,0,1,0,0,1,1,0,0,1,0,1,1,0,
11       1,0,0,1,0,1,1,0,0,1,1,0,1,0,0,1,0,1,1,0,1,0,0,1,1,0,0,1,0,1,1,0,
12       0,1,1,0,1,0,0,1,1,0,0,1,0,1,1,0,1,0,0,1,0,1,1,0,0,1,1,0,1,0,0,1,
13       1,0,0,1,0,1,1,0,0,1,1,0,1,0,0,1,0,1,1,0,1,0,0,1,1,0,0,1,0,1,1,0,
14       0,1,1,0,1,0,0,1,1,0,0,1,0,1,1,0,1,0,0,1,0,1,1,0,0,1,1,0,1,0,0,1,
15       0,1,1,0,1,0,0,1,1,0,0,1,0,1,1,0,1,0,0,1,0,1,1,0,0,1,1,0,1,0,0,1,
16       1,0,0,1,0,1,1,0,0,1,1,0,1,0,0,1,0,1,1,0,1,0,0,1,1,0,0,1,0,1,1,0,
17    };
```
/p16ecc/applications/USB2uart/uart.c   Code-18.13

The following, as illustrated in Code-18.14a and Code-18.14b, is a task called from the main loop in `main()` that controls the UART communication flow.

Send a byte to UART:

```
56   void UART_task(void)
57   {
58       BYTE i, len, tail;
59
60       if ( USBDeviceState < CONFIGURED_STATE || USBSuspendControl == 1 )
61           return;
62
63       if ( (uartRxPos < uartRxLen) && (PIR1bits->TXIF || TXSTAbits->TRMT) )    // anything to send to UART?
64       {
65           BYTE c = uartRxBuffer[uartRxPos];
66           BYTE parity = parityTable[c];
67           uartRxPos++;
68
69           switch ( uartFormat )
70           {
71               case 1: // Odd
72                   TXSTAbits->TX9D = 1;          // TXSTA |= 1;
73                   if ( parity ) TXSTAbits->TX9D = 0;
74                   break;
75
76               case 2: // Even
77                   TXSTAbits->TX9D = 0;          // TXSTA &= 0xfe;
78                   if ( parity ) TXSTAbits->TX9D = 1;
79                   break;
80           }
81
82           TXREG = c;
83       }
84
85       if ( uartRxPos >= uartRxLen )
86           uartRxPos = uartRxLen = 0;
87
88       // anything from USB(PC host)?
89       uartRxLen += getUSBData(&uartRxBuffer[uartRxLen], UART_BUFFER_SIZE - uartRxLen);
```
`/p16ecc/applications/USB2uart/uart.c`   Code-18.14a

- #65: get the byte from the buffer, to send it through UART Tx port.
- #69~80: set parity bit for sending if required (8/E/1 or 8/O/1).
- #85~89: get (new) input data bytes from USB receiving.

Receive a byte from UART:

```
91        if ( PIR1bits->RCIF )
92        {
93            uartTxBuffer[uartTxTail] = RCREG;
94            tail = (uartTxTail + 1) & (UART_BUFFER_SIZE - 1);
95            if ( tail != uartTxHead ) uartTxTail = tail;
96        }
97
98        // anything to USB(PC host)?
99        if ( USBUSARTIsTxTrfReady() && uartTxTail != uartTxHead )
100       {
101           len = UART_BUFFER_SIZE;
102           if ( uartTxTail > uartTxHead ) len = uartTxTail;
103           len -= uartTxHead;
104
105           putUSBUSART(&uartTxBuffer[uartTxHead], len);
106           uartTxHead += len;
107           uartTxHead &= (UART_BUFFER_SIZE - 1);
108       }
109
110       CDCTxService();
111   }
```
`/p16ecc/applications/USB2uart/uart.c`   Code-18.14b

- #91~96: receiving a byte from UART Rx port.
- #99~108: prepare sending the data to USR.
- #110: handle USB transmission.

18.2.3 Compile the project

Compiling from the terminal:

```
F:\p16ecc\applications\USB2uart>make
cc16e main.c
parse ... main.c
as16e main.asm
assembling 'main.asm' ...
cc16e usb_descriptors.c
parse ... usb_descriptors.c
as16e usb_descriptors.asm
assembling 'usb_descriptors.asm' ...
cc16e usb_function_cdc.c
parse ... usb_function_cdc.c
as16e usb_function_cdc.asm
assembling 'usb_function_cdc.asm' ...
cc16e usb_device.c
parse ... usb_device.c
as16e usb_device.asm
assembling 'usb_device.asm' ...
cc16e uart.c
parse ... uart.c
as16e uart.asm
assembling 'uart.asm' ...
cc16e tool.c
parse ... tool.c
as16e tool.asm
assembling 'tool.asm' ...
lk16e main.obj usb_descriptors.obj usb_function_cdc.obj usb_device.obj uart.obj tool.obj -o usb2uart
PIC16E linker, v0.1.7
linking 'main.obj' ...
linking 'usb_descriptors.obj' ...
linking 'usb_function_cdc.obj' ...
linking 'usb_device.obj' ...
linking 'uart.obj' ...
linking 'tool.obj' ...
parse ... _tmp_/crt0.c
assembling '_tmp_/crt0.asm' ...
linking '_tmp_/crt0.obj' ...
parse ... _tmp_/crt1.c
assembling '_tmp_/crt1.asm' ...
linking '_tmp_/crt1.obj' ...
RAM used: 471 bytes (46.73%)
ROM used: 3837 words (46.84%)
rm usb_descriptors.asm uart.asm main.asm usb_device.asm usb_function_cdc.asm tool.asm
```

The result shows the project costs 471 bytes of RAM and 3837 words of ROM.

18.2.4 Conclusions and considerations:

- A voltage shifter at the UART might be needed when interfacing to the target running at a voltage other than 5V.
- With modifications, this design can derive some other projects (USB to SPI, USB to I^2C, etc.).
- This design limits the data package to 64 bytes. A larger data package from the host to the USB may result in data bytes being lost.

STM8 MCUs are developed by STMicroelectronics NV in Europe. It's an 8-bit microprocessor series that runs up to 16/24 MHz. Compared with PIC16Fxxxx family MCUs, the STM8 MCU series is more powerful and provides better C-language support, thanks to its CISC instruction set and a complete set of addressing modes. Moreover, the STM8 MCU includes hardware support for 8-bit multiplication and division, as well as some 16-bit operations.

The design of the STM8 compiler tool is very similar to that of the PIC16F compiler tool. For this reason, this part just briefly introduces the design process.

Chapter-19

STM8 Compiler Tools

The design of the STM8S compilation tools is pretty much the same as for PIC16Fxxxx. The tool set contains four executable files as follows:

- `st8cpp1.exe`   - stm8 preprocessor
- `st8cc.exe`   - stm8 compiler
- `st8as.exe`   - stm8 assembler
- `st8lk.exe`   - stm8 linker

In addition, the library file, crt0.c/crt0x.c, and the header files for the specific MCUs should be provided. The project is organized and located under the folder "/stm8scc" as follows:

```
F:\stm8scc>ls -lsa
 4 drwxr-xr-x 20 mengj Administrators 4096 Dec 15 14:29 .
12 drwxr-xr-x 27 mengj Administrators 12288 Dec 11 21:17 ..
 4 drwxr-xr-x 7 mengj Administrators 4096 Dec 16 09:15 applications
 8 drwxr-xr-x 2 mengj Administrators 8192 Dec 14 16:37 as_source
 0 drwxr-xr-x 2 mengj Administrators    0 Oct 11 07:54 bin
40 drwxr-xr-x 3 mengj Administrators 40960 Dec 7 15:22 cc_source
 4 drwxr-xr-x 2 mengj Administrators 4096 Dec 9 10:29 cpp1_source
 4 drwxr-xr-x 2 mengj Administrators 4096 Dec 9 15:14 doc
 4 drwxr-xr-x 2 mengj Administrators 4096 Dec 14 15:41 include
 0 drwxr-xr-x 2 mengj Administrators    0 Dec 15 14:29 lib
12 drwxr-xr-x 2 mengj Administrators 12288 Dec 14 08:31 lk_source
```

19.1 STM8 MCU Architecture

Table-19.1 below lists the significant differences between PIC16F and STM8 MCUs.

	PIC16Fxxxx MCU	STM8 MCU
Instruction system	RISC	CISC
Memory organization	Harvard	Von Neumann
Data bytes order	Little endian	Big endian
Interrupt vector	Single interrupt vector	Multi-interrupt vectors
Maximum ROM size	32 Kword	16 Mbyte
Maximum RAM size	2~4 Kbyte	16 Kbyte
Maximum VCC voltage	5.5V	3.6V

Table-19.1

Among the differences, "Big endian" vs. "Little endian" has a significant impact on the compiler design.

Table-19.2 below shows the STM8 programming environment/structure.

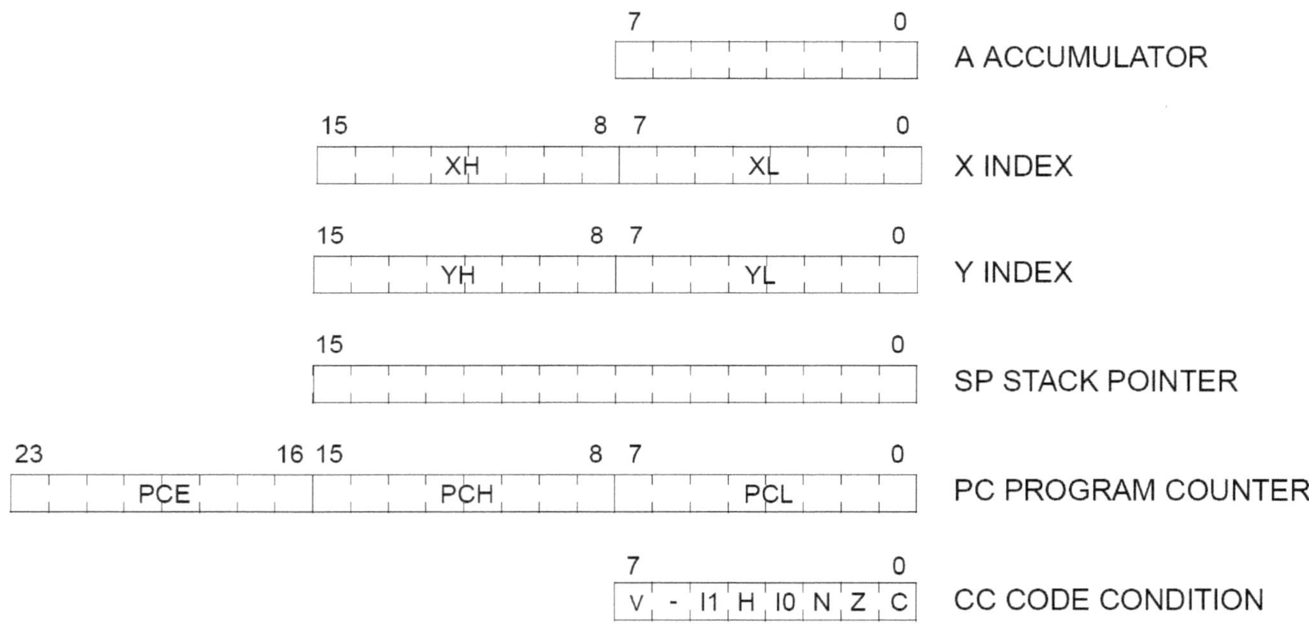

Table-19.2

The following lists the key parts that guide the compiler design:

- RAM starts at address 0x0000. The first 256 bytes of RAM are considered bank0 (or page0) and can be accessed using 1-byte address instructions.
- The first 8 bytes in bank0 are reserved for the compiler (0x0000~0x0003 is used for the virtual accumulator).
- ROM starts at address 0x8000. The ROM area from 0x8000 to 0x8080 is the interrupt vector area. Each vector consumes 4 bytes, including a 24-bit ISR target address. The first vector (starting at

384

0x8000) is always used as the reset vector. The vector is indicating an interrupt service routine (IRS) start address in memory, and it has the structure as shown in Table-19.3:

First byte	Second byte	Third byte	Fourth byte
0x82	ISR address bit23~16	ISR address bit15~8	ISR address bit7~0

Table-19.3

- When the target STM8 MCU has no more than 32 Kbytes of ROM on chip, the compiler tool runs in 'standard' (default) mode; otherwise, it runs in 'extended' mode.

19.1.1 STM8 instructions and addressing modes

In file /stm8/stm8s_inst.h, it lists all the basic STM8 instructions, plus the pseudo instructions (or directives).

```
 1  #ifndef ST8INST_H
 2  #define ST8INST_H
 3
 4  #define _ADD      (char*)"add"
 5  #define _ADDW     (char*)"addw"
 6  #define _ADC      (char*)"adc"
 7  #define _AND      (char*)"and"
 8  #define _BCCM     (char*)"bccm"
 9  #define _BCP      (char*)"bcp"
10  #define _BCPL     (char*)"bcpl"
11  #define _BRES     (char*)"bres"
12  #define _BSET     (char*)"bset"
13  #define _BTJF     (char*)"btjf"
14  #define _BTJT     (char*)"btjt"
15  #define _CALL     (char*)"call"
16  #define _CALLF    (char*)"callf"
17  #define _CALLR    (char*)"callr"
18  #define _CCF      (char*)"ccf"
19  #define _CLR      (char*)"clr"
20  #define _CLRW     (char*)"clrw"     // clears index reg.
21  #define _CP       (char*)"cp"       // compare
22  #define _CPW      (char*)"cpw"
23  #define _CPL      (char*)"cpl"
24  #define _CPLW     (char*)"cplw"
25  #define _DEC      (char*)"dec"
26  #define _DECW     (char*)"decw"
27  #define _DIV      (char*)"div"
28  #define _DIVW     (char*)"divw"
29  #define _EXG      (char*)"exg"
30  #define _EXGW     (char*)"exgw"
31  #define _HALT     (char*)"halt"
32  #define _INC      (char*)"inc"
33  #define _INCW     (char*)"incw"
34  #define _IRET     (char*)"iret"
35  #define _JP       (char*)"jp"       // 16-bit jump
36  #define _JPF      (char*)"jpf"      // 24-bit jump
37  #define _JRA      (char*)"jra"      //  8-bit jump (relative)
38  #define _JRC      (char*)"jrc"
39  #define _JREQ     (char*)"jreq"
40  #define _JRF      (char*)"jrf"
41  #define _JRH      (char*)"jrh"
42  #define _JRIH     (char*)"jrih"
43  #define _JRIL     (char*)"jril"
44  #define _JRM      (char*)"jrm"
45  #define _JRMI     (char*)"jrmi"
46  #define _JRNC     (char*)"jrnc"
47  #define _JRNE     (char*)"jrne"

               . . .
```

```c
101    #define _TNZ        (char*)"tnz"
102    #define _TNZW       (char*)"tnzw"
103    #define _TRAP       (char*)"trap"
104    #define _WFE        (char*)"wfe"
105    #define _WFI        (char*)"wfi"
106    #define _XOR        (char*)"xor"
107
108    #define __JP        (char*)".jp"
109    #define __JBZ       (char*)".jbz"
110    #define __JBNZ      (char*)".jbnz"
111
112    #define _SEGMENT    (char*)".segment"
113    #define _DEVICE     (char*)".device"
114    #define _INVOKE     (char*)".invoke"
115    #define _RS         (char*)".rs"
116    #define _VECT       (char*)".vect"
117    #define _EQU        (char*)".equ"
118    #define _DB         (char*)".db"      // 1-byte data
119    #define _DW         (char*)".dw"      // 2-byte data
120    #define _DT         (char*)".dt"      // 3-byte data
121    #define _DD         (char*)".dd"      // 4-byte data
122    #define _END        (char*)".end"
123
124    #endif
```

`/stm8scc/cc_source/stm8/stm8s_inst.h`

Code-19.1

There are four (pseudo-)instructions in the list above that are specific to the design. The following table, Table-19.4, gives the explanations:

Instruction	Assembly format	Function
.jp	.jp *lbl*	Make a jump to *lbl* unconditionally.
	.jp *c, lbl*	$c = 0$, mimic the instruction "jreq *lbl* ", $c = 1$, mimic the instruction "jrne *lbl* ", $c = 2$, mimic the instruction "jrc *lbl* ", $c = 3$, mimic the instruction "jrnc *lbl* ", $c = 4$, mimic the instruction "jrpl *lbl* ", $c = 5$, mimic the instruction "jrmi *lbl* ", $c = 6$, mimic the instruction "jrsgt *lbl* ", $c = 7$, mimic the instruction "jrsge *lbl* ", $c = 8$, mimic the instruction "jrslt *lbl* ", $c = 9$, mimic the instruction "jrsle *lbl* ", $c = 10$, mimic the instruction "jrugt *lbl* ", $c = 11$, mimic the instruction "jruge *lbl* ", $c = 12$, mimic the instruction "jrult *lbl* ", $c = 13$, mimic the instruction "jrule *lbl* ".
.jbz	.jbz *addr, bit, lbl*	Minic the instruction "btjf *addr, #bit, lbl*".
.jbnz	.jbnz *addr, bit, lbl*	Minic the instruction "btjt *addr, #bit, lbl*".
.vect	.vect *addr, ISR_lbl*	Generate an item in the vector table.

Table-19.4

Where the jump instructions remain unsolved until the linking procedure.

STM8 instruction set supplies a variety of addressing modes, listed in Table-19.5, to access the memory. Thus, it delivers more options for memory access.

Addressing mode	Example	Accessing range
SP register offset indirect	`LD %A, (10, %SP)`	0x000000 ~ 0x00FFFF
Short direct	`LD %A, 100`	0x000000 ~ 0x0000FF
Long direct	`LD %A, 1000`	0x000000 ~ 0x00FFFF
Extended direct	`LDF %A, 1000000`	0x000000 ~ 0xFFFFFF
Register indirect	`LD %A, (%X)`	0x000000 ~ 0x00FFFF
Register short offset indirect	`LD %A, (10, %X)`	0x000000 ~ 0x0100FE
Register long offset indirect	`LD %A, (1000, %X)`	0x000000 ~ 0x01FFFE
Register extended offset indirect	`LDF %A, (1000000, %X)`	0x000000 ~ 0xFFFFFF
Short pointer indirect	`LD %A, [100]`	0x000000 ~ 0x00FFFF
Long pointer indirect	`LD %A, [1000]`	0x000000 ~ 0x00FFFF
Long pointer extended indirect	`LDF %A, [1000]`	0x000000 ~ 0xFFFFFF
Short pointer + register indirect	`LD %A, ([100], %X)`	0x000000 ~ 0x01FFFE
Long pointer + register indirect	`LD %A, ([1000], %X)`	0x000000 ~ 0x01FFFE
Long pointer + extended register indirect	`LDF %A, ([1000], %X)`	0x000000 ~ 0xFFFFFF

Table-19.5

19.1.2 Stack operation of STM8 MCU

The STM8 MCU architecture provides a full-function hardware stack. It improves the efficiency of parameter passing and access to local data variables. And all the function-local data is located at the top of the stack. Moreover, the hardware stack can support recursive function calls (a function can call itself).

The stack pointer, SP, is always pointing to the first available byte location on the top of the stack. At the start of an application (after reset), SP should be initialized to the address of the last byte of RAM.

Here is an example of a program in which the function `caller()` calls `callee()`:

C source code (ch15_t1.c)	STM8 assembly code

<pre>int callee(int x)
{
    return x + 1;
}
 	
void caller()
{
    int n, r;
    r = callee(n);
}</pre>
	<pre> .segment CODE2 (REL)
	; callee_$_x: offset = 3
	callee::
	; :: ch15_t1.c #3: return x + 1;
	        ldw %X, (3, %SP)
	        addw %X, #1
	        ldw 0, %X
	        ret
	
	        .segment CODE2 (REL)
	; caller_$1_n: offset = 1
	; caller_$2_r: offset = 3
	caller::
	        sub %SP, #4
	; :: ch15_t1.c #9: r = callee(n);
	        ldw %X, (1, %SP)
	        pushw %X
	        call callee
	        popw %X
	        ldw %X, 0
	        ldw (3, %SP), %X
	        add %SP, #4
	        ret</pre>

The following (Table-19.6) shows the changes to SP during the function call described above.

Before the function call		Passing parameter **x**		Enter function `callee(int)`	
SP →		SP →		SP →	
SP + 1	Variable **n**	SP + 1	Parameter **x**	SP + 1	Return address
SP + 3	Variable **r**	SP + 3	Variable **n**	SP + 3	Parameter **x**
		SP + 5	Variable **r**	SP + 5	Variable **n**
				SP + 7	Variable **r**

Table-19.6

19.2 STM8 Compiler - st8cc.exe

<u>Project directory</u>: /stm8scc/cc_source

The infrastructure of st8cc.exe is the same as that for p16cce.exe. It also involves the following processes:

- Pre-processor
- Pre-scan
- Optimization
- Assembly code generation.

19.2.1 Compiling option to support 'extended' mode

To run in 'extended' mode, to support MCUs that have more than 32KB ROM on chip, the compiling option '**-E**' should be applied in the compiling command from the terminal, which is detected in the file option.cpp, as illustrated in Code-19.2.

```
41   bool Option :: add(char *str)
42   {
43       char *p;
44       const char *prompt = "unknown/unsupported option";
45       int length = strlen(str);
46
47       switch ( str[0] )
48       {
49           case '?': case 'h':
50               printf("\nFormat:\n\n\tst8cc [options] file1.c file2.c ...\n\n");
51               return false;

                                 . . .

112          case 'E':   // extended mode
113              extendedMode = true;
114              return true;
115      }
116
117      printf("%s - %s\n", prompt, str);
118      return false;
119  }
```
/stm8scc/cc_source/option.cpp Code-19.2

- #112~114: set up 'extended' mode.

In 'extended' mode, the assembly code generated will use instructions of 'callf' and 'retf' for function calls and return operations. In addition, the pointers that access data in ROM space will take 3 bytes, as seen in Code-19.3, which determines the size of a pointer.

```
13    int Sizer :: size(attrib *ap, SIZER_OPTION opt)
14    {
15        int n = 0;
16
17        if ( opt == INDIR_SIZE )
18        {
19            if ( ptrWeight(ap) > 0 )
20            {
21                attrib *attr = cloneAttr(ap);
22                reducePtr(attr);
23                n = size(attr, TOTAL_SIZE);
24                delAttr(attr);
25                return n;
26            }
27        }
28        if ( opt == SUBDIM_SIZE )
29        {
30            attrib *attr = cloneAttr(ap);
31            decDim(attr);
32            n = size(attr, TOTAL_SIZE);
33            delAttr(attr);
34            return n;
35        }
36
37        int dim_size = (opt == ATTR_SIZE)? 1: dimScale(ap);
38
39        if ( ap->ptrVect || ap->isFptr )
40        {
41            n = 2;
42            if ( EXT_MODE && (ap->isFptr || (ap->ptrVect && ap->ptrVect[1] == CONST)) )
43                n = 3;
44        }
```
/stm8scc/cc_source/sizer.cpp   Code-19.3

- #37~44: determine the size of a pointer.

19.2.2 Lexer and parser

Two changes or modifications have happened to the lexer/parser for st8cc.exe:

1) Replace the storage modifier '_linear_' with '_bank0_', in both cc.l and cc.y, as illustrated in Code-19.4a and Code-19.4b.

```
162   "struct"          { addSrcCode(); return STRUCT; }
163   "typedef"         { addSrcCode(); return TYPEDEF; }
164   "..."             { addSrcCode(); return ELIPSIS; }
165   "_bank0_"         { addSrcCode(); yylval.i = BANK0;    return MEM_BANK; }
166
167   {identifier}      { NameList *dp = searchName(newNameList, yytext);
168                       addSrcCode();
169                       if ( dp == NULL || (ignoreTypedef & 1) )
170                       {
171                          yylval.s = dupStr(yytext);
172                          return IDENTIFIER;
173                       }
174                       yylval.a = newAttr(TYPEDEF_NAME);
175                       yylval.a->typeName = dupStr(dp->name);
176                       return TYPEDEF_NAME;
177                     }
```
/stm8scc/cc_source/cc.l   Code-19.4a

- #165: identify the keyword '_bank0_'.

```
  32   %token CHAR INT SHORT LONG UCHAR UINT USHORT ULONG VOID
  33   %token FUNC_HEAD FUNC_DECL DATA_DECL
  34   %token _IF _ELSE _ENDIF _IFDEF _IFNDEF DEFINE UNDEF PRAGMA
  35   %token BREAK CASE CONST CONTINUE DEFAULT DO ELSE ENUM EXTERN
  36   %token FOR GOTO IF REGISTER RETURN SIZEOF STATIC SWITCH
  37   %token VOLATILE WHILE INTERRUPT UNION STRUCT TYPEDEF ELIPSIS
  38   %token BANK0
  39   %token ADD_ASSIGN SUB_ASSIGN MUL_ASSIGN DIV_ASSIGN MOD_ASSIGN
  40   %token LEFT_ASSIGN RIGHT_ASSIGN AND_ASSIGN XOR_ASSIGN OR_ASSIGN
  41   %token EQ_OP NE_OP LE_OP GE_OP LEFT_OP RIGHT_OP INC_OP DEC_OP
  42   %token OR_OP AND_OP PTR_OP FPTR CAST CALL
  43   %token POST_INC POST_DEC PRE_INC PRE_DEC POS_OF ADDR_OF NEG_OF
  44   %token EOL LABEL
  45
  46   %token <i> CONSTANT L_CONSTANT U_CONSTANT UL_CONSTANT C_CONSTANT MEM_BANK
  47   %token <i> RELATIONAL_OP EQUALITY_OP SHIFT_OP ASSIGN_OP
  48   %token <s> IDENTIFIER IDENTIFIER_ STRING AASM
  49   %token <a> TYPEDEF_NAME
```

`/stm8scc/cc_source/cc.y`   Code-19.4b

- #38: enumerate the token `BANK0`.

2) Modify the parsing rule for interrupt service routine syntax in cc.y, as illustrated in Code-19.5.

```
 506   func_definition
 507       : func_declarator
 508         func_body                { $$ = oprNode(FUNC_DECL, 2, $1, $2);
 509                                    $$->opr.attr = newAttr(INT);
 510                                  }
 511       | declaration_specifiers
 512         func_declarator2         { if ( $1->type == 0 )
 513                                    {
 514                                        if ( $2->id.attr && $2->id.attr->ptrVect )
 515                                            yyerror("function pointer type not defined!");
 516                                        else
 517                                            $1->type = INT;
 518                                    }
 519                                  }
 520         func_body                { $$ = oprNode(FUNC_DECL, 2, $2, $4);
 521                                    $$->opr.attr = $1;
 522
 523                                    if ( $2->id.attr )
 524                                    {
 525                                        $$->opr.attr->ptrVect = $2->id.attr->ptrVect;
 526                                        $2->id.attr->ptrVect = NULL;
 527                                        delAttr($2->id.attr);
 528                                        $2->id.attr = NULL;
 529                                    }
 530                                  }
 531       | INTERRUPT identifier
 532         '(' conditional_expr ')' { CLR_SRC(); }
 533         compound_statement       { node *np = oprNode(FUNC_HEAD, 1, $2);
 534                                    $$ = oprNode(FUNC_DECL, 2, np, $7);
 535                                    $$->opr.attr = newAttr(INTERRUPT);
 536                                    $$->opr.attr->atAddr = $4;
 537                                  }
 538       ;
```

`/stm8scc/cc_source/cc.y`   Code-19.5

- #531~537: the parsing rule for interrupt service routine syntax. Where the item 'conditional_expr' (i.e. $4 in action part) is the vector address, which should be a constant with value 0x8000~0x8080.

19.2.3 Assembly code generation – start/end of a function (P-Code FUNC_BEG & FUNC_END)

Assembly code generation is handled in the STM8S class.

P-code FUNC_BEG is the operation for the start of a function. First of all, if it's the main() function or an interrupt service routine, it needs to create a vector and calculate the total cost or size of local data (including temporary variables), as shown in Code-19.6.

```cpp
31    void STM8S :: funcBeg(Pnode *pnp)
32    {
33        char *fname = curFunc->name;
34        char func_dname[4096];
35        char *s = STRBUF();
36        attrib *attr = curFunc->attr;
37

                                       . . .

59        if ( strcmp(fname, "main") == 0 )
60        {
61            // reset vector init.
62            ASM_CODE1(_SEGMENT, "CODE0 (ABS, =0x8000)");
63            ASM_CODE1(_VECT, fname);
64            ASM_OUTP('\n');
65
66            // create static data init. link
67            ASM_CODE1(_SEGMENT, "CODEi (REL, BEG)");
68            ASM_OUTP("_$init$:\n");
69            ASM_OUTP_LN();
70
71            ASM_CODE1(_SEGMENT, "CODEi (REL, END)");
72            ASM_CODE0(EXT_MODE? _RETF: _RET);
73            ASM_OUTP_LN();
74        }
75        else if ( attr->type == INTERRUPT )
76        {
77            char *s = STRBUF();
78            sprintf(s, "CODE0 (ABS, =0x%X)", vectorAddr(attr));
79            ASM_CODE1(_SEGMENT, s);
80            ASM_CODE1(_VECT, fname);
81            ASM_OUTP_LN();
82        }
```
/stm8scc/cc_source/stm8/stm8s1.cpp   Code-19.6

- #62~63: create a vector for main().
- #78~80: create a vector for an ISR.

Some pre-running operation needs to be done if it's the main() function or an ISR, as shown in Code-19.7:

```cpp
84        int data_size = listData(curFunc, pnp);
85
86        ASM_CODE1(_SEGMENT, "CODE2 (REL)");
87        ASM_LABL(fname, true);
88        if ( !attr->isStatic && attr->type != INTERRUPT ) ASM_OUTP(':');
89        ASM_OUTP_LN();
90
91        if ( strcmp(fname, "main") == 0 )
92        {
93            // set Stack Poiner(SP), and clean up whole RAM...
94            ASM_CODE2(_LDW, REG_X, stackAddr, '#');
95            ASM_CODE2(_LDW, REG_SP, REG_X);
96            ASM_CODE1(_CLR, "(%X)");
97            ASM_CODE1(_DECW, REG_X);
98            ASM_CODE1(_JRPL, ".-4");
99
100            // call system init routine
101            ASM_CODE1(EXT_MODE? _CALLF: _CALL, (char*)"_$init$");
102        }
103        else if ( attr->type == INTERRUPT )
104        {
105            for (int addr = 0; addr < accSave; addr++)
106            {
107                if ( (accSave - addr) >= 2 )
108                {
109                    ASM_CODE2(_LDW, REG_X, addr);    // save ACC contents
110                    ASM_CODE1(_PUSHW, REG_X);
111                    addr++;
112                }
113                else
114                    ASM_CODE1(_PUSH, addr);
115            }
116        }
```
/stm8scc/cc_source/stm8/stm8s1.cpp   Code-19.7

- #84: calculate the memory cost for all types of data. It generates the following data size results:
  `totalTmpSize` – total memory cost for temporary variables.
  `totalVarSize` – total memory cost for local variables.
  `totalParSize` – total memory cost for parameters.
- #91~102: zero out the RAM space if it's the `main()` function.
- #103~116: save the (virtual) accumulator if it's an ISR.

Finally, it creates the data space on top of the stack, as shown in Code-19.8.

```
118         switch ( data_size )
119         {   // create local memory on stack
120             case 0: break;
121             case 1: ASM_CODE1(_PUSH, REG_A);      break;
122             case 2: ASM_CODE1(_PUSHW, REG_X);     break;
123             default:ASM_CODE2(_SUB, REG_SP, data_size, '#');
124         }
125         frameOffset = 0;
126  }
```
/stm8scc/cc_source/stm8/stm8s1.cpp   Code-19.8

- #118~124: adjust SP to create the data space.

P-code `FUNC_END` is the operation for the end of a function. It involves the following operations:
- Remove the local data space from the stack.
- Restore the (virtual) accumulator if it's an ISR.
- Append a returning instruction (IRET, RETF, or RET), depending on the situation.

```
128  void STM8S :: funcEnd(void)
129  {
130      int data_size = totalTmpSize + totalVarSize;
131      switch ( data_size )
132      {   // remove local memory on stack (if any)
133          case 0: break;
134          case 1: ASM_CODE1(_POP, REG_A);      break;
135          case 2: ASM_CODE1(_POPW, REG_X);     break;
136          default:ASM_CODE2(_ADD, REG_SP, data_size, '#');
137      }
138
139      if ( curFunc->attr->type == INTERRUPT )
140      {
141          for (int addr = accSave; addr > 0;)
142          {
143              if ( addr & 1 )
144              {
145                  addr--;
146                  ASM_CODE1(_POP, addr);         // restore ACC contentsa
147              }
148              else
149              {
150                  addr -= 2;
151                  ASM_CODE1(_POPW, REG_X);
152                  ASM_CODE2(_LDW, addr, REG_X);
153              }
154          }
155          ASM_CODE0(_IRET);
156      }
157      else if ( strcmp(curFunc->name, "main") == 0 )
158      {
159          ASM_OUTP(" _$iret$::\n");
160          ASM_CODE0(_IRET);
161      }
162      else
163          ASM_CODE0(EXT_MODE? _RETF: _RET);
164
165      // list all function called ...
166      if ( curFunc->fcall )
167          ASM_OUTP((char*)"; function(s) called:\n");
168
169      for (NameList *fcall = curFunc->fcall; fcall; fcall = fcall->next)
170          ASM_CODE2((char*)".fcall", curFunc->name, fcall->name);
171
172      delLocalData(&localDataList);
173  }
```
/stm8scc/cc_source/stm8/stm8s1.cpp   Code-19.9

- #130~137: remove the local data space from the stack.
- #141~154: restore the (virtual) accumulator for an ISR.

19.2.4 Assembly code generation – generic assignments (P-Code '=')

The assembly code generation for P-code '=' has the following form

$$\{\text{'='}, [X, Y]\}$$

and is implemented in file /stm8/stms2.cpp, shown in Code-19.10a~Code-19.10d.

```
21   void STM8S :: mov(Item *ip0, Item *ip1)
22   {
23       int size0 = ip0->acceSize();
24       int size1 = ip1->acceSize();
25
26       if ( IS_BF_VAR(ip0->attr) )
27       {
28           moveToBF(ip0, ip1);
29           return;
30       }
31       if ( IS_BF_VAR(ip1->attr) )
32       {
33           moveFromBF(ip0, ip1);
34           return;
35       }
```
/stm8scc/cc_source/stm8/stm8s2.cpp   Code-19.10a

- #26~35: handle the situations in which either X or Y is a bit-field variable.

```
46       if ( (ip0->type == ACC_ITEM || ip0->type == TEMP_ITEM) && related(ip0, ip1) && size1 >= size0 )
47       {   // special move operation for big-endian structure (moving starts at higher byte)
48           for (int i = size0; i > 0; i--) {
49               if ( i >= 2 ) { // move 2 bytes at a time
50                   ASM_CODE2(_LDW, REG_X, acceItem(ip1, i-2, 2));
51                   ASM_CODE2(_LDW, acceItem(ip0, i-2, 2), REG_X);
52                   i--;
53               }
54               else if ( ip0->type != ACC_ITEM ) {
55                   ASM_CODE2(_LD, REG_A, acceItem(ip1, i-1, 1));
56                   ASM_CODE2(_LD, acceItem(ip0, i-1, 1), REG_A);
57               }
58               else
59                   ASM_CODE2(_MOV, acceItem(ip0, i-1, 1), acceItem(ip1, i-1, 1));
60           }
61           return;
62       }
```
/stm8scc/cc_source/stm8/stm8s2.cpp   Code-19.10b

- #46~62: if X and Y overlap in memory (e.g., ACC), then the assignment should be done from the higher byte to the lower (start from the lower address). The following example code illustrates the situation.

```
long f1()                        .segment CODE2 (REL)
{                            f1::
  return 0x12345678;         ; :: ch15_t2.c #3: return 0x12345678;
}                                    ldw    %X, #22136
                                     ldw    2, %X
                                     ldw    %X, #4660
short f2()                           ldw    0, %X
{                                    ret
  return f1();
}                                    .segment CODE2 (REL)
                             f2::
                             ; :: ch15_t2.c #8: return f1();
                                     call   f1
                                     ldw    %X, 1
                                     ldw    0, %X
                                     mov    2, 3
                                     ret
```

```cpp
70          if ( isDirect(ip0) && isDirect(ip1) && size0 == 1 )
71          {
72              ASM_CODE2(_MOV, acceItem(ip0, 0, 1), acceItem(ip1, 0, 1));
73              return;
74          }
75
76          bool sign1 = ip1->acceSign();
77          bool flag0 = (size0 <= 2 && size0 <= size1);
78          bool flag1 = (size1 <= 2 && size0 >= size1);
79          char *reg0 = flag0? setPtr2(ip0, 0)           : setPtr(ip0, 0);
80          char *reg1 = flag1? setPtr2(ip1, reg0? 1: 0): setPtr(ip1, reg0? 1: 0);
81
82          if ( farAcce(ip1) )
83          {
84              for (int i = 0; i < size0; i++)
85              {
86                  if ( i >= size1 )
87                  {
88                      if ( sign1 )
89                      {
90                          if ( size1 == i ) call((char*)"_extendRegA");
91                          ASM_CODE2(_LD, acceItem(ip0, i, 1, reg0), REG_A);
92                      }
93                      else
94                          ASM_CODE1(_CLR, acceItem(ip0, i, 1, reg0));
95                  }
96                  else
97                  {
98                      ASM_CODE2(_LDF, REG_A, acceItem(ip1, i, 1, reg1));
99                      ASM_CODE2(_LD , acceItem(ip0, i, 1, reg0), REG_A);
100                 }
101             }
102             return;
103         }
```

`/stm8scc/cc_source/stm8/stm8s2.cpp`   Code-19.10c

- #70~74: if both *X* and *Y* are direct addressable and *X* is a 1-byte data, then use the MOV instruction for that.
- #82~103: when *Y* is 'far' data (in memory space $\geq$ 0x10000), it needs to use the LDF instruction to fetch.

Otherwise, the assignment is performed from the lower byte(s) to the higher byte (s).

```cpp
105         for (int i = 0; i < size0; i++)
106         {
107             if ( i >= size1 )
108             {
109                 if ( size1 == i && sign1 )
110                     call((char*)"_extendSign");
111
112                 if ( sign1 )
113                     ASM_CODE2(_LD, acceItem(ip0, i, 1, reg0), REG_A);
114                 else
115                     ASM_CODE1(_CLR, acceItem(ip0, i, 1, reg0));
116             }
117             else if ( (size0 - i) > 1 && (size1 - i) > 1 && !(reg0 && reg1) )
118             {
119                 char *r16 = (reg1 || reg0)? REG_Y: REG_X;
120                 if ( reg1 ) {
121                     if ( (i+2) >= size1 || (i+2) >= size0 )
122                         r16 = reg1;
123                     else
124                         ASM_CODE2(_LDW, r16, reg1);
125
126                     ASM_CODE2(_LDW, r16, acceItem(ip1, i, 2, r16));
127                 }
128                 else
129                     ASM_CODE2(_LDW, r16, acceItem(ip1, i, 2, reg1));
130
131                 ASM_CODE2(_LDW, acceItem(ip0, i, 2, reg0), r16);    i++;
132             }
133             else if ( (size0 - i) == 2 && (size1 - i) > 1 && (reg0 && reg1) && ip1->bias == 0 )
134             {
135                 ASM_CODE2(_LDW, reg1, acceItem(ip1, i, 2, reg1));
136                 ASM_CODE2(_LDW, acceItem(ip0, i, 2, reg0), reg1);    i++;
137             }
138             else
139             {
140                 ASM_CODE2(_LD, REG_A, acceItem(ip1, i, 1, reg1));
141                 ASM_CODE2(_LD, acceItem(ip0, i, 1, reg0), REG_A);
142             }
143         }
```

`/stm8scc/cc_source/stm8/stm8s2.cpp`   Code-19.10d

Note:
- The 'big-endian' structure and CISC instruction system make the design/implementation more complex.
- Use 16-bit operation instructions through the index registers X and Y to improve the quality of the code generated.
- The design can be easier if it does not support 24-bit integer data.

The following, Table-19.7, shows a test example that reveals the differences between 'standard' and 'extended' modes when compiling the same code.

Test code (ch15_t3.c)	Standard mode	Extended mode

<pre>typedef struct {
    int x;
    char y;
} Data;
 	
const Data data = {1000, 100};
const Data *const dptr = &data;
 	
void foo()
{
    int n;
    char m;
  n = dptr->x;
  m = dptr->y;
}</pre>
	Command: `st8cc ch15_t3.c`
	<pre>; foo_$1_n: cffset = 1
	; foo_$2_m: cffset = 3
	    .segment CODE2 (REL)
	foo::
	    sub %SF, #3
	; :: ch15_t3.c #13: n = dptr->x;
	    ldw %X, [dptr]
	    ldw (1, %SP), %X
	; :: ch15_t3.c #14: m = dptr->y;
	    ldw %X, dptr
	    ld %A, (2, %X)
	    ld (3, %SP), %A
	    add %SF, #3
	    ret
	
	    .segment CONSTi (REL)
	data::
	    .dw 10C0
	    .db 10C
	
	    .segment CONSTi (REL)
	dptr::
	    .dw data</pre>
		Command: `st8cc -E ch15_t3.c`
		<pre>; foo_$1_n: offset = 1
		; foo_$2_m: offset = 3
		    .segment CODE2 (REL)
		foo::
		    sub %SP, #3
		; :: ch15_t3.c #13: n = dptr->x;
		    ldf %A, dptr
		    ld 4, %A
		    ldf %A, dptr+1
		    ld 5, %A
		    ldf %A, dptr+2
		    ld 6, %A
		    clrw %X
		    incw %X
		    ldf %A, ([4], %X)
		    ld (2, %SP), %A
		    decw %X
		    ldf %A, ([4], %X)
		    ld (1, %SP), %A
		; :: ch15_t3.c #14: m = dptr->y;
		    ldf %A, dptr
		    ld 4, %A
		    ldf %A, dptr+1
		    ld 5, %A
		    ldf %A, dptr+2
		    ld 6, %A
		    clrw %X
		    incw %X
		    incw %X
		    ldf %A, ([4], %X)
		    ld (3, %SP), %A
		    add %SP, #3
		    retf
		
		    .segment CONSTi (REL)
		data::
		    .dw 1000
		    .db 100
		
		    .segment CONSTi (REL)
		dptr::
		    .dt data</pre>

Table-19.7

The file /stm8/stm8s0.cpp provides all the service routines/functions that help generate assembly code. The following lists two routines that are used most frequently:

`char *STM8S::setPtr(Item *ip, int index)` –
It sets up an index register when an indirect memory access occurs, where the parameter `index` determines which register to use (`index` = 0: use register X; `index` != 0: use register Y).

`char *STM8S::itemStr(Item *ip, int offset, int acce_size, char *reg)` –
Generating the string of an operand in the format of assembly code.

19.2.5 Assembly code generation – addition (P-Code '+')

The assembly code generation for the addition P-code has the form:

$$\{`+`, [Z, X, Y]\}$$

And it's implemented in /stm8/stm8s6.cpp, as illustrated in Code-19.11a~Code-19.11b.

STM8 MCUs include some 16-bit instructions for addition operations, which will speed up execution when used. In some situations (e.g., when either X or Y is stored in ROM and 'extended' mode is applied, or when the sizes of X and Y are matched, …), the operation must be modified or split into two steps. Or, the addition must be done byte by byte.

```
22    void STM8S :: add(Item *ip0, Item *ip1, Item *ip2)
23    {
24        int size0 = ip0->acceSize();
25        int size1 = ip1->acceSize();
26        int size2 = ip2->acceSize();
27
28        if ( (farAcce(ip2) && !farAcce(ip1)) || (LOCAL_ADDR(ip2) && !LOCAL_ADDR(ip1)) ) {
29            add(ip0, ip2, ip1);
30            return;
31        }
32        if ( (size2 < size0 && ip2->acceSign()) || farAcce(ip2) || LOCAL_ADDR(ip2) ) {
33            mov(ip0, related(ip0, ip1)? ip1: ip2);
34            add(ip0, related(ip0, ip1)? ip2: ip1);
35            return;
36        }
37        if ( LOCAL_ADDR(ip1) || (usePtr(ip0, ip1, ip2) && size0 > 1) ) {
38            mov(ip0, related(ip0, ip2)? ip2: ip1);
39            add(ip0, related(ip0, ip2)? ip1: ip2);
40            return;
41        }
42        if ( farAcce(ip1) || usePtr(ip1) || usePtr(ip2) ) {
43            if ( (size0 <= size1 || !ip1->acceSign()) && (size0 <= size2 || !ip2->acceSign()) )
44                add8(ip0, ip1, ip2);
45            else {
46                mov(ip0, related(ip0, ip1)? ip1: ip2);
47                add(ip0, related(ip0, ip1)? ip2: ip1);
48            }
49            return;
50        }
```
/stm8scc/cc_source/stm8/stm8s6.cpp   Code-19.11a

- #28~31: swap the operands X and Y:
  $$\{`+`, [Z, X, Y]\} \rightarrow \{`+`, [Z, Y, X]\}.$$
- #32~41: implement the operation in two steps:
  $$\{`+`, [Z, X, Y]\} \rightarrow \{`=`, [Z, X]\} \ \& \ \{`+=`, [Z, Y]\}$$
  or
  $$\{`+`, [Z, X, Y]\} \rightarrow \{`=`, [Z, Y]\} \ \& \ \{`+=`, [Z, X]\}.$$
- #42~50: if X is located in ROM or both X and Y are accessed indirectly, then the operation wouldn't use any 16-bit addition instruction (#44). Or, the same solution applies as above.

```cpp
53          char *r16  = reg0? REG_Y: REG_X;
54          bool const1= ip1->type == CON_ITEM || (ip1->type == IMMD_ITEM && !LOCAL_ADDR(ip1));
55          bool const2= ip2->type == CON_ITEM || (ip2->type == IMMD_ITEM && !LOCAL_ADDR(ip2));
56          for (int i = 0; i < size0; i++) {
57              if ( (i+1) < size0 && ((i+1) < size1 || const1) && ((i+1) < size2 || const2) )
58              {
59                  ASM_CODE2(_LDW, r16, acceItem(ip1, i, 2));
60                  if ( i > 0 ) {
61                      ASM_CODE1(_JRNC, (strcmp(r16, REG_Y) == 0)? ".+2": ".+1");
62                      ASM_CODE1(_INCW, r16);
63                  }
64                  ASM_CODE2(_ADDW, r16, acceItem(ip2, i, 2));
65                  ASM_CODE2(_LDW, acceItem(ip0, i++, 2, reg0), r16);
66              }
67              else
68              {
69                  char *inst = i? _ADC: _ADD;
70                  if ( i < size1 )
71                      ASM_CODE2(_LD,  REG_A, acceItem(ip1, i, 1));
72                  else
73                      ASM_CODE1(_CLR,  REG_A);
74
75                  if ( i < size2 )
76                      ASM_CODE2(inst, REG_A, acceItem(ip2, i, 1));
77                  else
78                      ASM_CODE2(inst, REG_A, 0, '#');
79                  ASM_CODE2(_LD, acceItem(ip0, i, 1, reg0), REG_A);
80              }
81          }
82      }
```

```
/stm8scc/cc_source/stm8/stm8s6.cpp
```
Code-19.11b

- #57~66: use 16-bit addition instruction(s).
- #68~80: use 8-bit addition instruction(s).

The following, Table-19.8, is a test example for this section.

C source code (ch15_t4.c)	Assembly code output

<pre>long n, m, r, *p;
 	
void foo()
{
  r = n + m;
  r = n + *p;
}</pre>
	<pre> .segment BANK (REL)
	n:: .rs 4
	        .segment BANK (REL)
	m:: .rs 4
	        .segment BANK (REL)
	r:: .rs 4
	        .segment BANK (REL)
	p:: .rs 2
	
	        .segment CODE2 (REL)
	foo::
	; :: ch15_t4.c #5: r = n + m;
	        ldw %X, n+2
	        addw %X, m+2
	        ldw r+2, %X
	        ldw %X, n
	        jrnc .+1
	        incw %X
	        addw %X, m
	        ldw r, %X
	; :: ch15_t4.c #6: r = n + *p;
	        ldw %X, p
	        ld %A, n+3
	        add %A, (3, %X)
	        ld r+3, %A
	        ld %A, n+2
	        adc %A, (2, %X)
	        ld r+2, %A
	        ld %A, n+1
	        adc %A, (1, %X)
	        ld r+1, %A
	        ld %A, n
	        adc %A, (%X)
	        ld r, %A
	        ret</pre>

Table-19.8

19.3 STM8 Assembler - st8as.exe

<u>Project directory</u>: /stm8scc/as_source

The assembler's design is relatively straightforward. Generally speaking, it simply translates the mnemonics of the assembly code generated by st8cc into machine or binary code.

The following, Table-19.9, lists the instruction set of STM8 MCUs. The instruction length varies from 1 to 5 bytes because the STM8 is a CISC architecture.

add	call	decw	jra	jrnh	jrule	or	rlwa	sllw	wfi
addw	callf	div	jrc	jrnm	jrult	pop	rrc	sra	trap
adc	callr	divw	jreq	jrnv	jrv	popw	rrcw	sraw	wfe
and	ccf	exg	jrf	jrpl	ld	push	rrwa	srl	wfi
bccm	clr	exgw	jrh	jrsge	ldf	pushw	rvf	srlw	xor
bcp	clrw	halt	jrih	jrsgt	ldw	rcf	sbc	sub	
bcpl	cp	inc	jril	jrsle	mov	ret	scf	subw	
bres	cpw	incw	jrm	jrslt	mul	retf	sim	swap	
bset	cpl	iret	jrmi	jrt	neg	rim	sla	swapw	
btjf	cplw	jp	jrnc	jrugt	negw	rlc	slaw	tnz	
btjt	dec	jpf	jrne	jruge	nop	rlcw	sll	tnzw	

Table-19.9

The output (*file*.obj files) uses the same format as in designing the assembler for the PIC16F MCU, with a few changes or expansions. Table-19.10 lists the output record types.

Record type	Usages
'B'	ROM content (items), see the following descriptions.
'E'	.equ instruction.
'G'	Global label
'I'	.invoke instruction.
'J'	.jp instruction.
'L'	Local label.
'M'	.dblank instruction.
'N'	.cblank instruction.
'P'	.device instruction.
'R'	.rs instruction (RAM).
'S'	.segment instruction.
'U'	.jbz instruction.
'V'	.jbnz instruction.

Table-19.10

Each type 'B' record can have multiple data value items. Each value item is 1 to 4 bytes based on its format:

value_exp	1 byte in length
[*value_exp*]	2 bytes in length
{*value_exp*}	3 bytes in length
<*value_exp*>	4 bytes in length

The assembler lexer, asm.l, contains an expansion that matches/identifies all register names, as illustrated in Code-19.12.

```
 54    %x OPCODE
 55    %x OPERAND
 56
 57    %%
 58
 59    ^{sp}*";".*                        {   append_str();
 60                                           if ( memcmp(yytext, "; :: ", 5) == 0 )
 61                                           {
 62                                               yylval.name = dupStr(yytext);
 63                                               return S_COMMENT;
 64                                           }
 65                                       }
 66    ^{sp}*{symbol}:(:)?                {   int i = 0;
 67                                           append_str();
 68                                           while ( yytext[i] == ' ' || yytext[i] == '\t' ) i++;
 69                                           BEGIN(OPCODE);
 70                                           yylval.name = dupStr(&yytext[i]);
 71                                           return LABEL;
 72                                       }
 73    ^{sp}+                             {   append_str();
 74                                           BEGIN(OPCODE); }
 75
 76    \n                                 {   append_str();
 77                                           return yytext[0];
 78                                       }

       .  .  .

108    <OPERAND>"%A"                      {   append_str(); yylval.value = REG_A;   return REG; }
109    <OPERAND>"%X"                      {   append_str(); yylval.value = REG_X;   return REG; }
110    <OPERAND>"%Y"                      {   append_str(); yylval.value = REG_Y;   return REG; }
111    <OPERAND>"%SP"                     {   append_str(); yylval.value = REG_SP;  return REG; }
112    <OPERAND>"%XH"                     {   append_str(); yylval.value = REG_XH;  return REG; }
113    <OPERAND>"%XL"                     {   append_str(); yylval.value = REG_XL;  return REG; }
114    <OPERAND>"%YH"                     {   append_str(); yylval.value = REG_YH;  return REG; }
115    <OPERAND>"%YL"                     {   append_str(); yylval.value = REG_YL;  return REG; }
116    <OPERAND>"%CC"                     {   append_str(); yylval.value = REG_CC;  return REG; }
```
/stm8scc/as_source/asm.l   Code-19.12

- #108~116: rules that identify register names in operand fields.

The assembler parser, asm.y, parses the input assembly lines and generates a parsing tree. Among the parsing rules, the following parts are 'new' for STM8 MCUs, which focus on operand syntax, as shown in Code-19.13a and Code-19.13b.

```
116    operands
117        :   operand                               {   $$ = $1; }
118        |   operands ',' operand                  {   $$ = appendItem($1, $3); }
119        ;
120
121    operand
122        :   exp                                   {   $$ = $1; }
123        |   '#' exp                               {   $$ = newItem('#'); $$->left = $2; }
124        |   '[' exp ']'                           {   $$ = newItem(TYPE_INDIR);
125                                                      $$->left = $2;
126                                                  }
127        |   '(' exp ',' REG ')'                   {   $$ = newItem(TYPE_INDEX);
128                                                      $$->left = $2;
129                                                      $$->right = regItem($4);
130                                                  }
131        |   '(' '[' exp ']' ',' REG ')'           {   $$ = newItem(TYPE_INDIR_INDEX);
132                                                      $$->left = $3;
133                                                      $$->right = regItem($6);
134                                                  }
135        |   '(' REG ')'                           {   $$ = newItem(TYPE_INDEX);
136                                                      $$->right = regItem($2);
137                                                  }
138        |   REG                                   {   $$ = regItem($1); }
139        ;
```
/stm8scc/as_source/asm.y   Code-19.13a

- #121~139: rules for parsing operand types.

```
227   exp
228       :    or_exp                              {   $$ = $1; }
229       |    '*' or_exp                          {   $$ = $2; $$->bank0 = 1; }
230       ;
/stm8scc/as_source/asm.y                                              Code-19.13b
```

- #229: the rule and action that identifies the operand accessing 'bank0'.

Note, (1) the other parts in both **asm.l** and **asm.y** are pretty much the same as those in designing the assembler for PIC16F MCUs; (2) parsing only applies to the grammar level.

19.3.2 Operation of assembling

The assembler's operation starts in main() of **main.cpp**, as shown in Code-19.18.

```
14   void asm0(char *filename);
15
16   ////////////////////////////////////////////////////////////////////////////
17   int main(int argc, char *argv[], char *env[])
18   {
19       printf("ST8AS Assembler, %s\n", VERSION);
20
21       for (int i = 1; i < argc; i++)
22       {
23           char *p = argv[i];
24           int  l = strlen(p);
25           if ( l > 4 && strcmp(&p[l-4], ".asm") == 0 )
26               asm0(p);   // parse source file
27       }
28
29       return error_cnt;
30   }
31
32   ////////////////////////////////////////////////////////////////////////////
33   /*
34       parse the input source file, and generate the raw file in
35       'line_t' format.
36   */
37   ////////////////////////////////////////////////////////////////////////////
38   void asm0(char *filename)
39   {
40       cur_file = filename;
41       printf("assembling '%s' ...\n", filename);
42
43       // using Lex and Bison to parse the source ...
44       error_cnt  = 0;
45       line_t *lp = _main(filename);
46
47       if ( error_cnt > 0 )
48           exit(1);
49
50       parse(lp);
51   }
/stm8scc/as_source/main.cpp                                           Code-19.14
```

- #23~26: get input file, start assembling.
- #45: run the lexer/parser. Results are buffered and pointed by `lp`.
- #50: syntax parsing and generating the outputs.

Function `parse()` is located in **parse.cpp**. It reorganizes the input lines into segments, parses them, and generates the output (**.obj** and **.lst**) files, as shown in Code-19.19. In there, multi-scan will occur on the input lines.

```
21    int parse(line_t *lptr)
22    {
23        int st8_dev = 0;
24
25        del_symbol_list();      // clean up the symbol list
26        clearupSegments();      // clean up the segment list

          .  .  .

102       switch ( st8_dev )
103       {
104           case 0:
105           case STM8:
106               // PASS1 - log macros (EQU)
107               for (curSeg = segList; curSeg != NULL; curSeg = curSeg->next)
108                   ST8_asm(curSeg, ASM_PASS1);
109
110               if ( error_cnt > 0 )
111                   return 1;
112
113               // PASS2 - log all labels
114               for (curSeg = segList; curSeg != NULL; curSeg = curSeg->next)
115                   ST8_asm(curSeg, ASM_PASS2);
116
117               if ( error_cnt > 0 )
118                   return 1;
119
120               // PASS3 - generate outputs
121               for (curSeg = segList; curSeg != NULL; curSeg = curSeg->next)
122                   ST8_asm(curSeg, ASM_PASS3);
123               break;
124
125           default:
126               yyerror("unknown device descriptor!");
127       }
128
129       return error_cnt;
130   }
```
`/stm8scc/as_source/parse.cpp`   Code-19.15

- #107~108: the first scan – log names of macros.
- #114~115: the second scan – log labels.
- #121~122: the third scan – generate outputs.

Function `ST8_asm()` is located in **st8asm.cpp**, and it fulfills the following things:
  (1) Log the macro symbols and labels.
  (2) Instruction syntax checking or validation.
  (3) Translate the input lines and generate the output files.

A translating table is created in **st8inst.h/st8inst.cpp**, seen in Code-19.16 and Code-19.7.

```c
 4    enum {
 5        ST8_REGA = 1,
 6        ST8_REGX,
 7        ST8_REGY,
 8        ST8_REGSP,
 9        ST8_REGXH,
10        ST8_REGXL,
11        ST8_REGYH,
12        ST8_REGYL,
13        ST8_REGCC,
14        ST8_EXP,           // mem
15        ST8_EXP0,          // *mem
16        ST8_INDIR,         // [mem]
17        ST8_INDIR0,        // [*mem]
18        ST8_INDEX_X,       // (X)
19        ST8_INDEX_Y,       // (Y)
20        ST8_INDIR_X,       // ([mem], X)
21        ST8_INDIR_Y,       // ([mem], Y)
22        ST8_INDIR0_X,      // ([*mem], X)
23        ST8_INDIR0_Y,      // ([*mem], Y)
24        ST8_OFFSET_X,      // (n, X)
25        ST8_OFFSET_Y,      // (n, Y)
26        ST8_OFFSET0_X,     // (n, X)
27        ST8_OFFSET0_Y,     // (n, Y)
28        ST8_OFFSET0_SP,    // (n, SP)
29        ST8_IMMD,          // #exp
30    };

32    typedef struct {
33        unsigned int symbl;        // instruction token
34        unsigned int code_size;    // bytes of opcode
35        unsigned int code;         // instruction code
36        unsigned int inst_size;    // whole instruction size (bytes)
37        unsigned int operands;     // operand types
38    } ST8_inst_t;
39
40    int  operCnt(const ST8_inst_t *);
41    int  pickOpr(const ST8_inst_t *, int i);
42    bool operandIsReg(const ST8_inst_t *p, int i);
43
44    extern const ST8_inst_t ST8_instCodeTbl[];
45
```

`/stm8scc/as_source/st8inst.h`   Code-19.16

- #4~30: operand types.
- #32~38: data structure/format in translating table.

```c
11    ///////////////////////////////////////////////////////////////////////
12    const ST8_inst_t ST8_instCodeTbl[] = {
13        {ADC,   1,  0xa9,     2,  (ST8_REGA << 8)|ST8_IMMD},
14        {ADC,   1,  0xb9,     2,  (ST8_REGA << 8)|ST8_EXP0},
15        {ADC,   1,  0xc9,     3,  (ST8_REGA << 8)|ST8_EXP},
16        {ADC,   1,  0xf9,     1,  (ST8_REGA << 8)|ST8_INDEX_X},
17        {ADC,   1,  0xe9,     2,  (ST8_REGA << 8)|ST8_OFFSET0_X},
18        {ADC,   1,  0xd9,     3,  (ST8_REGA << 8)|ST8_OFFSET_X},
19        {ADC,   2,  0x90f9,   2,  (ST8_REGA << 8)|ST8_INDEX_Y},
20        {ADC,   2,  0x90e9,   3,  (ST8_REGA << 8)|ST8_OFFSET0_Y},
21        {ADC,   2,  0x90d9,   4,  (ST8_REGA << 8)|ST8_OFFSET_Y},
22        {ADC,   1,  0x19,     2,  (ST8_REGA << 8)|ST8_OFFSET0_SP},
23        {ADC,   2,  0x92c9,   3,  (ST8_REGA << 8)|ST8_INDIR0},
24        {ADC,   2,  0x72c9,   4,  (ST8_REGA << 8)|ST8_INDIR},
25        {ADC,   2,  0x92d9,   3,  (ST8_REGA << 8)|ST8_INDIR0_X},
26        {ADC,   2,  0x72d9,   4,  (ST8_REGA << 8)|ST8_INDIR_X},
27        {ADC,   2,  0x91d9,   3,  (ST8_REGA << 8)|ST8_INDIR0_Y},
28
29        {ADD,   1,  0xab,     2,  (ST8_REGA << 8)|ST8_IMMD},
30        {ADD,   1,  0xbb,     2,  (ST8_REGA << 8)|ST8_EXP0},
31        {ADD,   1,  0xcb,     3,  (ST8_REGA << 8)|ST8_EXP},
32        {ADD,   1,  0xfb,     1,  (ST8_REGA << 8)|ST8_INDEX_X},
33        {ADD,   1,  0xeb,     2,  (ST8_REGA << 8)|ST8_OFFSET0_X},
34        {ADD,   1,  0xdb,     3,  (ST8_REGA << 8)|ST8_OFFSET_X},
35        {ADD,   2,  0x90fb,   2,  (ST8_REGA << 8)|ST8_INDEX_Y},
36        {ADD,   2,  0x90eb,   3,  (ST8_REGA << 8)|ST8_OFFSET0_Y},
37        {ADD,   2,  0x90db,   4,  (ST8_REGA << 8)|ST8_OFFSET_Y},
38        {ADD,   1,  0x1b,     2,  (ST8_REGA << 8)|ST8_OFFSET0_SP},
39        {ADD,   2,  0x92cb,   3,  (ST8_REGA << 8)|ST8_INDIR0},
40        {ADD,   2,  0x72cb,   4,  (ST8_REGA << 8)|ST8_INDIR},
41        {ADD,   2,  0x92db,   3,  (ST8_REGA << 8)|ST8_INDIR0_X},
42        {ADD,   2,  0x72db,   4,  (ST8_REGA << 8)|ST8_INDIR_X},
43        {ADD,   2,  0x91db,   3,  (ST8_REGA << 8)|ST8_INDIR0_Y},
44        {ADD,   1,  0x5b,     2,  (ST8_REGSP<< 8)|ST8_IMMD},
```

```
623        {XOR,   1, 0xa8,      2, (ST8_REGA << 8)|ST8_IMMD},
624        {XOR,   1, 0xb8,      2, (ST8_REGA << 8)|ST8_EXP0},
625        {XOR,   1, 0xc8,      3, (ST8_REGA << 8)|ST8_EXP},
626        {XOR,   1, 0xf8,      1, (ST8_REGA << 8)|ST8_INDEX_X},
627        {XOR,   1, 0xe8,      2, (ST8_REGA << 8)|ST8_OFFSET0_X},
628        {XOR,   1, 0xd8,      3, (ST8_REGA << 8)|ST8_OFFSET_X},
629        {XOR,   2, 0x90f8,    2, (ST8_REGA << 8)|ST8_INDEX_Y},
630        {XOR,   2, 0x90e8,    3, (ST8_REGA << 8)|ST8_OFFSET0_Y},
631        {XOR,   2, 0x90d8,    4, (ST8_REGA << 8)|ST8_OFFSET_Y},
632        {XOR,   1, 0x18,      2, (ST8_REGA << 8)|ST8_OFFSET0_SP},
633        {XOR,   2, 0x92c8,    3, (ST8_REGA << 8)|ST8_INDIR0},
634        {XOR,   2, 0x72c8,    4, (ST8_REGA << 8)|ST8_INDIR},
635        {XOR,   2, 0x92d8,    3, (ST8_REGA << 8)|ST8_INDIR0_X},
636        {XOR,   2, 0x72d8,    4, (ST8_REGA << 8)|ST8_INDIR_X},
637        {XOR,   2, 0x91d8,    3, (ST8_REGA << 8)|ST8_INDIR0_Y},
638        {VECT,  1, 0x82,      4, ST8_EXP},
639
640        {0, 0, 0}
641   };
```

`/stm8scc/as_source/st8inst.cpp` Code-19.17

The validation and translation are done based on the table above. Code-19.18 below illustrates the operation.

```
39    int ST8_asm(Cseg *sp, int pass)
40    {
41        line_t *lp = sp->lines;
42        int v, cur_addr = 0;
43
44        if ( pass == ASM_PASS3 )
45        {
46            if ( createOutputFiles(sp) )
47                return 1;
48        }
49
50        while ( lp != NULL )
51        {
52            int addr_inc = 0;
53            const ST8_inst_t *st8inst = NULL;
54
55            if ( pass == ASM_PASS2 && lp->label != NULL && lp->inst != EQU )
56            {
57                symbol_t *p = add_symbol(sp, lp, lp->label, valItem(cur_addr), LABEL);
58
59                if ( p != NULL )
60                {
61                    if ( sp->isABS() )
62                        p->isABS = 1;
63
64                    p->global = lp->global_lbl? 1: 0;
65                }
66            }
67
68            switch ( lp->inst )
69            {
    .  .  .
133                case VECT:  // vector ...
134                case ADD:    case ADDW:  case ADC:   case AND:    case BCCM:   case BCP:    case BCPL:
135                case BRES:   case BSET:  case BTJF:  case BTJT:   case CALL:   case CALLF:  case CALLR:
136                case CCF:    case CLR:   case CLRW:  case CP:     case CPW:    case CPL:    case CPLW:
137                case DEC:    case DECW:  case DIV:   case DIVW:   case EXG:    case EXGW:   case HALT:
138                case INC:    case INCW:  case IRET:  case JP:     case JPF:    case JRA:    case JRC:
139                case JREQ:   case JRF:   case JRH:   case JRIH:   case JRIL:   case JRM:    case JRMI:
140                case JRNC:   case JRNE:  case JRNH:  case JRNM:   case JRNV:   case JRPL:   case JRSGE:
141                case JRSGT:  case JRSLE: case JRSLT: case JRT:    case JRUGE:  case JRUGT:  case JRULE:
142                case JRULT:  case JRV:   case LD:    case LDF:    case LDW:    case MOV:    case MUL:
143                case NEG:    case NEGW:  case NOP:   case OR:     case POP:    case POPW:   case PUSH:
144                case PUSHW:  case RCF:   case RET:   case RETF:   case RIM:    case RLC:    case RLCW:
145                case RLWA:   case RRC:   case RRCW:  case RRWA:   case RVF:    case SBC:    case SCF:
146                case SIM:    case SLL:   case SLLW:  case SRA:    case SRAW:   case SRL:    case SRLW:
147                case SUB:    case SUBW:  case SWAP:  case SWAPW:  case TNZ:    case TNZW:   case TRAP:
148                case WFE:    case WFI:   case XOR:
149                    if ( pass != ASM_PASS1 )
150                    {
151                        st8inst = ST8asmParse(sp, lp);
152
153                        if ( st8inst == NULL )
154                            my_yyerror(lp, "operand(s) error!");
155                        else
156                            addr_inc = st8inst->inst_size;
157                    }
158                    break;
    .  .  .
```

```
179         if ( pass == ASM_PASS3 )
180             ST8asmOut(sp, lp, st8inst, cur_addr);
181
182         if ( pass == ASM_PASS3 && lp->next == NULL && sp->next == NULL )     // end of file
183         {
184             obj_file.flush();
185             lst_file.flush();
186
187             fclose(f_obj);
188             fclose(f_lst);
189             return 0;
190         }
191
192         cur_addr += addr_inc;
193         lp = lp->next;
194     }
```
/stm8scc/as_source/st8asm.cpp

Code-19.18

- #133~158: find the item from the translation table by calling `ST8asmParse()`.
- #179~180: during the third pass, generate the outputs by calling `ST8asmOut()`.
- #182~190: reach to the end of the input, flush and close the output files.

The following table, Table 19.9, shows the test results.

Assembly code (ch15_t3.asm)	Object output (ch15_t3.obj)

<pre>; foo_$1_n: offset = 1
; foo_$2_m: offset = 3
        .segment CODE2 (REL)
foo::
        sub   %SP, #3
; :: ch15_t3.c #13: n = dptr->x;
        ldw   %X, [dptr]
        ldw   (1, %SP), %X
; :: ch15_t3.c #14: m = dptr->y;
        ldw   %X, dptr
        ld   %A, (2, %X)
        ld   (3, %SP), %A
        add   %SP, #3
        ret
 	
        .segment CONSTi (REL)
data::
        .dw   1000
        .db   100
 	
        .segment CONSTi (REL)
dptr::
        .dw   data</pre>
	<pre>P stm8s
	S CODE2 REL
	G foo
	B 0x52 0x03
	; ; :: ch15_t3.c #13: n = dptr->x;
	B 0x72 0xCE [dptr] 0x1F 0x01
	; ; :: ch15_t3.c #14: m = dptr->y;
	B 0xCE [dptr] 0xE6 0x02 0x6B 0x03 0x5B 0x03 0x81
	S CONSTi REL
	G data
	B 0x03 0xE8 0x64
	S CONSTi REL
	G dptr
	B [data]</pre>

List output (ch15_t3.lst)

<pre>   00009 ; foo_$1_n: offset = 1
              00010 ; foo_$2_m: offset = 3
              00011   .segment CODE2 (REL)
00000:   00012 foo::
00000: 52 03   00013   sub   %SP, #3
              00014 ; :: ch15_t3.c #13: n = dptr->x;
00002: 72 CE ?? ??   00015   ldw   %X, [dptr]
00006: 1F 01   00016   ldw   (1, %SP), %X
              00017 ; :: ch15_t3.c #14: m = dptr->y;
00008: CE ?? ??   00018   ldw   %X, dptr
0000B: E6 02   00019   ld   %A, (2, %X)
0000D: 6B 03   00020   ld   (3, %SP), %A
0000F: 5B 03   00021   add   %SP, #3
00011: 81   00022   ret
              00023
              00024   .segment CONSTi (REL)
00000:   00025 data::
00000: 03E8   00026   .dw   1000
00002: 64   00027   .db   100
              00028
              00029   .segment CONSTi (REL)
00000:   00030 dptr::
00000: ????   00031   .dw   data</pre>

Table-19.9

19.4 STM8 Linker - st8lk.exe

`Project directory: /stm8scc/lk_source`

The design of st8lk.exe is relatively straightforward compared to lk16e.exe, since all functions' local variables are stored on the stack and no memory sharing is needed.

19.4.1 Lexer and parser for the linker

The following, as shown in Code-19.19, is the lexer (rule part) for the linker.

```
38    %%
39
40    ^"."{sp}+.*                      {    char *p = yytext;
41                                          appendStr();
42                                          for(p++; *p == ' '; p++);
43                                          p[strlen(p)-1] = 0;
44                                          yylval.syml = dupStr(p+1);
45                                          return COMMENT;
46                                     }
47    ^.{sp}+                          {    appendStr();
48                                          if ( isAnOpcode(yytext[0]) )
49                                          {
50                                              BEGIN(OPERAND);
51                                              yylval.value = yytext[0];
52                                              return TYPE;
53                                          }
54                                          BEGIN(ESCAPE);
55                                     }
56
57    <ESCAPE>.*                       {    /*ignore the line */ }
58    <ESCAPE>\n                       {    appendStr();
59                                          BEGIN(INITIAL);
60                                     }
61
62    <ITEMS>{symbol}                  {    appendStr();
63                                          yylval.syml = dupStr(yytext);
64                                          return SYMBOL;
65                                     }
66    <ITEMS>{string}                  {    int len = strlen(yytext);
67                                          appendStr();
68                                          yytext[len-1] = 0;
69                                          yylval.syml = dupStr(yytext+1);
70                                          return STRING;
71                                     }
72    <ITEMS>{hex} |
73    <ITEMS>{dec}                     {    appendStr();
74                                          yylval.value = convertNum((char *)yytext);
75                                          return NUMBER;
76                                     }
77    <ITEMS>">>"                      {    appendStr(); return RSHIFT; }
78    <ITEMS>"<<"                      {    appendStr(); return LSHIFT; }
79
80    <ITEMS>\n                        {    appendStr();
81                                          BEGIN(INITIAL);
82                                          return yytext[0];
83                                     }
84    <ITEMS>[-+%^:&|()*/~]            |
85    <ITEMS>"{"                       |
86    <ITEMS>"}"                       |
87    <ITEMS>"["                       |
88    <ITEMS>"]"                       |
89    <ITEMS>"<"                       |
90    <ITEMS>">"                       {    appendStr();    return yytext[0]; }
91    <ITEMS>{sp}+                     {    appendStr(); }
92    <ITEMS>"."                       {    appendStr();    return yytext[0]; }
93
94    {sp}+                            {    appendStr(); }
95    \n                               {    appendStr();
96                                          BEGIN INITIAL;
97                                          return yytext[0];
98                                     }
99    .                                {    appendStr();
100                                         yyerror("unknown character!");
101                                    }
```

/stm8scc/lk_source/lnk.l   Code-19.19

- #47~55: get the record type letter from an .obj file.
- #85~90: rules that identify the special letters.

Code-19.20 is the parser (rule part) for the linker.

```
29   %%
30   prog
31       : lines                              {   linePtr = $1; }
32       ;
33
34   lines
35       : line                               {   $$ = $1; }
36       | lines line                         {   $$ = $1;
37                                                appendLine(&$1, $2);
38                                            }
39       ;
40
41   line
42       : '\n'                               {   $$ = newLine(0, NULL);
43                                                $$->src = dupStr(__yyline);
44                                                $$->lineno = yylineno;
45                                            }
46       | obj_line '\n'                      {   $$ = $1;
47                                                $$->src = dupStr(__yyline);
48                                                $$->lineno = yylineno;
49                                            }
50       | COMMENT '\n'                       {   $$ = newLine(';', strItem($1));
51                                                free($1);
52                                                $$->src = dupStr(__yyline);
53                                                $$->lineno = yylineno;
54                                            }
55       ;
56
57   obj_line
58       : TYPE items                         {   $$ = newLine($1, $2); }
59       | TYPE                               {   $$ = newLine($1, NULL); }
60       ;
61
62   items
63       : item                               {   $$ = $1; }
64       | items item                         {   $$ = appendItem($1, $2); }
65       ;
66
67   item
68       : inclusive_or_expr                  {   $$ = $1;
69                                                $$->size = 1;
70                                            }
71       | '[' inclusive_or_expr ']'          {   $$ = $2;
72                                                $$->size = 2;    // 2-byte value
73                                            }
74       | '{' inclusive_or_expr '}'          {   $$ = $2;
75                                                $$->size = 3;    // 3-byte value
76                                            }
77       | '<' inclusive_or_expr '>'          {   $$ = $2;
78                                                $$->size = 4;    // 4-byte value
79                                            }
80       ;
```
/stm8scc/lk_source/lnk.y   Code-19.20

- #67~80: data items in a record line. Note that each item has a size indication (1~4 bytes).

<u>19.4.2 Operation of linking</u>

The following (Code-19.21a~Code-19.21c) illustrates the overall operation of the linking process.

```
36    int main(int argc, char *argv[], char *env[])
37    {
38        bool started = false;
39        printf("STM8S linker, %s\n", VERSION);
40
41        for (int i = 1; i < argc; i++)
42        {
43            if ( argv[i][0] == '-' )
44            {
45                if ( strcmp(argv[i], "-E") == 0 )
46                {
47                    extendedMode = true;
48                }
```
/stm8scc/lk_source/main.cpp   Code-19.21a

- ##41: process all the .obj files, one by on.
- #45~48: set the working mode ('standard' vs. 'extended').

```
62            }
63            else
64            {
65                if ( !started )
66                {
67                    dataMem = new Memory(DATA_MEMORY);
68                    codeMem = new Memory(CODE_MEMORY);
69                    libList = new Symbol(extendedMode? (char*)"crt0x": (char*)"crt0");
70                    started = true;
71                }
72
73                link0(argv[i], false);   // parse source file
74            }
75        }
```
/stm8scc/lk_source/main.cpp   Code-19.21b

- #67~68: create the memory management classes fcr both RAM and ROM.
- #69: log in the default library file that supports the generic math operations.
- #73: start processing the input file.

```cpp
 77        if ( started )
 78        {
 79            Path path(env);
 80            libPath = path.get();
 81
 82            std::string tmp_folder = TEMP_FOLDER;
 83            system(("mkdir " + tmp_folder).c_str());        // create a temp folder
 84            for (Symbol *sp = libList; sp; sp = sp->next)
 85            {
 86                char *ssp = getLibFile(sp->name);
 87                if ( ssp ) link0(ssp, true);
 88            }
 89            system(("rm -r " + tmp_folder).c_str());        // remove temp folder & files
 90
 91            ST8link linker(dataMem, codeMem);
 92            linker.scanInclusion();                         // code including process
 93            linker.assignSegmentsAddress();                 // assign addresses for segments
 94
 95            if ( linker.errorCount == 0 )
 96            {
 97                char *output = new char[strlen(outputFile)+10];
 98                sprintf(output, "%s.hex", outputFile);
 99                linker.outputHex(output);
100                sprintf(output, "%s.map", outputFile);
101                linker.outputMap(output, dataMem);
102                delete [] output;
103            }
104
105            deleteSegments(libSegGroup);
106            deleteSegments(codeSegGroup);
107            deleteSegments(dataSegGroup);
108            deleteSegments(miscSegGroup);
109
110            delete dataMem;
111            delete codeMem;
112        }
```
/stm8scc/lk_source/main.cpp

Code-19.21c

- #82~89: process the library files.
- #91: instantiate the class ST8link that fulfils the linking process.
- #92: remove the codes that are not used.
- #93: allocate the segments in memory (assign addresses for all the code in RAM and ROM).
- #95~103: generate output (.hex & .map) files.

In function link0(), as seen in Code-19.22, the input .obj file(s) are read and go through the lexer and parser. And finally, all the .obj record lines are grouped into segments.

There are some pseudo-instructions that are not translated into machine code and are left over from the compiling procedure. They need to be translated into actual STM8 instructions or machine code. The following table, Table-19.10, shows the translations.

Object (.obj) line	Short form translation	Long form translation	
		'standard' mode	'extended' mode
J lbl	jra lbl	jp lbl	jpf lbl
J C lbl	jrxx lbl	jrxx' .+3 jp lbl	jrxx' .+4 jpf lbl
U addr B lbl	btjf addr, #b, lbl	btjt addr, #b, .+3 jp lbl	btjt addr, #b, .+4 jpf lbl
V addr B lbl	btjt addr, #b, lbl	btjf addr, #b, .+3 jp lbl	btjf addr, #b, .+4 jpf lbl

Table-19.10

The object lines, listed in the table above, involve a jump operation. The 'short' form translation will be used if the jump distance is less than 127 bytes, which is likely to happen most of the time; the 'long' form translation will be adopted otherwise.

In function `link0()`, all these instructions are initialized to use the short form for translation after being parsed in, as illustrated in Code-19.22 below.

```
121   void link0(char *filename, bool lib_f)
122   {
123       if ( searchSymbol(fileList, filename) )
124           return;
125
126       addSymbol(&fileList, new Symbol(filename));
127       curFile = filename;
128
129       printf("linking '%s' ...\n", filename);
130
131       // using Lex and Bison to parse the source ...
132       errorCnt = 0;
133       line_t *lp = _main(filename);
134

139       while ( lp )
140       {
141           bool save_line = false;
142           bool add_seg   = false;
143           item_t *ip0, *ip1;
144
145           switch ( lp->type ) // input line type?
146           {

183               case 'U':   // variable bit test instruction('0') - ".jbz"  (btjf)
184               case 'V':   // variable bit test instruction('1') - ".jbnz" (btjt)
185                   lp->insert = 5; // init. size = 5 bytes
186                   save_line = true;
187                   break;
188
189               case 'J':   // variable branch instruction - ".jp" (jra or jrxx)
190                   lp->insert = 2; // init. size = 2 bytes
191                   save_line = true;
192                   break;
```

/stm8scc/lk_source/main.cpp Code-19.22

- #133: parse the input file.

- #183~192: initialize the size or translation form for the instructions.

The actual determination of the form translation for these instructions occurs during memory allocation/assigning addresses for the code.

```
170  static int updateVariableInstruction(Segment *segp, Symbol *symblist)
171  {
172      for (Segment *sp = segp; sp; sp = sp->next)
173      {
174          int addr = sp->addr;    // start address for the segment
175          for (line_t *lp = sp->lines; lp; lp = lp->next)
176          {
177              int w, rel_addr;
178              item_t *ip;
179              switch ( lp->type )
180              {
181                  case 'J':    // variable jump instruction
182                  case 'U':    // variable branch instruction
183                  case 'V':    // variable branch instruction
184                      ip = itemPtr(lp, itemCount(lp) - 1);
185                      expValue(ip, sp, symblist, &w, addr + lp->insert);
186                      rel_addr = w - (addr + lp->insert); // relative range
187                      if ( (rel_addr < -128) || (rel_addr > 127) )    // -128 ~ 127 ?
188                      {
189                          int new_insert = (lp->type == 'J')? 2: 5;    // base size

191                          if ( itemCount(lp) == 1 )                    // simple jump
192                              new_insert += (w > 0xffff)? 2: 1;
193                          else                                         // conditional branch
194                              new_insert += (w > 0xffff)? 4: 3;

196                          if ( lp->insert < new_insert )
197                          {
198                              lp->insert = new_insert;    // update instruction size.
199                              return 1;
200                          }
201                      }
202              }
203              addr += lineSize(lp);
204          }
205      }
206      return 0;
207  }
```
`/stm8scc/lk_source/st8link1.cpp`　　　　　　Code-19.23

19.4.4 STM8 library files

All the library files should be located in the "/stm8scc/lib" folder. Among the files, crt0.c and crt0x.c supply the basic service routines to support math operations as listed in the table below.

_extendRegA()	Extend the sign in register A.
_extendSign()	Extend the sign to register A, based on flags in register CC.
_mul16Ptr()	16-bit multiplication (referenced by index register X).
_mul24Ptr()	24-bit multiplication (referenced by index register X).
_mul32Ptr()	32-bit multiplication (referenced by index register X).
_divmod24Ptr()	24-bit division/modulation (referenced by index register X).
_divmod32Ptr()	32-bit division/modulation (referenced by index register X).
_divmod24()	24-bit division/modulation.
_divmod32()	32-bit division/modulation.
_leftShift16()	16-bit left-shift.
_leftShift24()	24-bit left-shift.

`_leftShift32()`	32-bit left-shift.
`_negPtrX24()`	Get the 24-bit negative value, referenced by index register X.
`_negPtrX32()`	Get the 32-bit negative value, referenced by index register X.
`_decodeZflag8()`	Decode and update the Z flag in register CC (8-bit value).
`_decodeZflag16()`	Decode and update the Z flag in register CC (16-bit value).
`_decodeZflag24()`	Decode and update the Z flag in register CC (24-bit value).

Note,
- (1) that the routines in both crt0.c and crt0x.c are highly optimized using inserted assembly instructions. And they will be automatically involved during the linking process.
- (2) Users can add or create other library files (*file*.c source code) to the folder. But it is necessary to add/create the corresponding header file in the "/stm8scc/include" folder.

19.5 Application Examples Using STM8 Compiler Tools

Under the folder "/stm8scc/applications", there are a few application projects based on STM8 MCUs that use the tools described in this part. They also illustrate how to use the STM8 compiler tools.

Among them, "/stm8scc/applications/mp3" is a project that accomplishes an MP3 player. This project is based on the STM8S-Discovery Kit from STMicroelectronics, which features an STM8S105C6 MCU (2KB RAM and 32KB ROM) on the board. A daughter board is added that has the following parts mounted:

- An MP3 decoder, VS1011E.
- An SD card (socket) that is used to store MP3 files.
- Three buttons that operate the player.

The project's makefile is shown in Table 19.11.

```
CC=st8cc
AS=st8as
LK=st8lk
RM=rm

HEX=mp3.hex
OBJ=main.obj spi.obj define.obj fat.obj key.obj vs1011.obj play.obj mmcspi.obj timer.obj sound.obj

$(HEX):$(OBJ) makefile
    $(LK) $(OBJ) -o mp3

%.obj: %.asm
    $(AS) $<

%.asm: %.c
    $(CC) $<

clean:
    $(RM) *.obj
    $(RM) $(HEX)
    $(RM) *.lst
```

Table-19.11

The following is the project's schematic.

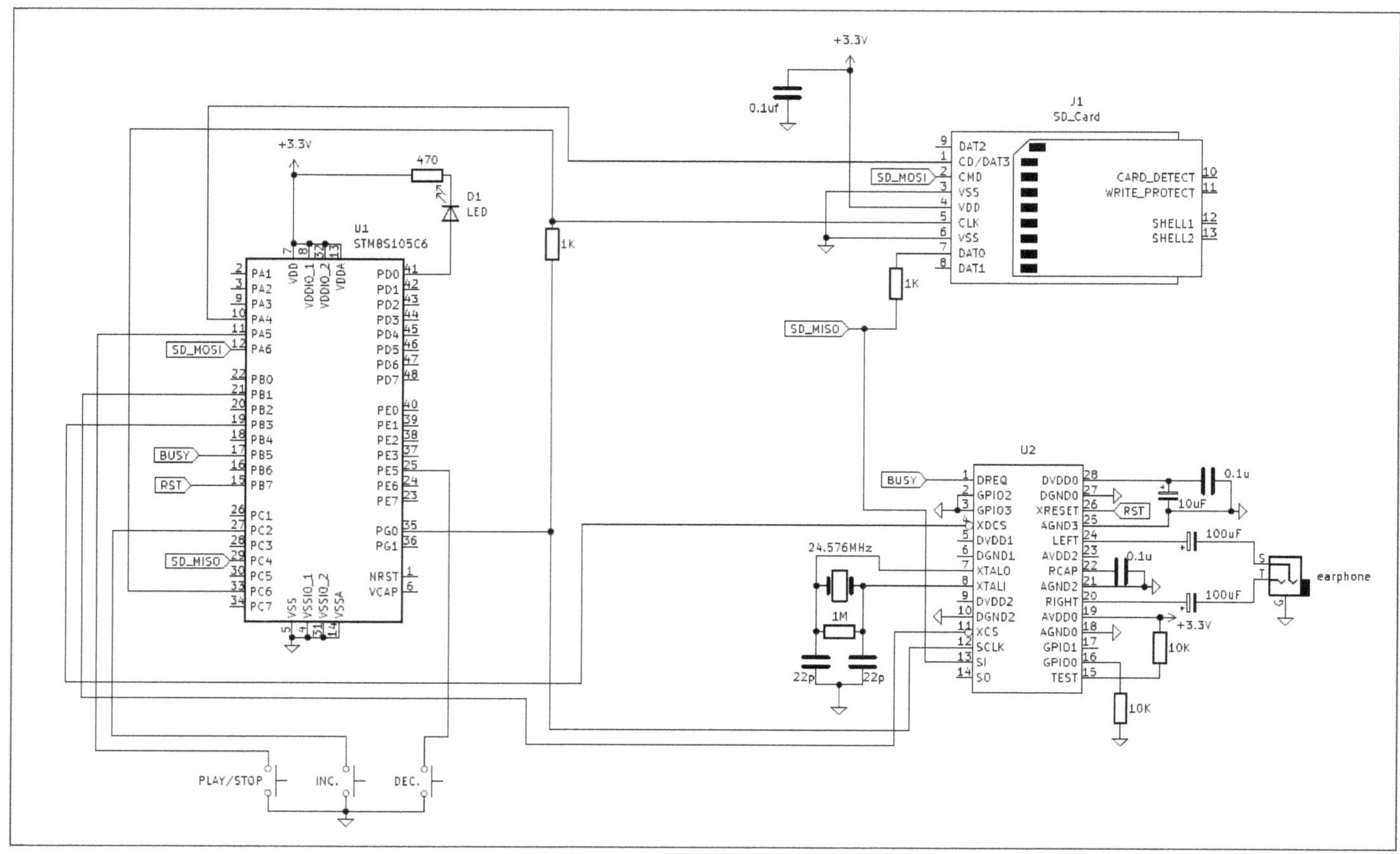

- Memory cost for the project: 94 bytes of RAM, 5877 bytes of ROM.
- This MP3 player can only support SD card (< 2GB capacity).
- It support both FAT16 and FAT32 file systems on SD card.